Fungal Biology

Fungal biology has an integral role to play in the development of the biotechnology and biomedical sectors. It has become a subject of increasing importance as new fungi and their associated biomolecules are identified. The interaction between fungi and their environment is central to many natural processes that occur in the biosphere. The hosts and habitats of these eukaryotic microorganisms are very diverse; fungi are present in every ecosystem on Earth. The fungal kingdom is equally diverse, consisting of seven different known phyla. Yet detailed knowledge is limited to relatively few species. The relationship between fungi and humans has been characterized by the juxtaposed viewpoints of fungi as infectious agents of much dread and their exploitation as highly versatile systems for a range of economically important biotechnological applications. Understanding the biology of different fungi in diverse ecosystems as well as their interactions with living and non-living is essential to underpin effective and innovative technological developments.

This series will provide a detailed compendium of methods and information used to investigate different aspects of mycology, including fungal biology and biochemistry, genetics, phylogenetics, genomics, proteomics, molecular enzymology, and biotechnological applications in a manner that reflects the many recent developments of relevance to researchers and scientists investigating the Kingdom Fungi. Rapid screening techniques based on screening specific regions in the DNA of fungi have been used in species comparison and identification, and are now being extended across fungal phyla. The majorities of fungi are multicellular eukaryotic systems and therefore may be excellent model systems by which to answer fundamental biological questions.

A greater understanding of the cell biology of these versatile eukaryotes will underpin efforts to engineer certain fungal species to provide novel cell factories for production of proteins for pharmaceutical applications. Renewed interest in all aspects of the biology and biotechnology of fungi may also enable the development of "one pot" microbial cell factories to meet consumer energy needs in the 21st century. To realize this potential and to truly understand the diversity and biology of these eukaryotes, continued development of scientific tools and techniques is essential. As a professional reference, this series will be very helpful to all people who work with fungi and should be useful both to academic institutions and research teams, as well as to teachers, and graduate and postgraduate students with its information on the continuous developments in fungal biology with the publication of each volume.

Arti Gupta
Editor

Prospects of Fungal Biotechnologies for Livestock Volume 1

Fungal Bioengineering in Livestock Health Management

Springer

Editor
Arti Gupta
Department of Zoology
Shri Avadh Raj Singh Smarak Degree College
Bishunpur, Uttar Pradesh, India

ISSN 2198-7777 ISSN 2198-7785 (electronic)
Fungal Biology
ISBN 978-3-031-90993-1 ISBN 978-3-031-90994-8 (eBook)
https://doi.org/10.1007/978-3-031-90994-8

This Springer imprint is published by the registered company Springer Nature Switzerland AG
The registered company address is: Gewerbestrasse 11, 6330 Cham, Switzerland

Preface

The interplay between biotechnology and livestock health has unlocked vast potential in recent years, reshaping the landscape of agricultural practices. This first volume delves into the innovative ways in which fungal biotechnologies are being harnessed to improve livestock health. By utilizing probiotics and biocontrol agents, we can not only enhance the well-being of animals but also promote sustainable agricultural practices.

In this comprehensive exploration, I also examine the significant economic advantages these biotechnologies can offer to farmers and stakeholders, potentially transforming the agricultural industry at large. The application of assisted reproduction technologies, which have played a pivotal role in elevating livestock production from a domestic endeavor to a thriving commercial enterprise, is also discussed in depth.

I hope that this volume serves as both a valuable resource and a source of inspiration for researchers, practitioners, and enthusiasts in the field of livestock biotechnology. As we stand at the forefront of a new era in agricultural innovation, we must continue to explore and embrace these biotechnologies to ensure a sustainable and prosperous future for all.

The first chapter, authored by Poonam and colleagues, discusses parasitism in farm animals. The second chapter, written by Shiwani, emphasizes the role of fungi as feed additives in animal nutrition and health.

In Chapter 3, Nahid and co-authors explore the effects of fungal feed additives on animals. Chapter 4, presented by John, highlights the benefits of fungal feed additives for animals and their future prospects.

Chapter 5, written by Susheela, addresses mycotoxins in food and feed and their relevance to human and animal health. In Chapter 6, Nahid and colleagues discuss the sustainable utilization of fungal resources and their applications in animal health.

Chapter 7, authored by Marian and colleagues, covers aquaculture production and management practices for farmed fish in Nigeria. Chapter 8, presented by Lily, highlights nematophagous fungi and their role in the biocontrol of parasites as well as biotechnological advances.

In Chapter 9, Abel discusses pathways for improving human and animal health using natural products derived from legumes.

Finally, in Chapter 10, Marian and coauthors outline disease management strategies for sustainable aquaculture in Nigeria, focusing on enhancements, innovations, and future directions.

Bishunpur, Uttar Pradesh, India Arti Gupta

Acknowledgments

First and foremost, I would like to thank God for his never-ending grace, mercy, and provision during what ended up being one of the toughest times of my life.

I am indebted to the many people who helped bring this book to light. I wish to thank series editors Dr. Vijai Kumar Gupta, Assistant Editor Natalie Muñoz, and Springer Nature for their generous assistance, constant support, and patience in initializing the series volume. I am very thankful to Vice Chancellor of Faizabad University, Ayodhya, for their kind support and constant encouragement. Special thanks are due to well-wishers, family members, and friends.

I owe a debt of gratitude to my wonderful husband, Saket Agrawal, and our son, Reyansh. Their unrelenting support and motivation have been pivotal in inspiring me to undertake this endeavor, overcome all obstacles, and assemble all the pieces into a coherent whole. I also give special thanks to all faculty colleagues and senior faculty members of our team.

Shri Avadh Raj Singh Smarak Degree College
Bishunpur, Uttar Pradesh, India

Arti Gupta

Contents

Author and Contributors

About the Author

Arti Gupta obtained her Ph.D. from Mahatma Jyotiba Phule Rohilkhand University, Bareilly, India, in 2010 in Animal Science. Her current research interests are animal biotechnology, molecular plant biotechnology, molecular animal biotechnology, bioprocess technology, and microbiology. From July 2014 to till date, she has been serving at Shri Avadh Raj Singh Smarak Degree College, Gonda, Uttar Pradesh, India, and has been engaged in some editorial work with Springer Nature, CRC Press, and De Gruyter. She has 13 years of teaching and research experiences in Plant and Animal Biotechnology, Microbial Biotechnology. She has worked as external examiner for central evaluation in 2023–24 at MJP Rohilkhand University, Bareilly.

She has received the following awards: University Topper (Gold Medal), M.Sc. (Biotechnology) by Ch.C.S. University, Meerut; Young Scientist Award (Gold Medal) by the Zoological Society of India, Lucknow; Best Poster Presenter by *Asian Journal of Experimental Science*, Jaipur; Best Poster Presenter by International Consortium of Contemporary Biologists (ICCB) and Madhawai-Shayam Educational Trust, Ranchi; Fellowship Award by International Consortium of Contemporary Biologist (FICCB) and Madhawi Shyam Educational Trust (FMSET); and Dr. V.P. Agarwal (Gold Medal) by D.A.V. (P.G.) College,

Muzaffarnagar. She has been involved with the following publications: 2 monographs, 18 edited books with Springer Nature, Switzerland; CRC Press, Taylor & Francis, USA; De Gruyter and GmBH Germany; 10 book chapters; and around 50 national and international research papers. She has presented many abstracts in national and international symposia/seminars/conference/workshops. She has lifetime membership of the Indian Science Congress Association, Biotech Research Society of India, Zoological Society of India, and International Consortium of Contemporary Biologist. She has been serving as a member on several editorial boards. Previously, she served as a member of Teaching Personnel, G.B. Pantnagar University, Pantnagar, Uttarakhand, India, and Visiting Academic Researcher, Biorefining and Advanced Material Research Center (BAMRC), Scotland's Rural College (SRUC), University of Edinburgh, Scotland, United Kingdom.

Contributors

Grace Abe Department of Aquatic Bio-resources, Bio-resources Development Centre, Kano, Nigeria

Fran Ehizokhale Abhulimen North Somerset Council, Weston-super-Mare, UK

Charles Oluwaseun Adetunji Department of Microbiology, Edo State University Uzairue, Iyahmo, Edo State, Nigeria

Temidayo Olanrewaju Aduloju Department of Quantity Surveying, Faculty of Environmental Science, Lagos State University, Lagos, Nigeria

Marian Onwude Agbugui Department of Biological Sciences, Edo State University Uzairue, Uzairue, Edo State, Nigeria

Alexander Ikechukwu Ajai Department of Chemistry, Federal University of Technology, Minna, Minna, Niger, Nigeria

Adamu Aliyu Department of Medicinal Plant Research and Traditional Medicine, NIPRD, Idogwari, Faidna, Nigeria

Okewu Liberty Apeh Department of Polymer and Textile Technology, Yaba College of Technology, Lagos, Nigeria

Yakubu Azeh Department of Chemistry, Ibrahim Badamasi Babangida University, Lapai, Lapai, Niger State, Nigeria

Esther Bernard Department of Chemical Engineering, Nassarawa State University, Keffi, Keffi, Nigeria

Kamlesh Kumar Bhatt Department of Science, Modern Institute of Technology, Rishikesh, Uttarakhand, India

Susheela Biranjia-Hurdoyal Department of Health Sciences, Faculty of Medicine and Health Sciences, University of Mauritius, Reduit, Mauritius

Isah Mohammed Chado Department of Biology, Ibrahim Badamasi Babangida University, Lapai, Lapai, Niger State, Nigeria

Tanishka Chauhan Department of Life Sciences, Parul Institute of Applied Sciences, Parul University, Vadodara, Gujarat, India

Koushalya Dangwal Department of Science, Modern Institute of Technology, Rishikesh, Uttarakhand, India

Oritseseyigbemi Gregory Edema Department of Physics with Electronics, Federal Polytechnic, Auchi, Edo State, Nigeria

Jonathan Hussaini Department of Chemistry, Ibrahim Badamasi Babangida University, Lapai, Lapai, Niger State, Nigeria

Abel Inobeme Department of Chemistry, Edo State University Uzairue, Edo State, Nigeria

Jonathan Inobeme Department of Geography, Ahmadu Bello University, Zaria, Nigeria

Poonam Jain Department of Life Sciences, Parul Institute of Applied Sciences, Parul University, Vadodara, Gujarat, India

Rahul Khanna Department of Microbiology, College of Basic Sciences and Humanities, Punjab Agricultural University, Ludhiana, Punjab, India

Gurvinder Singh Kocher Department of Microbiology, College of Basic Sciences and Humanities, Punjab Agricultural University, Ludhiana, Punjab, India

Hiral Krishnatray Amity Institute of Biotechnology, Amity University, Noida, Noida, Uttar Pradesh, India

Prabhanshu Kumar Amity Institute of Biotechnology, Amity University, Noida, Noida, Uttar Pradesh, India

Madhuri Kaushish Lily Department of Science, Modern Institute of Technology, Rishikesh, Uttarakhand, India

Gurumurthy Dummi Mahadevan Amity Institute of Biotechnology, Amity University, Noida, Noida, Uttar Pradesh, India

Amos Mamman Department of Chemistry, Ibrahim Badamasi Babangida University, Lapai, Lapai, Niger State, Nigeria

Vware Manawa Department of Biochemistry, Edo State University Uzairue, Iyamho, Edo State, Nigeria

Abdullahi Mann Department of Chemistry, Federal University of Technology, Minna, Minna, Niger, Nigeria

John Tsado Mathew Department of Chemistry, Ibrahim Badamasi Babangida University, Lapai, Lapai, Niger State, Nigeria

Prachi Mishra Department of Life Sciences, Parul Institute of Applied Sciences, Parul University, Vadodara, Gujarat, India

Musah Monday Department of Chemistry, Ibrahim Badamasi Babangida University, Lapai, Lapai, Niger State, Nigeria

Mohammed Jibrin Ndejiko Department of Microbiology, Ibrahim Badamasi Babangida University, Lapai, Lapai, Niger State, Nigeria

John Chiwuzulum Odozie Department of Agricultural and Development Economics, Ajayi Crowther University, Oyo Town, Oyo, Nigeria

John Temitope Ogbiti Department of Computer Science, Edo State University Uzairue, Uzairue, Edo State, Nigeria

Elizabeth Princess Okhamafe Department of Biological Sciences, Edo State University Uzairue, Uzairue, Edo State, Nigeria

Mathew Folaranmi Olaniyan Department of Medical Laboratory Science, Edo State University Uzairue, Iyahmo, Edo State, Nigeria

Igechi Gabriel Ozekhome Department of Chemistry, Ignatius Ajuru University, Port Harcourt, River State, Nigeria

Habibat Joy Ozemoka Department of Animal and Environmental Biology, University of Benin, Benin City, Edo State, Nigeria

Rohan Patel Department of Life Sciences, Parul Institute of Applied Sciences, Parul University, Vadodara, Gujarat, India

Ranjita Saikia Material Science Unit, Asian Institute of Technology, Klong Luang, Thailand

Elijah Yanda Shaba Department of Chemistry, Federal University of Technology, Minna, Lapai, Niger State, Nigeria

Bhavya Sharma Amity Institute of Biotechnology, Amity University, Noida, Noida, Uttar Pradesh, India

Nishu Sharma Department of Microbiology, College of Basic Sciences and Humanities, Punjab Agricultural University, Ludhiana, Punjab, India

Shiwani Guleria Sharma Department of Microbiology, College of Basic Sciences and Humanities, Punjab Agricultural University, Ludhiana, Punjab, India

Nahid Masood Siddiqui Amity Institute of Biotechnology, Amity University, Noida, Noida, Uttar Pradesh, India

Kshitij R. B. Singh Department of Chemistry, Govt, PG. Autonomous College, Durg, Chhattisgarh, India

Lavanya Singh Amity Institute of Biotechnology, Amity University, Noida, Noida, Uttar Pradesh, India

Suneel Kumar Singh Department of Science, Modern Institute of Technology, Rishikesh, Uttarakhand, India

Shambhavee Srivastav Amity Institute of Biotechnology, Amity University, Noida, Noida, Uttar Pradesh, India

Shailesh Srivastava Department of Zoology, Tripura University (A Central University), Agartala, Tripura West, India

Mary Ohunene Yakubu Department of Chemistry, Nigerian Defence Academy, Kaduna, Nigeria

Parasitism in Farm Animals

Rohan Patel, Tanishka Chauhan, Shailesh Srivastava, Prachi Mishra, and Poonam Jain

Abstract The chapter offers a detailed examination of the intricate dynamics of parasitism in agricultural animals, emphasizing its significant impact on animal well-being, productivity, and farming economics. It begins by providing a precise definition of parasitism and classifying the main types of parasites affecting livestock, which include helminths, and ectoparasites. It then explores the life cycles of these parasites, explaining how they adapt to and exploit their host organisms. In-depth discussions of key parasitic diseases, such as helminthiasis, and external infestations are caused by fleas, mites, and lice. These are included to illustrate the diverse impacts on various livestock species. The global cattle population, approximately 1.49 billion, is vulnerable to various external parasites, including insects, lice, mites, and ticks, which can significantly affect their health and productivity. Internal parasites, or endoparasites, such as tapeworms and flukes, inhabit the digestive or respiratory systems of their hosts, posing severe health risks. Helminths, particularly gastrointestinal nematodes, present major challenges to animal health and productivity in grazing livestock systems. Parasitic lungworms, including *Dictyocaulus, Muellerius*, and *Metastrongylus*, commonly infect ruminant animals and cause conditions such as verminous bronchitis. Trematodes, including liver flukes like *Fasciola hepatica* and Fascioloides magna, and the lancet liver fluke, *Dicrocoelium dendriticum*, significantly impact the health of livestock. Additionally, tapeworms such as *Moniezia benedeni* and zoonotic cestodes like those in the genus *Taenia*, contribute to severe illnesses in livestock and humans. Effective management and preventive measures are essential to mitigate the impact of these parasites on livestock health and productivity. By providing a comprehensive review of existing research and case studies, this section aims to equip veterinarians, farmers, and

R. Patel · T. Chauhan · P. Mishra · P. Jain (✉)
Department of Life Sciences, Parul Institute of Applied Sciences, Parul University, Vadodara, Gujarat, India
e-mail: poonam.jain26996@paruluniversity.ac.in

S. Srivastava
Department of Zoology, Tripura University (A Central University), Agartala, Tripura West, India

A. Gupta (ed.), *Prospects of Fungal Biotechnologies for Livestock Volume 1*, Fungal Biology, https://doi.org/10.1007/978-3-031-90994-8_1

researchers with a profound understanding of parasitism in farm animals. Several groups of fungi also show parasitism and bring bacterial infections. The immunocompromised livestock are often killed by the heavy infections of various infectious fungi such as *Candida, Aspergillus, Malassezia,* and so on. The objective is to empower stakeholders with the knowledge to effectively manage and mitigate parasitic infections, ultimately leading to improved animal health and farm productivity.

Keywords Ectoparasite · Endoparasite · Fungi · Helminthiasis · Livestock · Parasitism

1 Introduction

Cattle worldwide, about 1.49 billion, can be affected by various external parasites such as insects, lice, mites, and ticks. These parasites can significantly impact the well-being of cattle and food production (Pérez de León et al., 2020). Farm animals are at risk from internal and external parasites, which can affect their health. Farmers usually take preventative actions to protect their livestock. The health and general well-being of animals can be affected by these parasites, highlighting the necessity for farmers to put in place preventative measures to protect their livestock (Kumar et al., 2013). Ectoparasites are creatures that live on the outside of the skin or feathers of their host. They get their sustenance by feeding on the blood or other bodily fluids of the host (Hinkle & Corrigan, 2020). Ectoparasites are arthropod pests that can significantly impact the welfare and productivity of livestock by living on or burrowing into their host's epidermis, causing direct damage to the skin and other tissues (Colebrook & Wall, 2004). The small bugs known as lice, live as parasites and feed on the blood of warm-blooded creatures (Levot, 2000). Lice are external parasites with a strong preference for specific body areas showing high host specificity (Wall & Shearer, 2008). Lice are external parasites that must live on a host and show a strong preference for particular body areas, displaying a high level of host specificity. Sucking lice, known as Phthiraptera: Anoplura, are fascinating parasites that have formed enduring relationships with placental mammals (Light et al., 2010; Springer et al., 2023). Chewing lice, which are also called Mallophaga, are a common type of parasite that often affect various kinds of animals (Benelli et al., 2018). Fleas are tiny, flattened from side to side, flightless, and highly adaptive bugs (Bitam et al., 2010; Dobler & Pfeffer, 2011). With about 2500 species and subspecies, these small organisms are part of the order Siphonaptera (Mullen and Durden 2009). Fleas usually infest their host for a brief period before transferring to another host of the same species (Garedaghi, 2011).

Mange mites, also known as sarcoptic mites, are tiny wingless arthropods measuring less than 1 mm in size. They belong to the Arachnida group and the Acari subgroup (Ali et al., 2021). The northern fowl mite, *Ornithonyssus sylviarum*, is a widespread ectoparasite with a significant impact on poultry farming (McCulloch et al., 2020; Mullen and Durden 2009). Both immature and adult northern fowl mites need a blood meal to reproduce and can survive for up to 4 weeks without one in favorable conditions (Murillo & Mullens, 2020). Northern fowl mites need a blood meal to reproduce and can survive up to 4 weeks without one in favorable conditions (Murillo & Mullens, 2020). *Dermanyssus gallinae*, also known as the red mite is a common external parasite that mainly consumes the blood of laying hens, presenting a considerable threat to their health and welfare. *Ornithonyssus* sp. completes its entire life cycle on the host, usually inhabiting the down and feathers close to the cloaca area (Silva et al., 2023; Sparagano & Ho, 2020).

Ticks are fascinating creatures belonging to the suborder Ixodida of the order Parasitiformes. They are distinguished by their relatively large body size, ranging from 2 to 30 mm. This diverse group consists of about 820 species, further classified into two families: Argasidae (soft ticks) and Ixodidae (hard ticks) (Brites-Neto et al., 2015; Klompen et al., 1996; Sonenshine & Roe, 2014). Soft ticks can survive for a long time due to their slow metabolism and their ability to store bloodmeals without digesting them (Mans et al., 2022). Approximately 80% of the global cattle population is affected by ticks and tick-borne diseases, resulting in significant production losses (Burrow et al., 2019; Strydom et al., 2023). Tick bites may result in blood loss, harm to the skin, and the formation of wounds, which raises the likelihood of infections and diseases (Moyo & Masika, 2009). The order of flies, also known as Diptera, is important in transmitting various pathogens to humans and animals (Förster et al., 2009). Stomoxys flies can carry pathogens from the blood and skin of animal hosts, affecting livestock and sometimes humans (Baldacchino et al., 2013). Stable flies inflict painful bites, which can result in an increase in stress hormones and general irritation for horses and their owners (Machtinger et al., 2016; Cook, 2020). The common housefly, *Musca domestica* L., is known for spreading diseases and can pollute food and water with excrement (Banjo et al., 2005; Christensen, 1982).

Apart from these ectoparasites, endoparasites live inside the host's body. Tapeworms and flukes are examples of these types of parasites, and they are internal parasites that live within the digestive or respiratory systems of their hosts (Hinkle & Corrigan, 2020). Helminths have important traits that make them potential bioweapons. They are widely distributed, commonly found in domestic species, and easily obtainable (Kwak, 2016). Gastrointestinal nematodes are a major obstacle to animal health and productivity in grazing livestock systems (Szewc et al., 2021; Mendoza-de Gives et al., 2023).

The roundworms *Dictyocaulus*, *Muellerius,* and *Metastrongylus* are parasitic lungworms that commonly infect both domestic and wild ruminant animals. *Dictyocaulus* lungworms inhabit the air passages of ruminants. This specific nematode is known for causing verminous bronchitis, a respiratory condition that mainly impacts young cattle, particularly during their initial grazing season in areas with a moderate climate (Zafari et al., 2022; Ács et al., 2016; Lat-Lat et al., 2010). *Muellerius capillaris*, also called the nodular lungworm, is frequently present in sheep and goats all around the world (López & Martinson, 2017). Trematodes, a varied collection of parasites affecting the well-being of people and animals globally, originated from flatworms that were not parasitic (de Waal & Mehmood, 2021). The liver fluke, *Fasciola hepatica*, is a flat parasite that can grow up to 3 cm long and lives in the liver and bile ducts of infected cattle. Infestations can cause serious health problems, especially in sheep, and can lead to death (Lalor et al., 2021; Hayward et al., 2021). The giant liver fluke, or *Fascioloides magna*, is a parasite that affects North American cervids such as deer, elk, and caribou (Verocai et al., 2020). Livestock worldwide faces a significant threat from ruminant flukes, a parasitic disease caused by *Paramphistomum* spp. (Japa et al., 2020; Rizwan et al., 2022). The lancet liver fluke, known as *Dicrocoelium dendriticum*, lives in the bile ducts of animals such as sheep, cattle, pigs, and deer. Flukes from the *Paragonimus* genus cause Paragonimiasis in the lungs, with *Paragonimus westermani* being a significant species responsible for a foodborne parasitic zoonotic disease (Bellezza et al., 2002; Dharanesha et al., 2019). Tapeworms are parasitic flatworms that can cause severe illnesses in humans and other mammals. Part of the phylum Platyhelminthes and the class Cestoda (Deep et al., 2023). *Moniezia benedeni*, a type of tapeworm, resides in the small intestine of ruminant animals such as cattle, sheep, and goats (Irie et al., 2013).

Candidiasis is rarely observed in animals, but it is crucial to take it into account as a possible differential diagnosis when assessing bacterial infections. This fungal infection can also arise as a subsequent complication following a bacterial disease. It is essential for veterinarians to keep candidiasis in mind, particularly in cases where the host (Seyedmousavi et al., 2018). Likewise Candidiasis, *Aspergillus* species are common fungi that impact a range of organisms, including humans and animals, leading to diseases that vary from localized infections to severe systemic conditions (Seyedmousavi et al., 2015; Simões et al., 2023; Foley et al., 2014; Latgé, 1999; Knudtson & Kirkbride, 1992).

Livestock are commonly impacted by mycotoxins such as aflatoxins, zearalenone, and ergot alkaloids, which can result in carcinogenic, mutagenic, and immunosuppressive effects (Riet-Correa et al., 2024). These toxins can contaminate various food items and animal feed, threatening animal health and particularly affecting essential organs and reproductive capabilities (Zeece, 2020; Bennett & Klich, 2003; Moretti & Sarrocco, 2016; Wakenell, 2016). *Malassezia* are small yeasts (Hobi et al., 2022). These yeasts are typically found as commensals on the skin of warm-blooded animals (Sykes et al., 2014). Dermatophytosis, or ringworm, commonly affects dairy calves and can also impact adult cows, primarily caused by the fungus *Trichophyton verrucosum* (Scott, 2018).

2 Types of Parasites and Their Impact

Farm animals are at risk of various parasites, including internal parasites (endoparasites) and external parasites (ectoparasites). These parasites can impact the animals' health and overall wellness, underscoring the importance of farmers implementing preventive measures to safeguard their livestock.

2.1 Ectoparasites

Ectoparasites are organisms that reside on the exterior of the skin or feathers of their host. They may consist of various kinds of insects such as fleas, lice, and ticks (Fig. 1). These parasites acquire their nourishment by consuming the blood or other bodily fluids of the host (Hinkle & Corrigan, 2020). The majority of ectoparasites found on domestic animals are invertebrates, most of which belong to the arthropod group, including crustaceans, insects, and arachnids. Some ectoparasites play a significant role as carriers of pathogens (Hopla et al., 1994).

The global cattle population, numbering around 1.49 billion, is susceptible to a range of external parasitic organisms. The most economically impactful among these are arthropods, which include insects, lice, mites, and ticks. Severe infestations of these ectoparasites can have direct and negative effects on the health of cattle and the production of food (Pérez de León et al., 2020).

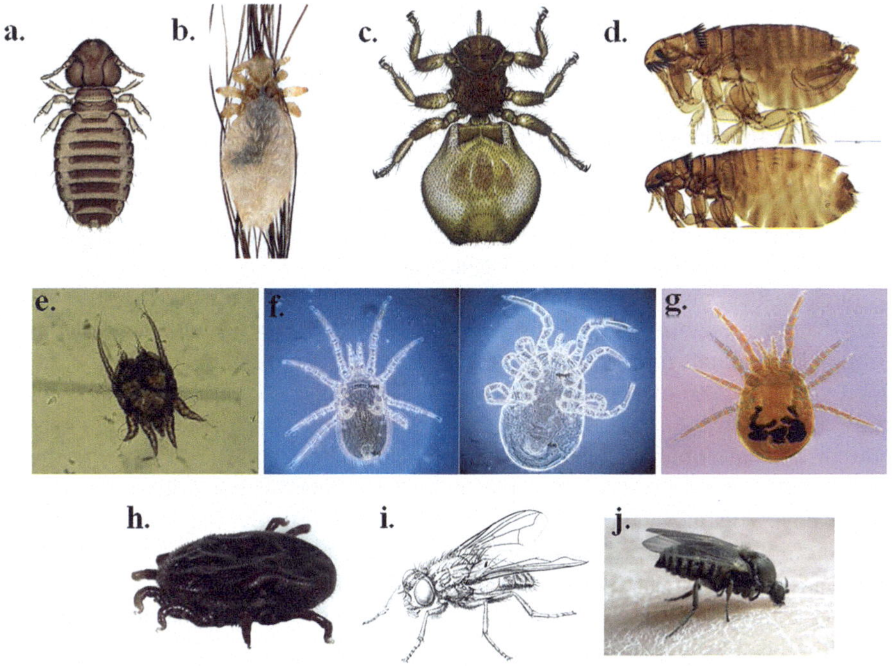

Fig. 1 The images showing several ectoparasites that cause infections in farm animals: (**a**) *Pediculus* or lice; (**b**) Chewing lice, known as *Mallophaga*; (**c**) Chewing lice, known as *Bovicola bovis*; (**d**) Fleas; (**e**) Mite; (**f**) red poultry mite known as *Dermanyssus gallinae*; (**g**) Tropical fowl mite known as *Ornithonyssus bursa*; (**h**) Tick; (**i**) Housefly; (**j**) Black flies

2.1.1 Lice

The small bugs, scientifically identified as Phthiraptera but commonly known as lice, have adapted to living as parasites and feeding on the blood of different warm-blooded creatures, such as household pets and humans (Fig. 1a). When they are present, the host often experiences continuous itching and discomfort, which can lead to frequent efforts to relieve these symptoms and potentially cause harm to the host's hair or fur (Levot, 2000). Lice are obligate external parasites that exhibit a distinct preference for specific anatomical regions, demonstrating a high degree of host specificity. Their life cycle is intricately linked to their host, as they remain on the host for their entire lifespan, only leaving to infest a new host. To thrive as permanent external parasites, lice have developed multiple adaptations that facilitate continuous close contact with their host (Wall & Shearer, 2008). Animals afflicted with lice suffer a deterioration in their physical well-being, leading to a lackluster and fragile appearance. Their fur takes on a greasy and discolored appearance as a result of the infestation (Sahito et al., 2017).

Lice experience an incomplete metamorphosis, implying they go through three stages during their life cycle: egg, nymph, and adult. The whole life cycle typically lasts about 4–5 weeks. Within 6–7 days, the eggs, also known as nits, emerge. Each female louse can lay anywhere from 20 to 50 (with the average being 30–40) eggs in her lifetime, placing them one by one on the hair close to the scalp. The incubation period for the eggs varies from 4 to 20 days before they hatch into nymphs (Egri, 2019). Lice, which are small insects belonging to the order Phthiraptera, are usually of tiny size, ranging from 2.5 to 3.0 mm, and have a flattened body. They can be classified into various groups based on their feeding habits, either blood-sucking or chewing on their hosts (Pérez de León et al., 2020).

Sucking lice, also called Phthiraptera: Anoplura, are interesting external parasites that have established a lasting connection with placental mammals. They can be found on animals from 12 out of the 29 recognized groups of mammals and infest approximately 20% of all mammal species. These blood-feeding bugs are specifically adapted to living on mammals. They lack wings, are flattened from top to bottom, and have tibiotarsal claws for grabbing onto host hair. In addition, they have piercing mouthparts for feeding (Light et al., 2010). Sucking lice, a kind of parasitic insect, have a distinct and exclusive association with particular host species. Known for their highly specialized nature, these lice are closely connected to and reliant on their chosen hosts for survival (Springer et al., 2023). Sucking lice possess a distinct method of managing the surplus water they ingest during their blood meals. In contrast to other blood-feeding insects, they do not excrete urine. Rather, they must figure out how to expel the excess water, as it could result in dehydration if they are unable to feed for only a few hours (Burgess, 2022).

Chewing lice, known as Mallophaga, are a prevalent form of parasite that commonly impacts different types of animals (Fig. 1b). They can be found on both terrestrial and aquatic birds, including pigeons, doves, swans, cormorants, and penguins. It is important to note that while Mallophaga have been observed on land mammals such as livestock and pets, they have not been discovered on marine mammals. These lice can cause damage and discomfort to affected animals, such as cattle, sheep, goats, horses, and poultry. Their bites can lead to itching and scratching, and may contribute to conditions like phthiriasis and dermatitis (Benelli et al., 2018).

Parasites collected in northeast Tunisia were examined using a stereomicroscope, and three species of lice were found: *Bovicola bovis*, *Linognathus vituli,* and *Haematopinus eurysternus* (Fig. 1c). Out of the observed animals, 32 were infested with the chewing lice *B. bovis* (4%), three with *H. eurysternus* (0.4%), and two with *L. vituli* (0.2%). The prevalence of infestation was higher in calves younger than 8 months (21.6%) compared to adults (0.8%) (Gharbi et al., 2013).

2.1.2 Fleas

Fleas are tiny, flattened from side to side, flightless, and highly adaptive bugs (Bitam et al., 2010). Fleas, which are part of the Order Siphonaptera, are intriguing insects recognized for their distinct adaptations and specialized feeding behaviors (Fig. 1d). With around 2500 species and subspecies, these small organisms are divided into 15 families and 220 genera, displaying their impressive variety and presence in different ecosystems (Mullen and Durden 2009). Fleas, which are wingless ectoparasites, are small creatures that sustain themselves by feeding on the blood of their hosts. They are present across the globe and can infest a wide variety of warm-blooded animals, including humans. The bites of fleas can result in irritation and discomfort, and in cases of severe infestation, they may cause blood loss and anemia in their host. Furthermore, fleas have the potential to transmit various pathogens, such as bacteria and parasites, posing a potential risk to health. They are commonly found in warmer climates and are known to be widely distributed geographically (Brasil et al., 2023). Fleas, which measure from 1 to 8 mm, are wingless insects. Their bodies are flat with a rounded abdomen, a compact thorax, and a blunt head. Fleas are usually dark brown, and they might have small eyes. Their bodies are covered in numerous spines, which can help identify different flea species (Dobler & Pfeffer, 2011).

Numerous species and subspecies of fleas (Siphonaptera) are known to infest domestic birds and mammals with around 20 species being particularly common. These highly mobile blood-sucking parasites can cause significant problems for adult hosts, including anemia, dermatitis, hypersensitivity, and the transmission of pathogens (Hopla et al., 1994). Fleas typically infect their host for a short period before moving on to another host of the same species. However, there are occasions when they may move to a host of a different species (Garedaghi, 2011). Fleas are known to play a key role in the transmission of pathogens in many regions globally. These tiny insects primarily feed on blood, which can potentially damage blood vessels. What is concerning is that fleas can carry pathogens and act as a natural vehicle for spreading them. They often transmit pathogens by regurgitating blood or by leaving contaminated fecal matter (Bitam et al., 2010). Fleas that commonly feed on dogs and cats can also infest farm animals and cause significant issues in livestock farming. Several types of fleas, including *C. felis*, *C. canis*, *P. irritans*, *T. penetrans*, and *X. cheopis*, can reproduce in poultry houses and other facilities where livestock are kept, especially in materials such as straw, sawdust, and sand used for bedding. Consequently, these infestations can affect the animals, farm workers, and owners (Oguntomole et al., 2018).

2.1.3 Mites

Mange mites, also referred to as sarcoptic mites, are minuscule wingless arthropods measuring less than 1 mm in size, making them extremely challenging to detect with the naked eye (Fig. 1e). They belong to the Arachnida group and the Acari (or Acarina) subgroup. Mange is a skin infestation in animals caused by these parasitic mites, resulting in symptoms such as itching (pruritus), hair loss (alopecia), crusted lesions, and redness (erythema). The intense itching often drives animals to vigorously scratch against surfaces such as walls and tree trunks, leading to skin damage and making them vulnerable to secondary viral and bacterial infections. This emphasizes the necessity of promptly and effectively treating animals affected by mange mites (Ali et al., 2021).

The discomfort caused by scabies mites to humans and domestic animals holds medical significance. The larger females are typically found in areas such as the spaces between the fingers and toes, in the groin, on external genitalia, and in axillary regions. They burrow into the epidermis and may even tunnel into the dermal tissue of their hosts, consuming blood and tissue fluids. This activity leads to irritations, inflammatory reactions, and hypersensitivity, resulting in clinical lesions, follicular papules, and intense itching, ultimately leading to scabies. The mites' bites and the resulting itching can cause restlessness and sleep disturbances (Ajobiewe & Ajobiewe, 2016). The various parasitic mites that impact farm animals, especially poultry, can be classified into acarine families. These families include Sarcoptidae (*Sarcoptes, Knemidokoptes*), Psoroptidae (*Psoroptes, Chorioptes*), Demodicidae (*Demodex*), Dermanyssidae (*Dermanyssus*), and Macronyssidae (*Ornithonyssus, Macronyssus, Liponyssus*). These mites can cause significant issues such as the formation of scabs, reduction in weight, poor growth rates, anemia, and even fatality in the affected animals. It is important to note that Sarcoptes mites have the ability to move from animal hosts to humans (Gabaj et al., 1992). Blood-feeding arthropods spreading diseases are receiving increased attention due to their ability to transmit various diseases affecting both animals and humans. Of particular note among these arthropods are poultry mites such as *Ornithonyssus sylviarum* (Northern Fowl Mite-NFM), *Dermanyssus gallinae* (Poultry Red Mite-PRM), and *O. bursa* (Tropical Fowl Mite-TFM). These mites, originally from Asia, can play a significant role in disease spread (Sparagano & Ho, 2020).

The northern fowl mite, scientifically named *O. sylviarum* Canestrini and Fanzago, is an ectoparasite found widely that has a significant effect on poultry farming. This small mite feeds on birds' blood, creating both economic and health risks for avian populations. It is capable of infesting various bird species, including wild birds, thereby increasing its potential impact on both poultry and wildlife (McCulloch et al., 2020). The northern fowl mite is commonly present in domestic chickens, particularly in caged laying hens or breeding stock. This is mainly due to the extended periods for which these populations are kept, leading to rapid reproduction of the mites and a substantial rise in infestations (Mullen and Durden 2009). It is important to keep in mind that wild birds can introduce mites to farms, and once these pests take hold, they often linger among the facilities and poultry for an extended period. They can easily spread among flocks via rodents, equipment, and human activity. Both immature and adult northern fowl mites require a blood meal from a host to develop and reproduce, completing their life cycle in as little as 5–12 days. While these mites typically cannot survive without a host for more than 1–2 weeks, they can endure up to 4 weeks without a blood meal in favorable conditions. High infestations of these mites can result in financial losses, including decreased egg production and reduced feed efficiency (Murillo & Mullens, 2020).

Dermanyssus gallinae, commonly referred to as the red poultry mite (Fig. 1f), is a prevalent external parasite that primarily feeds on the blood of laying hens, posing a significant risk to their health and well-being. This parasitic infestation can lead to issues such as anemia, skin irritation, behavioral changes, feather loss, increased pecking, and dermatitis in hens housed in aviaries year-round. Furthermore, this infestation adversely affects both the quantity and quality of egg production and has been associated with higher mortality rates among affected birds (Sparagano & Ho, 2020). *D. gallinae*, identified as the poultry red mite, primarily feeds on its avian host at night and hides in nests, cracks, and crevices throughout its life cycle. In contrast, *Ornithonyssus* sp. goes through its entire life cycle on the host, typically residing in the down and feathers near the cloaca region (Silva et al., 2023).

The *O. bursa*, also referred to as the tropical fowl mite, is a diminutive arachnid commonly found on both domestic and wild birds in tropical and subtropical areas (Fig. 1g). It is known to cause significant discomfort, severe dermatitis, and potentially even anemia in the birds it infests (Lareschi et al., 2017). The tropical fowl mite, *O. bursa*, is a minuscule organism that serves as an external parasite in the nesting materials used by birds. It mainly sustains itself by feeding on the blood of juvenile birds. The life cycle of *O. bursa* comprises five stages: egg, larva, protonymph, deutonymph, and adult. After consuming a blood meal, the adult mites rapidly become engorged, following which the fully developed female mite lays her eggs. Under favorable conditions, these eggs can hatch in as little as one day after being laid (Arce et al., 2018).

Cheyletiellosis, also known as Cheyletiella dermatitis or walking dandruff, is a contagious mite infestation that affects a wide range of animals, including domestic pets, wildlife, and humans. Its primary symptoms include skin scaling and crusting. This condition has been observed in various animals such as dogs, cats, rabbits, squirrels, poultry, foxes, and even humans (Kwochka, 1987).

2.1.4 Ticks

Ticks are intriguing creatures that belong to the suborder Ixodida of the order Parasitic forms. What distinguishes them from other Acari is their relatively large body size, which ranges from 2 to 30 mm, and their specialized mouth-parts known as hypostomes. Moreover, ticks have a unique and highly specialized arrangement of sensory structures (Fig. 1h). This diverse group consists of approximately 820 species, further classified into three families: Argasidae (soft ticks), Ixodidae (hard ticks), and Nuttalliellidae. The Nuttalliellidae family is particularly noteworthy as it is monotypic and shares characteristics of both Argasidae and Ixodidae, while also possessing many distinctive features of its own (Klompen et al., 1996).

The variety of relationships between ticks and their hosts, as well as tick adaptations, depends on the range of tick species, the diversity of potential host species, and the natural habitats in which these interactions take place (Sonenshine & Roe, 2014). Infestations of ticks raise important veterinary worries because these parasites can spread diseases, cause paralysis or toxic reactions, and inflict physical harm on livestock (Klompen et al., 1996). Ticks are important carriers of diseases affecting humans, livestock, and companion animals, more so than any other type of arthropod. They can lead to serious toxic conditions such as toxicosis, paralysis, allergies, and irritation in the host, while also transmitting harmful microorganisms such as Rickettsiae, viruses, protozoa, and spirochaetes (Mahlobo & Zishiri, 2021). Ticks extract blood from various animals such as reptiles, birds, and mammals. They go through a life cycle where larvae and nymphs target one set of host species, while adult ticks choose different host species (Sonenshine & Roe, 2014).

Approximately 80% of the global cattle population is impacted by ticks and tick-borne diseases, resulting in substantial losses in production. The resistance of cattle to ticks plays a crucial role in the economics of tick control, yet it is often overlooked in control programs due to the challenges and expenses of identifying variations in resistance among individual animals (Burrow et al., 2019). Ticks have the potential to cause significant economic harm in various ways. This includes direct impacts such as decreased productivity and income when ticks attach to animals, as well as the injection of toxins leading to "tick paralysis" and blood loss through feeding, resulting in economic losses. Additionally, ticks can indirectly contribute to economic losses by transmitting disease-causing pathogens. Heavy tick infestations in cattle can lead to "tick worry," causing discomfort and restlessness, ultimately resulting in significant energy loss and weight decrease. This not only affects production economics but also the welfare of the animals (Strydom et al., 2023). The presence of ticks can cause animals to lose condition, weaken, and become more susceptible to diseases, resulting in economic losses due to lower growth and production. Tick bites can also lead to blood loss, damage to skin and teats, and the development of wounds, increasing the risk of secondary bacterial infections and diseases (Moyo & Masika, 2009).

There is a specific period during which younger animals are more resistant than older animals to most tick-borne diseases. This means that younger animals are just as likely to get infected, but the illness is less severe. If the rate of infection, also known as the inoculation rate, is high enough to infect all young animals during this period, there will be little to no clinical disease because all older, potentially vulnerable animals are already immune. This creates a situation of endemic stability. The inoculation rate depends on the number of ticks present and their infection rate. In areas where the climate is marginal for the tick vector, or in farms and ranches where intense acaricidal tick control is implemented, the inoculation rate may drop below the level required for endemic stability. This could vary from year to year based on fluctuations in rainfall (Uilenberg, 1995). Females are more prone to infestation by ticks compared to males, as males exhibit greater resistance to certain parasites than females (Nasirian, 2023).

Ticks, which are commonly known as "hard ticks," can be recognized by their large anterodorsal sclerite and dorsal plate, also called scutum. They are quite big, ranging in size from 2 to 20 mm (Brites-Neto et al., 2015). Hard ticks are widely recognized as parasitic arachnids that rely on blood meals from domesticated ruminant hosts as part of their life cycle (Nasirian, 2023). Hard ticks consume blood from their hosts for long periods of time, which can range from several days to weeks. The duration of their feeding depends on factors such as their life stage, the type of host, and the tick species (Rajput et al., 2006). Soft ticks' remarkable survival capability over extended periods is due to their slow metabolism and their capacity to store bloodmeals without digesting them. A unique feature of soft ticks is their ability to feed multiple times without digesting previous bloodmeals, enabling the analysis of past hosts. This biological trait greatly enhances the identification of natural hosts of soft ticks, which is particularly challenging given their tendency to reside in burrows inhabited by multiple species of vertebrates, making host identification difficult (Mans et al., 2022).

2.1.5 Flies

Diptera, or the order of flies, holds great medical and veterinary significance as they are responsible for transmitting a wide variety of pathogens, including viruses, bacteria, fungi, protozoan, and metazoan parasites in both humans and animals. Although there are hundreds of fly species, only a few are common pests in and around human homes and farms (Fig. 1i). These insects are primarily important due to their frequent pathogenic properties. Some of the most common nuisance fly of the genus *Musca* species include the house fly (*M. domestica*), the face fly (*M. autumnalis*), the stable fly (*Stomoxys calcitrans*), and blowflies (e.g., *Lucilia* sp. and *Sarcophaga* sp.). These flies feed on and breed in animal dung, organic waste, and carcasses (blowflies), which can contain bacteria, viruses, and parasites that may cause diseases in both humans and animals (Förster et al., 2009).

The house flies are typically between 3 and 8 mm in length, have reddish eyes, sponging mouthparts, and are identifiable by four dark stripes on the upper side of the middle part of their bodies. The lower part of the abdomen is usually yellowish, especially on the sides, with males displaying more yellowing on the sides than females. There is a dark band running down the middle of the top side of the front parts of the abdomen. Female houseflies have eyes that are much farther apart than those of males. These flies are commonly found in places where animals are kept and also in urban areas where their larvae develop in a variety of decaying plant matter, feces, and household waste (Geden et al., 2021). The larvae exist as tissue parasites, causing myiasis in the host, and significantly affect the productivity and well-being of infested animals (Yadav et al., 2017). Adult house flies have sponging mouthparts and require liquid food or regurgitated crop contents to soften solid foods before consumption. The behavior of "bubbling" is commonly seen, during which flies expel a droplet of regurgitated substance, thought to eliminate excess water (Hendrichs et al., 1992).

Livestock producers are currently encountering considerable difficulties due to biting and nuisance flies. The types of flies causing problems vary depending on the time of year, geographic location, and cattle management practices. For example, stable flies (*Stomoxys calcitrans*) and houseflies (*M. domestica*) commonly present major issues when cattle are housed in stables or dry lots with organic debris accumulation and high cattle density (Craig, 2003).

Stomoxys flies act as mechanical carriers of pathogens found in the blood and skin of their animal hosts, primarily affecting livestock but occasionally humans. Their presence can lead to skin lesions, reduced food intake, stress, blood loss, and a general suppression of the immune system in livestock. Additionally, they prompt animals to gather for mutual protection while promoting the development and transmission of pathogens within the hosts. Their indirect impact involves the mechanical transmission of these pathogens (Baldacchino et al., 2013). Stable flies cause painful bites that can lead to a rise in stress hormones and overall irritation for horses and their owners (Machtinger et al., 2016). Increased exposure to stable flies led to elevated levels of the stress hormone cortisol in dairy cows, particularly when the cows displayed behaviors to deter the flies. Even as few as 20 flies can result in reduced weight gain for animals. The painful bites of stable flies can diminish milk production in dairy cows, lower weight gain in beef cattle, and affect feed efficiency (Cook, 2020). Adult stable flies need vertebrate blood to survive and reproduce, but they have limited contact with the host. The larvae of stable flies develop in manure, spilled feed, and decaying plant material. Proper sanitation is crucial for controlling stable flies by effectively managing these larval habitats (Foil & Hogsette, 1994). Stable flies take between 3 weeks and 1 month to transition from an egg to an adult, indicating that they have a longer life cycle (Christensen, 1982).

M. domestica, insects are often found around exposed items, leading to the potential contamination of drinking water, food, and dairy products with fecal matter. In poultry environments, houseflies are the main insect pests, thriving in conditions with organic waste and making large fly populations (Banjo et al., 2005). Houseflies have specialized sponging mouthparts that allow them to feed on a wide variety of foods. They liquefy solid foods with regurgitated fluids, a behavior that contributes to the transmission of disease-causing organisms. House flies breed in various moist, decomposing organic materials such as animal waste, accumulated manure, decomposing or damp hay, animal feed, silage, grass cuttings, and garbage (Christensen, 1982). Furthermore, *Musca domestica* is a species that is highly adaptable and can thrive in various environments including freshwater areas, waterlogged regions in mangrove forests, septic tanks, domestic waste sites, desert coolers, as well as indoor and outdoor locations with stagnant water. They are also commonly found in high-humidity environments, such as the exposed roots of aquatic plants and in moist, sheltered areas (Alsaad, 2023). The house fly has a complete metamorphosis with distinct egg, larval or maggot, pupal, and adult stages. The housefly overwinters in either the larval or pupal stage under manure piles or in other protected locations. Warm summer conditions are generally optimum for the development of the house fly, and it can complete its life cycle in as little as 7–10 days. However, under suboptimal conditions, the life cycle may require up to 2 months. As many as 10–12 generations may occur annually in temperate regions, while more than 20 generations may occur in subtropical and tropical regions (Capinera, 2008).

Face flies, or *M. autumnalis*, are so named because of their preference for the eyes and noses of livestock, particularly cattle. They can also be found on the withers, neck, sternum, and flank. Their mouthparts are specialized for absorbing saliva, tears, and mucus (Šlapeta, 2022). Male and female flies use their sponge-like proboscis, which is equipped with prestomal teeth, to feed on moisture around the mucus membranes on an animal's face and other body areas. While feeding, face flies can provoke defensive behaviors in livestock such as ear flapping, head shaking, seeking shelter in shade, and adjusting grazing patterns. Additionally, they can transmit eye-inhabiting nematodes and pathogenic bacteria, potentially leading to the development of pinkeye (Fryxell et al., 2021).

The face fly female only lays eggs in recently deposited cattle manure found in rangeland or pasture. The larvae of face flies undergo three developmental stages before locating their hosts. Female face flies' persistent feeding behavior leads to significant annoyance and irritation. Livestock react to fly feeding by gathering in groups, seeking shelter under trees, and occasionally standing in water to evade the flies (Boxler, 2015).

Black flies are tiny two-winged insects that lay their eggs in or close to flowing water (Fig. 1j). Despite being only 1–5 mm long, these insects that bite during the day can emerge in large groups, causing significant irritation and blood loss. Their effects can include poisoning, skin irritation, allergies, and, in cattle, sheep, and wild ruminants, death (presumably due to suffocation). The young stages of these insects live in the water. They are organisms that filter out organic matter from the water or actively consume materials found within the water. In regions with moderate climates, each species typically only has one generation per year, even though the sequence of species throughout a summer may cause it to seem like a single species is consistently present (Cortinas & Jones, 2006).

2.2 Endoparasites

Endoparasites inhabit the body of the host. Examples of endoparasites comprise tapeworms and flukes, which are internal parasites residing within the digestive or respiratory systems of their hosts (Fig. 2) (Hinkle & Corrigan, 2020). Endoparasites take up residence in an animal's bloodstream or tissues. Some organisms enter the animal's system when they consume contaminated food or water, while others burrow through the skin, make their way into the bloodstream, and settle in a favorable location to thrive and multiply. Endoparasites often interfere with the digestion and absorption of nutrients, resulting in hindered growth, temporary or permanent damage, or even death (National Agricultural Library et al., 2024). Endoparasites also affect other aspects of behavior. It has been shown that parasitized animals take

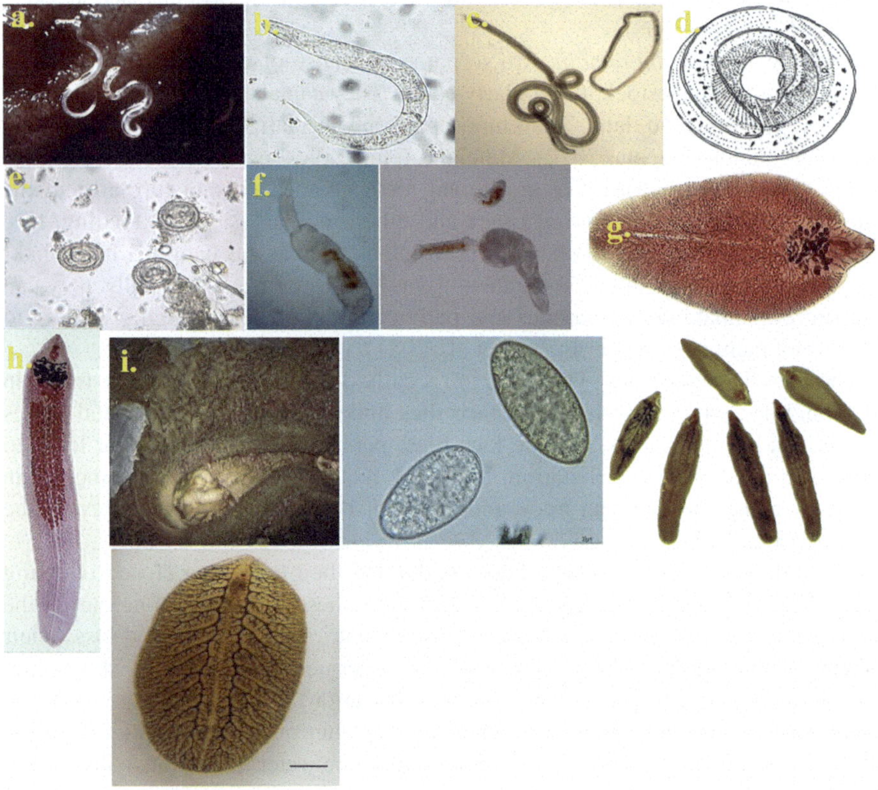

Fig. 2 The images showing several endoparasites that cause infections in farm animals: (**a**) Cestodes large parasite; (**b**) Gastrointestinal nematode; (**c**) lungworm *Dictyocaulus filaria*; (**d**) *Metastrongylus* spp. *of nematode parasite.*; (**e**) Microscopic image of *Metastrongylus* spp. (**f**) Trematode; (**g**) Leaf-shaped *Fasciola hepatica*; (**h**) Cylinder-shaped *Fasciola hepatica*; (**i**) *Paramphistomum* spp. *known as ruminant trematode*; (**j**) *Paramphistomum cervi.* Known as rumen fluke; (**k**) *Dicrocoelium dendriticum*, known as lancet liver fluke; (**l**) *Fascioloides magna* known as a giant liver fluke; (**m**, **n**, and **o**) *Paragonimus westermani* known as lung fluke; (**p**) *Taenia solium*

fewer steps, lie longer, and change posture less frequently than non-parasitized animals. This reduced activity may be a consequence of lethargy likely aimed at conserving energy, or as another parasite-avoidance strategy.

2.2.1 Helminths

Helminths possess several significant biological characteristics that make them potentially dangerous and effective as bioweapons. The likelihood of accidental exposure to these agents by individuals using them as weapons is much lower compared to more common biological agents, primarily because most helminthic infections result from consuming contaminated food. This group is taxonomically diverse, and many potentially hazardous helminths are widespread, often found in domestic or cosmopolitan species and easily collected. Several helminths are also known to cause a variety of distressing and often socially stigmatizing effects in those who are infected, including larval migrants and symptoms related to nerve damage. Diseases caused by many species also have a delayed onset, and the signs and symptoms are often difficult to distinguish from those of other more common diseases (Kwak, 2016). Helminths, or parasitic worms, are a diverse group of multicellular parasites categorized as nematodes, trematodes, and cestodes. Each category of parasites has unique life cycles and is found in different tissues or luminal organs of their hosts. These parasitic helminths present a considerable challenge to sustainable livestock production, often leading to significant economic burdens on the agricultural industry worldwide (Foster & Elsheikha, 2012).

The well-being of large livestock such as cattle and goats can be significantly affected by gastrointestinal helminths, which are parasitic worms that primarily reside in the intestines. It is essential to regularly monitor these animals for helminth infestations to effectively manage and control their impact on overall health (Income et al., 2021).

2.2.2 Gastrointestinal Nematodes

Gastrointestinal nematodes are the most significant group of internal parasites (Fig. 2a). They pose a major challenge to animal health, productivity, and profitability in grazing livestock production systems. Gastrointestinal nematodes (GIN) are a substantial concern across all animal production systems, especially in ruminants and horses that primarily graze on grass. The economic impact of GIN arises from production losses in animals with subclinical infections, the cost of treating clinical cases, and potential mortalities, which can have significant repercussions for both the farmer and the economy (Szewc et al., 2021).

The presence of tiny parasitic worms, or nematodes, can pose a significant threat to young grazing ruminants. These parasites can lead to severe malnutrition, weakness, anemia, reduced immune response, and even fatalities. The severity of these health issues depends on the harmfulness of the parasite infections and the quantity of worms present. Even in the absence of apparent signs of illness, the presence of these gastrointestinal nematode parasites can still diminish the animals' performance, resulting in long-term financial losses (Mendoza-de Gives et al., 2023). The illness often has financial repercussions due to a decreased ability to gain weight or produce milk at the expected rate. Severe cases can result in death or significant weight loss. Gastric parasitism may manifest in symptoms including reduced appetite, thickening of the stomach lining, impaired initial digestion, decreased production of stomach acid and digestive enzymes, indigestion, loss of protein in the blood, and bleeding (Craig, 2018).

Gastrointestinal nematodes (GINs) are parasitic worms that impact livestock (Fig. 2b). These worms follow a direct life cycle, which means that infection occurs when animals graze on contaminated pasture. In regions with cold winters, like temperate areas, cattle deposit parasite eggs at the beginning of the pasture season in the spring. As the summer progresses, the pasture becomes more contaminated with infective larvae, typically reaching its peak in early autumn. During the winter and early spring, when pastures are ungrazed, contamination levels decrease to their lowest points (Charlier et al., 2020). Gastrointestinal parasites can greatly affect the body. One of the primary impacts is a reduction in food consumption, which can be attributed to the pain and discomfort associated with the infection. Furthermore, hormonal changes resulting from disrupted gastrointestinal function may also play a role in this decrease. Particularly, helminths and other parasites can harm the gut tissue and interfere with mucoprotein secretion, exacerbating the detrimental effects on the body (Villalba et al., 2014).

Sheep, scientifically referred to as *Ovis aries*, are particularly susceptible to different gastrointestinal nematodes (GIN). These infections can be influenced by various factors such as the animals' age and breed, the specific parasite species involved, and the extent of the parasite infestation. Nematode infections have the potential to significantly affect animal welfare, leading to reduced productivity on farms, irrespective of the production system in use (Tachack et al., 2022).

2.2.3 Lung Worms

The lungworms from the groups *Dictyocaulus, Muellerius*, and *Metastrongylus* are parasitic roundworms that frequently infest domestic and wild ruminant animals. These parasites hold considerable veterinary and economic significance because of their effects on the health and productivity of animals (Zafari et al., 2022). Lungworms called *Dictyocaulus* are parasites that live in the air passages of ruminant animals. The level of worm infestation, also known as "worm burden," can vary from mild to severe, causing significant health problems in the host, a condition called "dictyocaulus." For example, *Dictyocaulus viviparus* can lead to severe and often fatal bronchitis and pneumonia in cattle, a condition known as "husk." This is a major concern for both the veterinary and agricultural industries due to its impact on animal well-being, decreased production, and the costs of treatment (Ács et al., 2016). *D. viviparus* is a pathogenic parasitic nematode, meaning it can cause disease. This particular nematode is notorious for causing verminous bronchitis, a respiratory condition that primarily affects young cattle, especially during their first grazing season in regions with a moderate climate (Lat-Lat et al., 2010).

The life cycle of *D. viviparus* follows a direct pattern. Adult worms lay eggs with thin shells measuring 82–88 mm by 32–38 mm, which then become embryonated. Upon being laid, the eggs swiftly hatch either in the airways or, if coughed up and swallowed, in the large intestine. The first-stage larvae are released in the host's feces into the environment, where they can mature into second and third-stage larvae within a week. The enveloped third-stage larvae are capable of infecting grazing cattle. Upon consumption, these larvae shed their outer layer in the abomasum, penetrate the intestine, and migrate to the mesenteric lymph nodes near the ileum, cecum, and proximal colon, where they reside for a duration of 3–8 days. Fourth-stage larvae can reach the lungs as early as 7 days after infection and are found in significant quantities by day 13 to 15 post-infection (Panuska, 2006).

The lungworm *Dictyocaulus* filaria is a parasitic nematode that commonly infects sheep all around the world (Fig. 2c). This type of infection can lead to a condition known as verminous pneumonia, which is characterized by inflammation and damage to the lungs. Infected sheep may exhibit symptoms such as persistent coughing, labored breathing, and in severe cases, weight loss and even fatalities (Prabhu et al., 2023). The adult worms inhabit the respiratory system of small ruminants, specifically the trachea, bronchi, and bronchioles. The host gets infected through ingestion, after which the larvae penetrate the intestinal wall and navigate through the lymphatic vessels to the mesenteric lymph nodes. It is there that they mature and undergo the third molting stage. Subsequently, they move through the lymphatic and blood vessels to the lungs, where they encroach upon the airways after getting stuck in the capillaries (Kırcalı Sevimli et al., 2011).

Muellerius capillaris, also known as the nodular lungworm, is commonly found in sheep and goats worldwide. It depends on slugs or snails as intermediate hosts. In affected sheep, multiple subpleural nodules develop in the dorsal areas of the caudal lung lobes. These nodules initially appear soft and hemorrhagic, but later become gray-green, hard, or even calcified. Under the microscope, a localized, eosinophilic, and granulomatous response occurs in the subpleural alveoli, where the adults, eggs, and coiled larvae are located. Clinical signs are typically not observable. Infections of *M. capillaris* in sheep and goats can result in reduced weight gain and potentially secondary bacterial infections (López & Martinson, 2017).

Metastrongylus spp. are a type of parasitic nematode that follows a complex life cycle involving suids, such as pigs, as their definitive hosts. The adult worms typically reside in the bronchi of the suid hosts, where the females deposit eggs. These eggs then travel through the airways to reach the suid's digestive tract. This type of nematode displays a heteroxenous nature, which means that it relies on multiple hosts to complete its life cycle (Spieler & Schnyder, 2021). The lung parasites of the *Metastrongylus* genus deposit mature eggs in the host's respiratory system. After the boar coughs or sneezes, the eggs are expelled into its mouth and then ingested and excreted in the feces. Once in the soil, the eggs hatch and release L1 larvae, which are then consumed by earthworms, the intermediate hosts. The final host becomes infected by consuming earthworms that have ingested the larvae. These earthworms are plentiful in the final host's environment (Poglayen et al., 2016).

Metastrongylus spp. can lead to breathing difficulties and gradual weight loss by causing damage to the interstitial tissues, resulting in obstruction and eventual consolidation of the lungs. Animals with moderate or severe immunosuppression are at a higher risk of experiencing these harmful effects (Fig. 2d). *Metastrongylus* species are parasitic roundworms that reside in the respiratory tract of pigs and have indirect life cycles involving several earthworm species (Gassó et al., 2014). Six different *Metastrongylus* species have been observed globally. These species are *M. apri* (also called M. elongatus), *M. asymmetricus, M. confusus, M. madagascariensis, M. pudendotectus, and M. salmi. M. apri, M. salmi*, and *M. pudendotectus* are the most commonly reported species worldwide (Fig. 2e) (Poglayen et al., 2016).

2.2.4 Trematodes

Trematodes, a diverse group of parasites that impact the health of both humans and animals worldwide, evolved from free-living flatworms (Fig. 2f). Over time, they developed close associations with mollusks and eventually transitioned into parasitic forms. These parasites can infest various organs, such as the intestines, lungs, liver, and vascular system, leading to significant economic losses in livestock production and a decline in human well-being (de Waal & Mehmood, 2021).

For centuries, both humans and animals have suffered from snail-borne trematodes (SBTs) (Trematoda: Digenea), with devastating consequences. These parasites have been known to induce life-threatening conditions, including cancer, leading to fatalities. Diseases caused by SBTs, such as fascioliasis, paramphistomiasis, and schistosomiasis, have profoundly disrupted profitable livestock production in various countries, particularly in the humid tropics and sub-tropics. The ramifications include reduced milk and meat production, poor weight gain, infertility, decreased vitality, heightened susceptibility to various diseases, and increased mortality (Yasin et al., 2018).

2.2.5 Liver Fluke

The liver fluke, scientifically known as *Fasciola hepatica*, is a flat leaf-shaped parasite that can grow up to 3 cm in length and resides within the liver and bile ducts of infected cattle (Fig. 2g). This parasitic organism has a complex life cycle that involves freshwater snails as intermediate hosts (Liver Fluke and Rumen Fluke, n.d.). Fascioliasis is a parasitic disease caused by liver fluke species of the genus *Fasciola*, with *F. hepatica* being the primary cause due to its widespread distribution across Europe, Africa, Asia, Oceania, and the Americas. Infections caused by the liver fluke species, *F. hepatica* and *F. gigantica*, belong to the most neglected zoonotic diseases, despite their global prevalence (Mas-Coma et al., 2009).

Liver flukes are extremely efficient parasites capable of infecting both humans and various ruminant animals, including sheep, cattle, goats, buffalo, camelids, and cervids. Additionally, these parasites can infect non-ruminant herbivores, albeit less frequently. Infection with Fasciola spp. in livestock can result in severe illness and mortality, leading to reduced productivity and fertility, and increased susceptibility to concurrent infections (Lalor et al., 2021).

The life cycle of *Fasciola* begins with the release of immature eggs in the biliary ducts and stool. These eggs then mature in water, giving rise to miracidia that invade a suitable snail intermediate host, such as *Galba, Fossaria,* and *Pseudosuccinea.* Inside the snail, the parasites undergo several developmental stages (sporocysts, rediae, and cercariae) before being released from the snail. The cercariae then encyst as metacercariae on aquatic vegetation or other surfaces. Mammals become infected by consuming vegetation containing metacercariae. Human infection occurs through ingesting metacercariae-containing freshwater plants, especially watercress. After ingestion, the metacercariae hatch in the duodenum and move through the intestinal wall, peritoneal cavity, and liver parenchyma into the biliary ducts, where they mature into adults (Tolan, 2011).

Liver fluke infestations can lead to significant health issues as a result of the migration of juvenile flukes within the host's body, particularly in sheep, where it can ultimately result in fatalities. Furthermore, adult liver flukes can induce chronic health complications by feeding on the host's blood, as they are capable of surviving for as long as the host does (Hayward et al., 2021). Sheep, cattle, horses, pigs, goats, alpacas, and deer can all serve as hosts for the sexual maturation of liver fluke. Additionally, kangaroos, wombats, and rabbits are capable of maintaining pasture contamination as reservoirs. Humans can become infected by ingesting watercress from naturally contaminated creeks (Boray et al., 2017).

The repercussions of livestock deaths extend beyond the immediate loss of animals. They encompass broader impacts, including decreased wool production and quality, reduced lambing rates, slower growth of lambs, higher expenses for replacing stock, diminished milk production and quality in dairy cattle, as well as lower growth and feed efficiency in fattening cattle (Boray et al., 2017) (Fig. 2h).

2.2.6 Giant Liver Fluke

The giant liver fluke, also known as the deer fluke and scientifically referred to as *Fascioloides magna*, is a parasite that impacts North American cervids such as deer, elk, and caribou (Fig. 2l). When passed in the feces of cervids, the eggs of this parasite resemble those of another parasite called *F. hepatica* and can be identified using specific sedimentation methods. Although the infection primarily affects wild cervids, domestic ruminants such as cattle, sheep, and goats can also become infected, potentially leading to fatal consequences. In cattle, the flukes develop into thick-walled cysts in the liver, while in sheep and goats, they migrate extensively and cause significant liver damage. Infections in domestic livestock can have serious and even fatal outcomes (Verocai et al., 2020). Fascioloides magna is highly pathogenic to infected wildlife and is therefore of great veterinary importance. The migration of young flukes and the encapsulation of adult trematodes in the liver cause mechanical damage to the tissue, leading to organ enlargement and potential liver failure. As a result, the presence of these flukes is linked to substantial reductions in body size and weight, as well as impaired reproduction in infected animals (Filip-Hutsch et al., 2022).

2.2.7 Rumen Fluke

The parasitic disease called ruminant flukes, which is induced by *Paramphistomum* spp., presents a major danger to livestock worldwide. Livestock afflicted by these flukes encounter reduced growth rates and diminished productivity, resulting in considerable financial losses for livestock farmers (Japa et al., 2020).

Paramphistomum is a genus of flukes that are part of the trematode flatworm family. They commonly infect cattle, sheep, goats, and other livestock, as well as various wild ruminants (Fig. 2i). Their presence is global, particularly in humid areas, and their prevalence varies based on climatic and ecological factors. In regions where they are endemic, their prevalence can reach remarkably high levels, with as many as 80% of animals in a herd being affected (Liver Fluke and Rumen Fluke, n.d.).

Paramphistomosis occurs in the forestomach of ruminants due to parasites called Paramphistomum spp. (Trematoda: Paramphistomatidae) in areas with higher temperatures. Snails act as the intermediary hosts for this parasite. Ruminants get infected by ingesting metacercaria found in water and on aquatic plants (Rizwan et al., 2022). The adult flukes are typically nonpathogenic for their hosts. However, the migration of immature worms in the duodenal mucosa can lead to severe enteritis, and possibly necrosis and hemorrhage. This migration is responsible for symptoms such as anorexia, polydipsia, unthriftiness, and severe diarrhea, and can even lead to mortality. Paramphistomosis has a high prevalence throughout tropical and subtropical regions, particularly in Africa, Asia, Australia, and Eastern Europe (Russia) (Khedri et al., 2015).

The specific species of parasite responsible for this condition varies by region. *Paramphistomum cervi*, commonly known as the rumen fluke, is considered to be one of the most significant species of *Paramphistomes* (Fig. 2j). Adult flukes are found in the rumen of ruminant animals, while immature flukes mainly reside in the small intestine, particularly in the duodenum. *P. cervi* follows a di-heteroxenous life cycle, involving snails (such as *Bulinus* spp., *Glyptanisus gilberti, Indoplanorbis exustus, Planorbis planorbis*, and *Lymnaea bulimoides*) as intermediate hosts and mammals as definitive hosts (Chaoudhary et al., 2015).

2.2.8 Lancet Fluke

Dicrocoelium dendriticum is the cause of liver fluke disease in ruminants and is both zoonotic and economically significant (Fig. 2k). It is prevalent in various regions across the globe. This parasite takes up residence in the liver of its definitive host, utilizing a gastropod snail as its first intermediate host and an ant as its second intermediate host during its life cycle. The larval stages of the parasite, comprising sporocysts and cercariae, inhabit the hepatopancreas of terrestrial snails, while metacercariae, also larval stages, are found in the abdomen and brain of ants (the second intermediate host). The parasite exhibits a certain level of specificity toward the definitive host, showing a preference for ruminants, primarily sheep and goats, and secondarily, large ruminants such as cattle (Chougar et al., 2019).

Dicrocoelium dendriticum, commonly known as the lancet liver fluke, inhabits the bile ducts of various animals including sheep, cattle, pigs, deer, cottontail rabbits, and woodchucks. The fluke eggs are excreted in the feces of the definitive hosts and are then consumed by terrestrial snails. Within the snails, cercariae develop, are released in mucus, and are subsequently ingested by ants. The definitive host then consumes the ants while grazing. The metacercariae travel up the common bile duct. In sheep and woodchucks, liver scarring and histologic lesions characteristic of perilobular cirrhosis are observed (Bellezza et al., 2002). Macroscopic liver injuries may present as bile duct inflammation, liver enlargement, and the formation of fibrous tissue. In studies with infected lambs, researchers noted diverse histopathological changes such as periductal fibrosis, ductal reaction, and the presence of infiltrating white blood cells (Piegari et al., 2021).

2.2.9 Lung Fluke

Paragonimiasis in the lungs is attributable to various flukes from the Paragonimus genus. Particularly, *Paragonimus westermani* is a significant species responsible for causing a potentially foodborne parasitic zoonotic disease (Fig. 2m, n, and o) (Dharanesha et al., 2019). Paragonimiasis, also known as lung fluke disease, is caused by various species of the trematode genus *Paragonimus.* This infection primarily affects the lungs and pleura of humans and animals. While the parasite can cause a pulmonary disease similar to tuberculosis when it infects the lungs, it can also invade other visceral organs such as the brain, spinal cord, and abdominal organs. The adult worms are relatively large, measuring 7–12 mm in length for *P. westermani*, and have a plump, coffee bean-like appearance. The eggs are yellowish-brown, operculated, and vary in size (ranging from 70 to 90 mm in length and 40 to 50 mm in width). The patterns of lobation of the ovary and testes are the most important morphological features used for species discrimination (Chai, 2013).

Small ruminant lungworm parasites are a significant challenge for livestock production and are prevalent in many tropical and subtropical regions worldwide. The severity of lungworm infections can be influenced by factors such as the number of larvae ingested, the strength of the animal's immune system, and the location of the lungworms within the respiratory tract. Despite the potential impact on the animals, the clinical signs of lungworm infection may be less apparent compared to other livestock diseases (Tessema et al., 2024).

P. westermani's natural definitive hosts, besides humans, include a wide variety of mammals. Adult worms are usually located in the lungs. Upon release with sputum and exposure to water, the eggs transform into miracidia. These miracidia then emerge and penetrate mollusk hosts, eventually developing into microcercous cercariae. Subsequently, the cercariae infest crustacean hosts, where they mature into metacercariae. Once ingested by the definitive host, the metacercariae excyst in the small intestine before traveling to the lungs through the peritoneal cavity. This intricate migration route allows the worms to settle and mature in atypical locations as well. Additionally, pigs serve as important paratenic hosts for *P. westermani*. It is commonly found in stream valleys in hilly and mountainous regions (Kim, 1984).

2.2.10 Cestodes

Tapeworms, which are parasitic segmented flatworms, are part of the phylum Platyhelminthes and the class Cestoda. They are elongated, segmented, and hermaphroditic. Cestodes are divided into two types: intestinal and tissue cestodes (Deep et al., 2023). Tapeworms are flatworms with a ribbon-shaped, segmented body. As adults, they reside exclusively in the human small intestine. The larvae can be found in the skin, liver, muscles, central nervous system, or other organs. Their life cycles include a unique process of survival and transmission to specific intermediate hosts, through which they can infect another human host. Each life cycle is distinct for a particular species of tapeworm (Heyneman, 1996). Cestode parasites play a significant role by causing tapeworm infections, which are flatworms with a flat, segmented body. When they reach adulthood, they exclusively reside in the human small intestine, while their larvae can be found in the skin, liver, muscles, central nervous system, or other organs. Their life cycles involve a unique process of survival and transmission to specific intermediate hosts, through which they can infect other human hosts. Each life cycle is unique to a specific tapeworm species. These parasites contribute to production losses, especially in the sheep, beef, and pig meat industries, and certain species can also cause serious diseases in humans as zoonotic parasites (Lightowlers, 1996).

Cestodes, which are commonly referred to as tapeworms, constitute a significant category of zoonotic parasites characterized by their ribbon-like structures. Their intricate life cycles encompass final and intermediate hosts, and can result in serious clinical conditions such as sparganosis, neurocysticercosis, and echinococcosis (Jeon et al., 2024). The adult flatworms, which reach sexual maturity inside the final host, are ribbon-shaped, multi-segmented, hermaphroditic organisms. Each segment contains both male and female reproductive systems. The worm has an anterior holdfast organ called the scolex, followed by a germinative portion known as the "neck," along with segments at progressively later stages of development.

Larvae form cysts in various tissues of the intermediate host, and these larval cysts contain one or multiple scoleces of future adult worms (Johnston & Taubert, 2001).

Tapeworms are a significant type of internal parasite, some of which have the potential to cause severe illnesses in humans and other mammals. Usually, the mature phase of tapeworms lives in the intestines of a definitive host, where sexual maturation takes place. The eggs produced through self- or cross-fertilization contain a zygote, which then progresses to a prelarval stage. This prelarval stage is consumed orally by an intermediate host, where the parasite targets specific compartments or organs and matures into the larval stage. The infective pre-adult parasites subsequently mature within the larval stage. The life cycle comes full circle when the definitive host consumes tissue from the infected intermediate host. In certain instances, two intermediate hosts are necessary to complete the life cycle (Siles-Lucas & Hemphill, 2002).

Tapeworms of the *Moniezia* spp. are parasitic organisms with intricate life cycles, mainly found in the digestive tracts of ruminant animals. By absorbing a significant portion of the host's nutrients, they cause monieziasis, resulting in economic challenges in animal husbandry. Despite their considerable impact on the economy, there has been a lack of comprehensive understanding regarding the molecular characterization and taxonomic classification of these parasites. In a recent study, cestodes were removed from the intestines of sheep and goats and then stained using Gower's carmine. A thorough examination of their morphological characteristics led to the recognition of two distinct species: *M. denticulata* and *M. expansa* (Kumar & Kaur, 2023). *M. expansa* and *M. benedeni* are two prevalent tapeworm species found in domestic ruminants worldwide. However, there is a lack of morphological and molecular data on these species due to limited specimen availability from a few countries (Tam et al., 2020).

M. benedeni is a type of tapeworm that inhabits the small intestine of ruminant animals, including cattle, sheep, and goats. This parasite is found all over the world and is particularly common among grazing animals. To complete its life cycle, the parasite requires an intermediate host, the oribatid mite. There are approximately 170 families, 1200 genera, and 9000 valid species and subspecies of oribatid mites that reside in soil and litter. It is generally believed that the pathogenicity of *M. benedeni* is low, as infected hosts typically do not exhibit specific or severe symptoms. However, infected calves may experience reduced growth efficiency. Additionally, some farmers have reported that calves expelling large proglottids may have diminished value at auction (Irie et al., 2013).

Hydatidosis is a common parasitic zoonosis worldwide caused by the larval stage of *Echinococcus granulosus* sensu lato, a small tapeworm in the Taeniidae family. Herbivores and canids act as the intermediate and definitive hosts, respectively, in the parasite's life cycle. The infection is primarily spread globally by livestock and dogs (Matini et al., 2019). The life cycles of the parasites involve two mammalian hosts. The adult tapeworm resides in the small intestine of a carnivorous animal (definitive host) and produces eggs containing infective oncospheres. The carnivore's intestinal tract releases either tapeworm segments (proglottids) with eggs or free eggs into the environment. When an intermediate host consumes the eggs, a larval stage known as the metacestode forms in its internal organs. Normally, the mature metacestode produces multiple protoscoleces, each capable of developing into an adult tapeworm when ingested by a suitable definitive host. On occasion, humans and other "aberrant" hosts unintentionally ingest the eggs, even though they are not part of the natural cycle (Eckert & Deplazes, 2004).

The genus *Taenia* consists of numerous zoonotic cestodes (tapeworms) that primarily infect cattle (*T. saginata*) and pigs (*T. solium*) among other animals (Fig. 2p) (Pradhan & Karanth, 2023). The larval stage of the pork tapeworm (*T. solium*) can invade the human nervous system and cause neurocysticercosis. This disease is a significant contributor to epileptic seizures in many developing nations and is on the rise in developed countries due to immigration from regions where the illness is prevalent. In places where pigs are bred for consumption, there is widespread occurrence of *T. solium* infection and the resulting neurocysticercosis. *T. solium* has a complex life cycle involving two hosts. Humans act as the definitive host and can carry the adult tapeworm (taeniasis), while both humans and pigs can act as intermediate hosts, harboring the larvae or cysticerci (García et al., 2003).

The life cycle of *T. solium* in nature involves humans as the primary hosts, where the adult tapeworm resides in the intestine, and pigs as the intermediate hosts, housing the larval stage, also known as metacestode. Consumption of raw or undercooked meat containing larval cysticerci leads to the development of adult tapeworms in the small intestine. While infections with adult tapeworms are generally not severe, infection with certain larval stages can lead to life-threatening conditions (Cuervo et al., 2024).

3 Fungal Parasites

3.1 Candidiasis

Candida exhibits several morphological forms, including yeast (blastoconidia), pseudohyphae, and true hyphae, although in some cases, only yeast forms may be observed (Fig. 3a, b). These organisms can be effectively stained using either Grocott's methenamine silver and they demonstrate a strong positive Gram stain. Occasionally, the blastoconidia of *Candida* species can be sufficiently large to resemble those of other fungal infections. *C. glabrata* (torula glabrata) features multiple budding yeasts measuring 2–5 μM, displaying amphophilic characteristics in H&E-stained sections. It can be distinguished from other *Candida* species by its lack of pseudo-hyphae or true hyphae (Kradin & Mark, 2018).

Candida species are classified under the Deuteromycetes, the Fungi Imperfecti and display dimorphism, existing in both yeast and mycelial forms. In liquid environments, the yeast form is predominant, whereas a combination of mycelial forms can be found in tissue. The presence of mycelial elements is often linked to tissue invasion. In horses, instances of *Candida* endometritis have been documented,

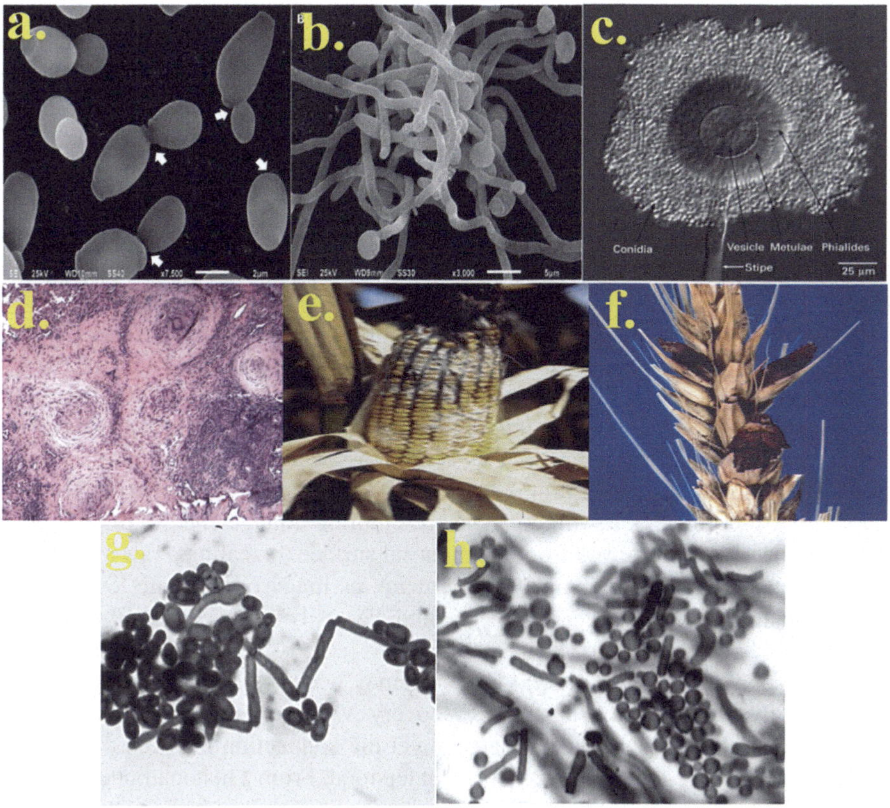

Fig. 3 The images showing several fungal-parasites that cause infections in farm animals: (**a**) and (**b**) Transmission electron microscopic (TEM) images of *Candida albicans*; (**c**) Stain image of *Aspergillus;* (**d**) Stain image of *Aspergillus terreus*; (**e**) Rotten maize by *Fusarium*; (**f**) Mature barley with large black sclerotia of *Claviceps purpurea*; (**g**) TEM image of *Malassezia furfur;* (**h**) TEM image of *Malassezia globosa*

which is thought to lead to early embryonic loss and infertility. Furthermore, alterations in the host's immune response during pregnancy are recognized to elevate the risk of vaginal candidiasis in expectant women (Foley & Schlafer, 1987).

C. albicans is a multifaceted and polymorphic fungus celebrated for its capacity to assume various morphological forms. It can take the shape of ovoid budding yeast, recognizable by its rounded, pear-like appearance. In addition, it may appear as elongated ellipsoid cells featuring constrictions at the septa, referred to as pseudohyphae. Moreover, *C. albicans* can develop into true hyphae, distinguished by parallel-walled filaments that enhance its intricate structure. Each of these forms plays a crucial role in its adaptability and potential for causing disease (Mayer et al., 2013). *C. albicans* is recognized as one of the most significant opportunistic pathogenic fungi. When the immune system is compromised, this yeast-like organism can transition from a typically harmless saprophyte to a highly virulent pathogen. Its remarkable ability to adapt to the varying conditions within host tissues enables it to thrive in environments that would normally be unsuitable. This adaptability poses a serious threat, as it can lead to infections in individuals with weakened immune systems, making *C. albicans* a major concern in the realm of infectious diseases (Ruiz-Herrera et al., 2006). *C. albicans* is widely recognized as one of the most prevalent pathogens associated with various diseases; however, it is not the only species of concern. Other species within the Candida genus also play a significant role in causing infections. Notably, *C. tropicalis* and *C. parapsilosis* are closely related to *C. albicans*, while *C. glabrata*, which is less closely associated, has emerged as an important contributor to candidiasis. Together, these organisms present considerable challenges in the field of infectious diseases (Calderone & Clancy, 2011). Some commensal fungi can cause infections in animals when the host's immune system is compromised. *Candida* species, especially *C. albicans*, may lead to oral and gastrointestinal *candidiasis* in various animals, including chickens, horses, cattle, dogs, cats, and pigs. This is often associated with antibiotic use or a weakened immune response (Simões et al., 2023). Common oral manifestations of *candidiasis* present as soft, white patches resembling cottage cheese. Importantly, these patches can be removed from the mucosal surface, revealing a raw and bleeding area underneath. This distinction is crucial, as many other white lesions that appear plaque-like, such as leukoplakia, cannot be easily removed from the mucosa (Greenwald & Brandt, 2010).

Bovine mastitis due to *Candida* infection may occur as an acute or temporary condition, frequently with spontaneous recovery. Various species of *Candida* have been associated with this form of mastitis, yet the connection between the fungus and the associated lesions remains poorly understood. From a histopathological perspective, lesions associated with *C. mastitis* can be categorized into two types. The first type exhibits marked suppurative inflammation with necrosis, displaying numerous fungal elements within the necrotic areas, along with a significant influx of neutrophils and granulation tissue, which indicates a robust inflammatory response. The second type is characterized by the presence of giant cells in the affected alveoli and intralobular ducts, representing a distinct immune response to the ongoing fungal infection and underscoring the complexity of the inflammatory reaction in this condition (Kitamura et al., 1990). The objective of this research was to isolate yeast *(Candida)* from the milk of cow udders across 37 dairy farms in Brazil and to identify the various species associated with mastitis. Milk samples were collected between October 2002 and February 2003, resulting in a total of 260 samples from cows diagnosed with clinical and subclinical mastitis. These samples were cultured on Blood agar, Mac Conkey agar, and Sabouraud dextrose agar. Among them, 45 samples (17.3%) tested positive for the *Candida* genus. The isolated species included *C. krusei* (44.5%), *C. rugosa* (24.5%), *C. albicans* (8.9%), *C. guilliermondii* (8.9%), and other species (13.2%) (Casia dos Santos & Marin, 2005).

Candidiasis is not commonly seen in animals, but it is important to consider it as a differential diagnosis when evaluating bacterial infections. This fungal infection can also occur as a secondary complication after a bacterial disease. Veterinarians need to consider candidiasis, especially in situations where the host shows little to no improvement with antibiotic treatment, as this may suggest an underlying fungal problem that needs to be addressed. Being aware of the possibility of candidiasis can facilitate more effective and focused treatment approaches (Seyedmousavi et al., 2018).

3.2 Aspergillosis

Aspergillus species are widely distributed around the world, affecting humans, a majority of domesticated animals and birds, as well as various wild species. This distribution leads to a range of diseases, from localized infections to severe systemic illnesses, along with allergic reactions triggered by inhaled conidia (Fig. 3c). Among the common forms of animal aspergillosis are life-threatening infections in sea fan corals, mummification due to stonebrood in honey bees, respiratory and air sac infections in birds, mycotic abortions and mammary gland infections in cattle, and guttural pouch mycosis in horses (Seyedmousavi et al., 2015). Aspergillosis, commonly referred to as brooder pneumonia, is a severe fungal disease that primarily affects young poultry and is often challenging to treat. This infection typically manifests as a respiratory issue, leading to significant respiratory distress in the affected birds. The disease is notorious for its potential to produce high mortality rates within poultry populations, posing a considerable risk to farmers and breeders. Birds that manage to survive frequently face lasting consequences, which often result in stunted growth and reduced size due to the prolonged impact on their cardiovascular health. The implications of this condition extend beyond mere survival, adversely affecting the overall health and productivity of the flock (MSD Veterinary Manual, 2023).

The *Aspergillus* genus includes over 300 characterized species; however, the majority of cases of animal aspergillosis are primarily associated with *Aspergillus fumigatus*. Incidences involving other species within this diverse genus occur infrequently (Heitman, 2011). Aspergillus spp. are primarily saprophytic fungi commonly found in soils and on various organic and inorganic surfaces. Their conidia (asexual spores) are hydrophobic and easily become airborne, enabling them to germinate under a wide range of conditions. These fungi can withstand high temperatures and thrive in environments ranging from 12 °C to over 50 °C, which has contributed to their establishment as widespread opportunistic pathogens in vertebrates (Foley et al., 2014). *Aspergillus fumigatus* is a saprophytic fungus that plays an essential role in the recycling of carbon and nitrogen in the environment. Its natural habitat is the soil, where it flourishes on organic matter. While this species is not the most common fungus worldwide, it ranks among the most widespread in terms of airborne conidia. The identification of *A. fumigatus* primarily depends on the morphology of its conidia and conidiophores. This organism is characterized by its green, echinulate conidia, which measure 2.5–3 μm in diameter and are produced in chains from greenish phialides measuring 6–8 by 2–3 μm (Latgé, 1999). The hyphae of *Aspergillus* species typically range from 2.5 to 4.5 µm in diameter and are characterized by frequent septation. These fungi exhibit a distinctive branching pattern, primarily at acute angles around 45°, which can be likened to the architecture of a branching limb. However, in cross-section, their structure may resemble yeasts, although the absence of budding is a key differentiator that aids in accurate identification. In areas of active mycelial growth, the hyphae can become entangled and exhibit swelling and distortion, complicating the morphological diagnosis without additional confirmatory methods (Kradin & Mark, 2018).

Fungal pneumonia in horses is a rare condition that typically occurs under debilitating circumstances, which allow for the invasion and proliferation of fungal organisms. Contributing factors may include exposure to high concentrations of fungi in the environment, housing horses in damp or dusty conditions, prolonged use of antimicrobials that disrupt the balance of microflora or inhibit vitamin production, and the presence of immunosuppressive conditions that can be either primary or secondary, arising from endocrine disorders, tumors, or long-term use of immunosuppressive medications (Melo et al., 2024). In horses, the fungus *A. fumigatus* is responsible for a serious condition known as equine guttural pouch mycosis, which can lead to severe complications if not treated promptly. In cattle, this pathogen is associated with various health issues, including mycotic pneumonia, an infection affecting the lungs and gastroenteritis, which causes inflammation of the gastrointestinal tract. It can also lead to mastitis, an infection of the udder, placentitis, impacting the placenta, and has been linked to cases of abortion. These conditions underscore the significant impact of *A. fumigatus* on the health of both equine and bovine species (Simões et al., 2023). Invasive pulmonary mycosis caused by *Aspergillus niger* and *Rhizopus stolonifer* has been reported in a 2-year-old horse, one of three that died after being kept in a disused, uncleaned stable. The lesions were characterized by thrombosis of blood vessels accompanied by hemorrhage and tissue necrosis. Fungal hyphae were observed in both the thrombosed vessels and the surrounding necrotic tissue. In culture media inoculated with lung samples and bedding hay samples, two types of colonies were identified as A. niger and R. stolonifera. This study represents the first documented case of equine pulmonary mucormycosis and identifies A. niger as an etiological agent in horses (Knudtson & Kirkbride, 1992).

Mycotic infections were identified in 6.8% of the 6858 cases of bovine abortion and stillbirth analyzed over a nine-year period. Aspergilli were associated with approximately 5% of all abortion cases and accounted for 71% of the 446 cases that underwent fungal culture and were confirmed as mycotic abortion. The most frequently isolated species was *A. fumigatus* (62%), followed by *A. terreus* (6.7%), Emericella (*Aspergillus*) nidulans (3.0%), and *A. flaws* (2.9%) (Knudtson & Kirkbride, 1992). This type of abortion occurs infrequently and may be influenced by the consumption of low-quality, contaminated feed collected during wet seasons. *A. fumigatus* can thrive in moisture-laden hay, inferior silage, and brewers' grains. The infection spreads to the uterus via the bloodstream, leading to placentitis and resulting in abortion during the later stages of pregnancy. Cows affected by this condition typically do not show signs of systemic illness. The intern cotyledonary regions of the placenta appear thickened and leathery, while the cotyledons exhibit signs of necrosis. Aborted fetuses may present with elevated skin plaques that resemble ringworm lesions (Journal of Animal Feed Science and Technology, 2019).

Bovine mycotic abortion is a sporadic illness found worldwide, impacting only a small percentage of animals within a herd. Typically, cows experience abortion during their second or third trimester of pregnancy. In the northern hemisphere, this condition is most common during the winter months when pregnant cows are housed in sheds and cubicles, consuming hay or silage that is often heavily contaminated with *Aspergillus* spores. Aspergillus species are frequently identified, with *A. fumigatus* accounting for the majority of cases. It is believed that infection occurs primarily through ingestion or inhalation of the spores (Tell, 2005). A 31-month-old Japanese Black cow (Bos taurus) experienced an abortion at 5 months of gestation, exhibiting no clinical signs prior to the event. Histopathological analysis of both the placenta and fetus revealed significant necrotizing placentitis (necrotizing lesions due to *Aspergillus*) characterized by the presence of numerous irregular degenerative fungi and inflammatory cells. Typical filamentous fungi were also identified, while no inflammatory response was observed in the fetal digestive and respiratory systems. Both types of fungi displayed aleuria conidia and septa within the placenta and fetal organs, and were positively stained with antibodies specific to Aspergillus spp. *A. terreus* was successfully isolated from the fetal lung and abomasal contents, as confirmed through mycological and molecular techniques (Fig. 3d). This marks the first documented instance of immunohistochemical, morphological, and molecular identification of *A. terreus* in the bovine placenta and aborted fetuses (Sato et al., 2022).

3.3 Mycotoxicosis

Mycotoxicosis are toxic condition caused by exposure to mycotoxins. Several mycotoxicoses that affect animals, including aflatoxins, zearalenone, and ergot alkaloids, can result in subclinical issues or decreased productivity (Table 1). Furthermore, these toxins may induce carcinogenic, mutagenic, teratogenic, immunosuppressive, and endocrine effects, which can be difficult to detect (Riet-Correa et al., 2024). Mycotoxins are a diverse group of harmful compounds produced by various fungi that can contaminate products such as wheat, corn, sorghum, rice, cocoa beans, figs, and certain spices, including ginger and nutmeg, as well as peanuts and tree nuts. Individuals can be exposed to mycotoxins by consuming foods derived from these contaminated grains and cereals (Zeece, 2020). Mycotoxins are byproducts produced by fungi in cereals and are frequently found as contaminants in animal feed within the livestock industry. Global research on feed mycotoxins has shown that various mycotoxins such as aflatoxin, ochratoxin *A. fumonisin*, zearalenone, and deoxynivalenol are present in a significant number of feed samples. Exposure to these mycotoxins can lead to immunotoxicity and negatively

Table 1 List of fungal species that are producing mycotoxin and seriously affected crops that are indirectly involved in the transmission of fungal-parasite to farm animals

Fungal species	Mycotoxin classification	Commodity
Aspergillus flavus, Aspergillus parasiticus	Aflatoxins	Corn/maize, peanuts, tree nuts, cottonseed, rice
Fusarium sp. *(F. graminearum, F. Poae, F. acuminatum, F. sporotrichiodes, F. equiseti)*	Trichothecenes	Wheat, maize, hay, straw, overwintered grains
Claviceps sp. *C. purpurea, C. fusiformis, C. paspali*	Ergot alkaloids	Rye, wheat, cereals, grasses
Aspergillus alutaceus var. alutaceus (formerly A.ochraceus), A. westerdijkiae and A. Steynii	Ochratoxins	Cereal grains

Reference: Mostrom (2016)

impact reproductive functions in farm animals. Additionally, the exposure of vital organs, including the kidneys, liver, and intestines, to mycotoxins can result in histopathological changes that may disrupt animal growth and survival (Yang, 2020). Mycotoxins present in contaminated feed or cereals may also be inhaled, posing risks to domestic animals. Although the harmful effects of mycotoxins through consumption are well documented, there is a scarcity of research focusing on their negative impact on the respiratory system. A recent study emphasized the harm that aflatoxins can inflict on respiratory epithelial cells in occupational environments (Prathap-Kumar, 2023). The five primary mycotoxins that negatively affect growth and reproductive efficiency in North America include aflatoxins, zearalenone, deoxynivalenol, ochratoxin, and ergot (Table 1). *Aflatoxins* are generated by the fungi *A. flavus* and *A. parasiticus* (Diekman & Green, 1992).

3.3.1 Aflatoxins

Aflatoxins are a class of carcinogenic furanocoumarins primarily produced by the fungi *A. flavus* and *A. parasiticus.* These saprophytic fungi are found worldwide. Aflatoxin production is likely in environments with a relative humidity of 85%, especially when corn has a moisture content between 15% and 28% and temperatures exceed 25 °C for more than 48 h. Under these or similar unfavorable conditions, aflatoxin levels can rise significantly; therefore, effective grain drying and proper storage are essential to mitigate risks to livestock and prevent related economic losses (Haschek & Voss, 2013). Aflatoxin is a highly toxic mycotoxin that poses significant health risks due to its carcinogenic and harmful properties affecting both humans and animals. The adverse effects of aflatoxin exposure are collectively referred to as aflatoxicosis. There are two primary forms of these conditions: acute aflatoxicosis, which can lead to rapid death, and chronic aflatoxicosis, associated with long-term health issues such as cancer, immune system suppression, and other chronic diseases. The liver is particularly vulnerable to the detrimental effects of aflatoxin, making it the organ most severely impacted. This toxicity has been observed in a variety of species, including birds, fish, rodents, and nonhuman primates, highlighting the widespread threat that aflatoxin poses in contaminated environments (Bennett & Klich, 2003).

3.3.2 Trichothecenes

The genus *Fusarium* (Ascomycetes, Hypocreales) encompasses numerous species that pose pathogenic threats to plants, resulting in diseases in several essential crops, especially cereals. A considerable number of these species are known to produce a wide variety of mycotoxins with significant chemical diversity (Fig. 3e). Among the mycotoxins produced by *Fusarium* species, only a select few are recognized as significant for human and animal health, with trichothecenes being one of the most notable (Moretti & Sarrocco, 2016). Trichothecenes (TCNs) are significant mycotoxins that cause substantial economic losses in the food supply each year. These toxins are naturally produced by specific filamentous fungi, including *Aspergillus, Penicillium, Fusarium, Spicellum*, and *Stachybotrys* with Fusarium being the primary source of toxins found in cereals, vegetables, and animal feed. Consumption of food contaminated with TCNs can negatively affect the gastrointestinal tract, kidneys, liver, and immune system, leading to health issues such as loss of appetite, nausea, abdominal pain, and cardiovascular problems (Liu et al., 2022).

3.3.3 Ergotism

Ergotism is caused by the fungus Claviceps, which is commonly found in cereal grains, especially rye, and is known to be one of the most ancient mycotoxins (Fig. 3f). Molds that develop in the feed transform the grain into a thick black material called ergot. During the Middle Ages, this ailment resulted in the deaths of many and was often referred to as St. Anthony's fire. The alkaloids generated by these toxins influence the nervous and vascular systems, leading to vasoconstriction and potential gangrene, as well as impacting the endocrine system. Signs of ergotism include reduced food consumption and growth, tissue death in the beak, comb, and toes, abnormal feather formation, and inflammation of the intestines. This condition mainly impacts leghorns and other breeds that are known for egg-laying, while its occurrence in meat is limited (Wakenell, 2016). Currently, modern grain cleaning techniques have successfully eliminated ergotism as a human disease; however, it remains a significant and detrimental issue in veterinary medicine, particularly affecting sheep, cattle, pigs, and chickens. In animals, the clinical signs of ergotism include gangrenous extremities, abortions, convulsions, agalactia, and ataxia (Craig et al., 2015).

3.3.4 Ochratoxins

Aflatoxins can lead to bleeding and hepatorenal syndrome, and they may also compromise immune system function; notably, aflatoxin B1 (AFB1) is recognized as a carcinogen. Meanwhile, ochratoxin A (OTA) possesses properties that are toxic to the kidneys, carcinogenic, and teratogenic, as well as effects that can suppress the immune system. Most poultry feeds are vulnerable to fungal contamination at various stages of the manufacturing process, including production, processing, transportation, and storage. Therefore, effective quality control of poultry feed necessitates the identification of sources of aflatoxins and ochratoxins (Fraga et al., 2007). Ochratoxin A (OTA) is recognized as one of the most significant mycotoxins due to its harmful effects on health. This particular mycotoxin is known to adversely affect the kidneys and may also play a role in the development of kidney cancer, rendering it a serious concern. OTA is produced by specific fungal species, primarily *Penicillium verrucosum*, which thrives in temperate and cooler climates. In warmer and tropical regions, various species of the Aspergillus genus, including *A. ochraceus*, *A. melleus*, *A. carbonarius*, and *A. niger*, are responsible for its production. The presence of these fungi in diverse environments highlights the widespread risk of ochratoxin A contamination in food products (Rosa et al., 2006).

3.4 *Malasseziosis*

Malassezia are small yeasts characterized by thick walls and may take on ovoid, ellipsoid, or cylindrical shapes. They are commonly found as commensals in warm-blooded vertebrates. The mycelial form of *Malassezia* species has been observed naturally in certain skin lesions and can be induced in specific culture media at an incubation temperature of 30 °C. These yeasts reproduce asexually through unipolar broad-based budding. Although the sexual form has yet to be documented, the mating-type locus region has been identified (Hobi et al., 2022). *Malassezia* spp., formerly referred to as *Pityrosporum* spp., are yeasts that thrive on lipids and typically exist in small quantities on the skin of both humans and animals (Fig. 3g, h). There are at least 13 recognized species of *Malassezia*. Most of these species are classified as lipid-dependent; when cultured in a laboratory environment, they require long-chain fatty acids for growth, which they utilize as their carbon source (Sykes et al., 2014). *Malassezia furfur* is a species of fungi that belongs to a single lineage typically found on the skin of humans and animals. These yeasts, which rely on lipids for their growth, generally constitute over 80% of the total fungal community on human skin. They can be found in both healthy individuals and those with various skin conditions (National Center for Biotechnology Information, n.d.).

3.5 Ringworm

Dermatophytosis, commonly known as ringworm, is prevalent among dairy calves and can also affect adult cows (Fig. 4). The main pathogen is *Trichophyton verrucosum* and *T. mentagrophytes* occur less frequently. Calves aged 2 months to a year are most susceptible, especially when managed in groups. These fungi can survive for months on surfaces, bedding, and soil even after cattle have been removed, leading to increased cases, particularly in colder months. Dermatophytes affect keratinized skin and hair, causing lesions that are usually oval or circular and can last from 1 to 3 months. Transmission occurs through contact, especially with contaminated items that irritate the skin (Scott, 2018). *T. verrucosum* is the main dermatophyte species associated with ringworm in cattle, though it can also affect other ruminants. Dermatophytosis is often characterized as an infection that resolves on its own; however, it has the potential to spread between pets, livestock, and humans, posing a zoonotic risk. In humans, infections are typically quite inflammatory and arise from direct contact with infected cattle or contaminated objects, affecting areas such as the scalp, beard, or other exposed skin (Spanamberg et al., 2023). Raised, crusted lesions of ringworm involving the facial and periocular region of a calf.

Multifocal ringworm lesions that appear as dry, crusted areas of alopecia of the head, neck, and shoulders in a calf. Dermatophytes are highly contagious, spreading primarily through direct contact between horses or through the use of contaminated equipment. Infections can easily proliferate, particularly via infected saddle girths, where the fungus can survive for up to 12 months. Additionally, insects have been noted as contributors to the transmission of this illness. The most common cause of dermatophytosis in horses is *T. equinum*, which has two identified varieties: *T. equinum and T. autotrophicum* (Marsella, 2014). *T. equinum* is a fungal species primarily affecting animals, with infrequent infections in humans. It is known to cause ringworm in horses and donkeys, and can occasionally impact dogs. Young horses and yearlings are especially susceptible, initially presenting with swellings that can be felt through their fur. These swellings may progress into small, inflamed sores with discharge, a condition commonly referred to as girth itch. As the lesions expand and become chronic, hair loss may occur, and crusts can form and shed from the

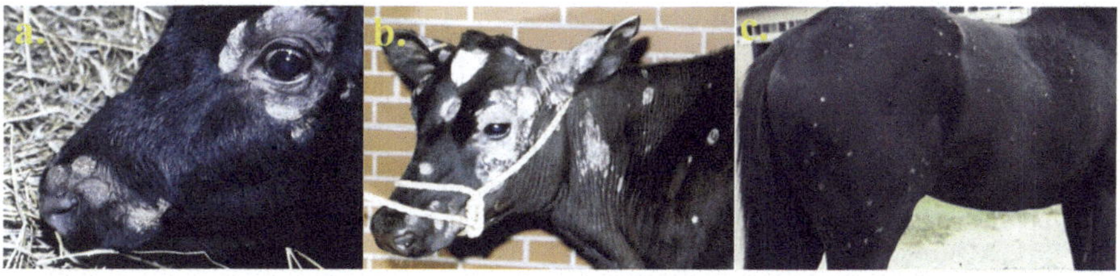

Fig. 4 The pictures of animals showing several fungal-parasites which cause infections in farm animals: (**a**) Raised, crusted lesions of ringworm involving the facial and periocular region of a calf; (**b**) Multifocal ringworm lesions that appear as dry, crusted areas of alopecia of the head, neck, and shoulders in a calf; and (**c**) Ringworm due to *Trichophyton equinum* in a horse

healed areas, leading to bald patches that resemble a moth-eaten appearance (Carter, 1990).

A live vaccine utilizing an attenuated strain of *T. verrucosum* was developed to address ringworm in cattle and was assessed through both controlled challenge-exposure tests and field studies. After analyzing data from approximately 400,000 vaccinated cattle over a five-year period, it was concluded that immunoprophylaxis can effectively manage trichophytosis in cattle. Adverse reactions associated with the vaccination are infrequent (Gudding & Naess, 1986).

4 Conclusion

These ectoparasites, living on or burrowing into the host's epidermis, can cause direct damage to the skin and other tissues, leading to substantial economic losses in the livestock industry. Key ectoparasites include lice, which display high host specificity, and fleas, which are known for their short infestation periods. Mange mites, particularly the northern fowl mite and the red poultry mite, pose significant threats to poultry health. Ticks, classified into families such as Argasidae and Ixodidae, are notable for their ability to cause significant production losses and health issues in cattle. Diptera, or flies, also play a crucial role in transmitting various pathogens to both humans and animals. In addition to this, the fungi-based parasitic is found as one of the dangerous parasitic attacks. The parasitic fungi of *Taenia, Candida, Aspergillus, Malassezia,* etc. cause skin infections and act as both ecto- and endoparasite to livestock animals.

References

Ács, Z., Hayward, A., & Sugár, L. (2016). Genetic diversity and population genetics of large lungworms (*Dictyocaulus*, Nematoda) in wild deer in Hungary. *Parasitology Research, 115*, 3295–3312. https://doi.org/10.1007/s00436-016-5088-0

Ajobiewe, F. H., & Ajobiewe, O. J. (2016). Differential mites infestation of domesticated animals and handlers dermatitis in Ijumu Nigeria. *Research in Zoology, 6*(1), 11–16. https://doi.org/10.5923/j.zoology.20160601.03

Ali, A., Hameed, K., Mohsin, M., Khan, W., Rafiq, N., Iqbal, M. A., Kabir, M., Hassan, H. U., Usman, T., & Kamal, M. (2021). Prevalence and risk factors assessment of mange mites in livestock of Malakand Division, Pakistan. *Saudi Journal of Biological Sciences, 28*(11), 6480–6487. https://doi.org/10.1016/j.sjbs.2021.07.001

Alsaad, R. (2023). Control study of *Musca domestica* (Diptera, Muscidae) in Misan Province. *F1000Research, 12*, 459. https://doi.org/10.12688/f1000research.132636.2

Arce, S. I., Manzoli, D. E., Saravia-Pietropaolo, M. J., Quiroga, M. A., Antoniazzi, L. R., Lareschi, M., & Beldomenico, P. M. (2018). The tropical fowl mite, Ornithonyssus bursa (Acari: Macronyssidae): Environmental and host factors associated with its occurrence in Argentine passerine communities. *Parasitology Research, 117*(11), 3521–3531. https://doi.org/10.1007/s00436-018-6025-1

Baldacchino, F., Muenworn, V., Desquesnes, M., Desoli, F., Charoenviriyaphap, T., & Duvallet, G. (2013). Transmission of pathogens by Stomoxys flies (Diptera, Muscidae): A review.

Parasite, 20, 26. https://doi.org/10.1051/parasite/2013026. PMCID: PMC3756335, PMID: 23985165.

Banjo, A. D., Lawal, O. A., & Adeduji, O. O. (2005). Bacteria and fungi isolated from housefly (Musca domestica L.) larvae. *African Journal of Biotechnology, 4*(8), 780–784. Available online at http://www.academicjournals.org/AJB

Bellezza, C. A., et al. (2002). Chapter 8 - Woodchucks as laboratory animals. In J. G. Fox, L. C. Anderson, F. M. Loew, & F. W. Quimby (Eds.), *Laboratory animal medicine* (2nd ed., pp. 309–328). Academic. https://doi.org/10.1016/B978-012263951-7/50011-9

Benelli, G., Caselli, A., Di Giuseppe, G., & Canale, A. (2018). Control of biting lice, Mallophaga - a review. *Acta Tropica, 177*, 211–219. https://doi.org/10.1016/j.actatropica.2017.05.031. Epub 2017 Jun 3.

Bennett, J. W., & Klich, M. (2003). Mycotoxins. *Clinical Microbiology Reviews, 16*(3), 497–516. https://doi.org/10.1128/CMR.16.3.497-516.2003

Bitam, I., Dittmar, K., Parola, P., Whiting, M. F., & Raoult, D. (2010). Fleas and flea-borne diseases. *International Journal of Infectious Diseases, 14*(8), e667–e676. https://doi.org/10.1016/j.ijid.2009.11.011

Boray, J. C., et al. (2017). *Liver fluke disease in sheep and cattle* (Primefact 446, 4th ed.). NSW Department of Primary Industries. Retrieved June 21, 2024, from https://www.dpi.nsw.gov.au/__data/assets/pdf_file/0004/114691/liver-fluke-disease-in-sheep-and-cattle.pdf

Boxler, D. J. (2015, February). *Face fly. Index: Insects and Pests Livestock. University of Nebraska–Lincoln*. Retrieved from https://entomology.unl.edu/livestock/facefly.pdf

Brasil, G. M., Riet-Correa, F., Mota, R. A., Silva, N. S., Soares, M. P., de Miranda Neto, E. G., de Almeida, M. O., & de Oliveira, D. M. (2023). Equine piroplasmosis in northeastern Brazil: Occurrence and risk factors. *Brazilian Journal of Veterinary Medicine, 45*(1), e03523. https://doi.org/10.29374/2527-2179.bjvm003523

Brites-Neto, J., Roncato Duarte, K. M., & Fernandes Martins, T. (2015). Tick-borne infections in human and animal population worldwide. *Veterinary World, 8*(3), 301–315. https://doi.org/10.14202/vetworld.2015.301-315. PMCID: PMC4774835, PMID: 27047089.

Burgess, I. F. (2022). Sucking lice and spiracular transpiration: Turning a liability into a benefit and a necessity. *The Journal of Experimental Biology, 225*(7), jeb244041. https://doi.org/10.1242/jeb.244041. Epub 2022 Apr 19.

Burrow, H. M., et al. (2019). Towards a new phenotype for tick resistance in beef and dairy cattle: A review. *Animal Production Science, 59*(8), 1401–1427. https://doi.org/10.1071/AN18487

Calderone, R. A., & Clancy, C. J. (Eds.). (2011). *Candida and candidiasis*. American Society for Microbiology Press. https://doi.org/10.1128/9781555817176

Capinera, J. L. (2008). *House fly, Musca domestica Linnaeus (Insecta: Diptera: Muscidae)*. Retrieved from https://entnemdept.ufl.edu/creatures/urban/flies/house_fly.htm

Carter, G. R. (1990). Dermatophytes and dermatophytoses. In G. R. Carter & J. R. Cole (Eds.), *Diagnostic procedure in veterinary bacteriology and mycology* (5th ed., pp. 381–404). Academic Press. https://doi.org/10.1016/B978-0-12-161775-2.50033-5

Casia dos Santos, R. d., & Marin, J. M. (2005). Isolation of Candida spp. from mastitic bovine milk in Brazil. *Mycopathologia 159*, 251–253. https://doi.org/10.1007/s11046-004-2229-2

Chai, J.-Y. (2013). Chapter 23 - Paragonimiasis. In H. H. Garcia, H. B. Tanowitz, & O. H. Del Brutto (Eds.), *Handbook of clinical neurology* (Vol. 114, pp. 283–296). Elsevier. https://doi.org/10.1016/B978-0-444-53490-3.00023-6

Chaoudhary, V., Hasnani, J. J., Khyalia, M. K., Pandey, S., Chauhan, V. D., Pandya, S. S., & Patel, P. V. (2015). Morphological and histological identification of Paramphistomum cervi (Trematoda: Paramiphistoma) in the rumen of infected sheep. *Veterinary World, 8*(1), 125–129. https://doi.org/10.14202/vetworld.2015.125-129

Charlier, J., Höglund, J., Morgan, E. R., Geldhof, P., Vercruysse, J., & Claerebout, E. (2020). Biology and epidemiology of gastrointestinal nematodes in cattle. *The Veterinary Clinics of North America. Food Animal Practice, 36*(1), 1–15. https://doi.org/10.1016/j.cvfa.2019.11.001

Chougar, L., Harhoura, K., & Aissi, M. (2019). First isolation of *Dicrocoelium dendriticum*among cattle in some Northern Algerian slaughterhouses. *Veterinary World, 12*(7), 1039–1045. https://doi.org/10.14202/vetworld.2019.1039-1045

Christensen, C. M. (1982). External parasites of dairy cattle. *Journal of Dairy Science, 65*(11), 2189–2193. https://doi.org/10.3168/jds.S0022-0302(82)82481-8

Colebrook, E., & Wall, R. (2004). Ectoparasites of livestock in Europe and the Mediterranean region. *Veterinary Parasitology, 120*(4), 251–274. https://doi.org/10.1016/j.vetpar.2004.01.012

Cook, D. (2020). A historical review of management options used against the stable fly (Diptera: Muscidae). *Insects, 11*(5), 313. https://doi.org/10.3390/insects11050313

Cortinas, R., & Jones, C. J. (2006). Ectoparasites of cattle and small ruminants. *The Veterinary Clinics of North America. Food Animal Practice, 22*(3), 673–693. https://doi.org/10.1016/j.cvfa.2006.06.003

Craig, T. M. (2003). Treatment of external and internal parasites of cattle. *The Veterinary Clinics of North America. Food Animal Practice, 19*(3), 661–678vi–vii. https://doi.org/10.1016/s0749-0720(03)00053-7

Craig, T. M. (2018). Gastrointestinal nematodes, diagnosis and control. *The Veterinary Clinics of North America. Food Animal Practice, 34*(1), 185–199. https://doi.org/10.1016/j.cvfa.2017.10.008

Craig, A. M., Klotz, J. L., & Duringer, J. M. (2015). Cases of ergotism in livestock and associated ergot alkaloid concentrations in feed. *Frontiers in Chemistry, 3*, 8. https://doi.org/10.3389/fchem.2015.00008

Cuervo, M. P., et al. (2024). Overview of biological hazards and foodborne diseases. In G. W. Smithers (Ed.), *Encyclopedia of food safety* (2nd ed., pp. 1–18). Academic. https://doi.org/10.1016/B978-0-12-822521-9.00113-1

de Waal, T., & Mehmood, K. (2021). Editorial: Trematode infection in ruminants. *Frontiers in Veterinary Science, 8*. https://doi.org/10.3389/fvets.2021.719577

Deep, A., et al. (2023). Chapter 14: Vermifuge and vermicide drugs. In P. C. Acharya & M. Kurosu (Eds.), *Medicinal chemistry of chemotherapeutic agents* (pp. 463–497). Academic. https://doi.org/10.1016/B978-0-323-90575-6.00011-9

Dharanesha, N. K., Saminathan, M., Mamta, P., Ramesh, K. R., Ananda, K. J., Giridhar, P., & Byregowda, S. M. (2019). Parasitic pneumonia caused by *Paragonimus* spp. in a wild Royal Bengal Tiger, Mysuru, South India. *Journal of Parasitic Diseases: Official Organ of the Indian Society for Parasitology, 43*(3), 528–533. https://doi.org/10.1007/s12639-019-01112-8

Diekman, M. A., & Green, M. L. (1992). Mycotoxins and reproduction in domestic livestock. *Journal of Animal Science, 70*(5), 1615–1627. https://doi.org/10.2527/1992.7051615x

Dobler, G., & Pfeffer, M. (2011). Fleas as parasites of the family Canidae. *Parasites & Vectors, 4*, 1–12.

Eckert, J., & Deplazes, P. (2004). Biological, epidemiological, and clinical aspects of echinococcosis, a zoonosis of increasing concern. *Clinical Microbiology Reviews, 17*. https://doi.org/10.1128/cmr.17.1.107-135.2004

Egri, B. (2019). *Louse infestation of ruminants*. IntechOpen. https://doi.org/10.5772/intechopen.79257

Filip-Hutsch, K., Pyziel-Serafin, A. M., Hutsch, T., Bulak, K., Czopowicz, M., Merta, D., Kobielski, J., & Demiaszkiewicz, A. W. (2022). The occurrence of *Fascioloides Magna* (Bassi, 1875) in the wild cervid population in the lower Silesian wilderness - epidemiological and pathological aspects. *Journal of Veterinary Research, 66*(3), 381–387. https://doi.org/10.2478/jvetres-2022-0042

Foil, L. D., & Hogsette, J. A. (1994). Biology and control of tabanids, stable flies and horn flies. *Revue scientifique et technique (International Office of Epizootics), 13*(4), 1125–1158. https://doi.org/10.20506/rst.13.4.821

Foley, G. L., & Schlafer, D. H. (1987). Candida abortion in cattle. *Veterinary Pathology, 24*(6), 532–536. https://doi.org/10.1177/030098588702400610

Foley, K., Fazio, G., Jensen, A. B., & Hughes, W. O. H. (2014). The distribution of Aspergillus spp. opportunistic parasites in hives and their pathogenicity to honey bees. *Veterinary Microbiology, 169*(3–4), 203–210. https://doi.org/10.1016/j.vetmic.2013.11.029

Förster, M., Klimpel, S., & Sievert, K. (2009). The house fly (Musca domestica) as a potential vector of metazoan parasites caught in a pig-pen in Germany. *Veterinary Parasitology, 160*(1-2), 163–167. https://doi.org/10.1016/j.vetpar.2008.10.087. Epub 2008 Oct 28.

Foster, N., & Elsheikha, H. M. (2012). The immune response to parasitic helminths of veterinary importance and its potential manipulation for future vaccine control strategies. *Parasitology Research, 110*, 1587–1599. https://doi.org/10.1007/s00436-012-2832-y

Fraga, M. E., Curvello, F., Gatti, M. J., et al. (2007). Potential aflatoxin and ochratoxin A production by Aspergillusspecies in poultry feed processing. *Veterinary Research Communications, 31*(4),343–353. https://doi.org/10.1007/s11259-006-3434-x

Fryxell, R., Moon, R., Boxler, D., & Watson, D. (2021). Face fly (Diptera: Muscidae)—Biology, pest status, current management prospects, and research needs. *Journal of Integrated Pest Management, 12*. https://doi.org/10.1093/jipm/pmaa020

Gabaj, M. M., Beesley, W. N., & Awan, M. A. Q. (1992). A survey of mites on farm animals in Libya. *Annals of Tropical Medicine & Parasitology, 86*(5), 537–542. https://doi.org/10.1080/00034983.1992.11812704

García, H. H., Gonzalez, A. E., Evans, C. A., Gilman, R. H., & Cysticercosis Working Group in Peru. (2003). Taenia solium cysticercosis. *Lancet (London, England), 362*(9383), 547–556. https://doi.org/10.1016/S0140-6736(03)14117-7

Garedaghi, Y. (2011). Flea infestation in farm animals and its zoonotic importance in East-Azerbaijan Province. *American Journal of Animal and Veterinary Sciences, 6*(4), 193–196. https://doi.org/10.3844/ajavsp.2011.193.196

Gassó, D., Rossi, L., Mentaberre, G., et al. (2014). An identification key for the five most common species of *Metastrongylus*. *Parasitology Research, 113*, 3495–3500. https://doi.org/10.1007/s00436-014-4001-y

Geden, C. J., Nayduch, D., Scott, J. G., Burgess, E. R., Gerry, A. C., Kaufman, P. E., Thomson, J., Pickens, V., & Machtinger, E. T. (2021). House fly (Diptera: Muscidae): Biology, pest status, current management prospects, and research needs. *Journal of Integrated Pest Management, 12*(1), 39. https://doi.org/10.1093/jipm/pmaa021

Gharbi, M., Ben Abdallah, H., Mbarek, Y., Jedidi, M., & Darghouth, M. A. (2013). Cross-sectional study of cattle lice infestation in the region of Nabeul in north-east Tunisia. *Revue Scientifique et Technique, 32*(3), 879–883. https://doi.org/10.20506/rst.32.2.2208

Greenwald, D. A., & Brandt, L. J. (2010). The Upper Gastrointestinal Tract. In H. M. Fillit, K. Rockwood, & K. Woodhouse (Eds.), *Brocklehurst's textbook of geriatric medicine and gerontology* (7th ed., pp. 608–625). W.B. Saunders. https://doi.org/10.1016/B978-1-4160-6231-8.10076-5

Gudding, R., & Naess, B. (1986). Vaccination of cattle against ringworm caused by Trichophyton verrucosum. *American Journal of Veterinary Research, 47*(11), 2415–2417

Haschek, W. M., & Voss, K. A. (2013). Mycotoxins. In Haschek and Rousseaux's handbook of toxicologic pathology (pp. 1187–1258). https://doi.org/10.1016/b978-0-12-415759-0.00039-x

Hayward, A. D., Skuce, P. J., & McNeilly, T. N. (2021). The influence of liver fluke infection on production in sheep and cattle: A meta-analysis. *International Journal for Parasitology, 51*(11), 913–924. https://doi.org/10.1016/j.ijpara.2021.02.006

Heitman J. (2011). Microbial pathogens in the fungal kingdom. *Fungal Biology Reviews, 25*(1), 48–60. https://doi.org/10.1016/j.fbr.2011.01.003

Hendrichs, J., Cooley, S. S., & Prokopy, R. J. (1992). Post-feeding bubbling behavior in fluid-feeding Diptera: Concentration of crop contents by oral evaporation of excess water. *Physiological Entomology, 17*(2), 153–161.

Heyneman, D. (1996). Cestodes. In S. Baron (Ed.), *Medical microbiology* (4th ed.). University of Texas Medical Branch at Galveston.

Hinkle, N. C., & Corrigan, R. M. (2020). External parasites and poultry pests. In D. E. Swayne, M. Boulianne, C. M. Logue, L. R. McDougald, V. Nair, D. L. Suarez, S. Wit, T. Grimes, D. Johnson, M. Kromm, T. Y. Prajitno, I. Rubinoff, & G. Zavala (Eds.), *Diseases of poultry*. https://doi.org/10.1002/9781119371199.ch26

Hobi, S., Cafarchia, C., Romano, V., & Barrs, V. R. (2022). Malassezia: Zoonotic implications, parallels and differences in colonization and disease in humans and animals. *Journal of Fungi (Basel, Switzerland), 8*(7), 708. https://doi.org/10.3390/jof8070708

Hopla, C. E., Durden, L. A., & Keirans, J. E. (1994). Ectoparasites and classification. *Revue Scientifique et Technique-Office international des epizooties, 13*(4), 985–1034. https://doi.org/10.20506/rst.13.4.815

Income, N., Tongshoob, J., Taksinoros, S., Adisakwattana, P., Rotejanaprasert, C., Maneekan, P., & Kosoltanapiwat, N. (2021). Helminth infections in cattle and goats in Kanchanaburi, Thailand, with focus on strongyle nematode infections. *Veterinary Sciences, 8*(12), 324. https://doi.org/10.3390/vetsci8120324

Irie, T., Sakaguchi, K., Ota-Tomita, A., Tanida, M., Hidaka, K., Kirino, Y., Nonaka, N., & Horii, Y. (2013). Continuous Moniezia benedeni infection in confined cattle possibly maintained by an intermediate host on the farm. *The Journal of Veterinary Medical Science, 75*(12), 1585–1589. https://doi.org/10.1292/jvms.13-0250

Japa, O., Siriwechviriya, P., & Prakhammin, K. (2020). Occurrence of fluke infection in beef cattle around Phayao Lake, Phayao, Thailand. *Veterinary World, 13*(2), 334–337. https://doi.org/10.14202/vetworld.2020.334-337

Jeon, H.-K., et al. (2024). Chapter 149: Cestodes and cestodiasis. In Y.-W. Tang, M. Y. Hindiyeh, D. Liu, A. Sails, P. Spearman, & J.-R. Zhang (Eds.), *Molecular medical microbiology* (3rd ed., pp. 2941–2963). Academic. https://doi.org/10.1016/B978-0-12-818619-0.00044-7

Johnston, M. J., & Taubert, A. (2001). Parasitic helminths. In S. Baron (Ed.), *Medical microbiology* (4th ed.). University of Texas Medical Branch at Galveston. Retrieved from https://www.ncbi.nlm.nih.gov/books/NBK8399/

Journal of Animal Feed Science and Technology Volume 7 Number 2, July – December 2019. https://doi.org/10.21088/jafst.2321.1628.7219.5

Khedri, J., Radfar, M. H., Borji, H., & Mirzaei, M. (2015). Prevalence and intensity of Paramphistomum Spp. in cattle from south-eastern Iran. *Iranian Journal of Parasitology, 10*(2), 268–272.

Kim, D. C. (1984). Paragonimus westermani: Life cycle, intermediate hosts, transmission to man and geographical distribution in Korea. *Arzneimittel-Forschung, 34*(9B), 1180–1183.

Kitamura, H., Anri, A., Fuse, K., Seo, M., & Itakura, C. (1990). Chronic mastitis caused by candida maltosa in a cow. *Veterinary Pathology, 27*(6), 465–466. https://doi.org/10.1177/030098589902700616

Kırcalı Sevimli, F., Kozan, E., & Doğan, N. (2011). Efficacy of eprinomectin pour-on treatment in sheep naturally infected with Dictyocaulus filaria and Cystocaulus ocreatus. *Journal of Helminthology, 85*(4), 472–475. https://doi.org/10.1017/S0022149X10000854

Klompen, J. S. H., Black, W. C., Keirans, J. E., & Oliver, J. H., Jr. (1996). Evolution of ticks. *Annual Review of Entomology, 41*(1), 141–161. https://doi.org/10.1146/annurev.en.41.010196.001041

Knudtson, W. U., Kirkbride, C. A. (1992). Fungi associated with bovine abortion in the northern plains states (USA). *Journal of Veterinary Diagnostic Investigation, 4*(2):181–185. https://doi.org/10.1177/104063879200400211

Kradin, R. L., & Mark, E. J. (2018). Pathology of pulmonary infection. In R. L. Kradin (Ed.), *Diagnostic pathology of infectious disease* (2nd ed., pp. 143–206). Elsevier. https://doi.org/10.1016/B978-0-323-44585-6.00008-4

Kumar, S., & Kaur, H. (2023). Molecular characterization of Moniezia denticulata (Rudolphi, 1810) and its distinction from M. expansa infecting sheep and goats raised in the north and north-western regions of India. *Parasitology, 150*(9), 831–841. https://doi.org/10.1017/S003118202300063X

Kumar, N., Rao, T. K. S., Varghese, A., et al. (2013). Internal parasite management in grazing livestock. *Journal of Parasitic Diseases, 37*, 151–157. https://doi.org/10.1007/s12639-012-0215-z

Kwak, M. (2016). Helminths as weapons of bioterrorism: An unrecognised threat. *Journal of Bioterrorism & Biodefense, 7*. https://doi.org/10.4172/2157-2526.1000148

Kwochka, K. W. (1987). Treatment of canine and feline demodicosis. *Veterinary Clinics of North America: Small Animal Practice, 17*(6), 1187–1204.

Lalor, R., Cwiklinski, K., Calvani, N. E. D., Dorey, A., Hamon, S., Corrales, J. L., & De Marco Verissimo, C. (2021). Pathogenicity and virulence of the liver flukes *Fasciola hepatica* and *Fasciola Gigantica* that cause the zoonosis Fasciolosis. *Virulence, 12*(1), 2839–2867. https://doi.org/10.1080/21505594.2021.1996520

Lareschi, M., Cicuttin, G. L., De Salvo, M. N., Ibañez, L., & Montalti, D. (2017). The tropical fowl mite Ornithonyssus bursa (Acari: Mesostigmata: Macronyssidae) parasitizing the European starling Sturnus vulgaris (Aves: Passeriformes: Sturnidae), an invasive bird in central

Argentina. An approach to the bacterial fauna of this mite. *Revista de Microbiologia, 49*(3), 322–329. https://doi.org/10.1016/j.rmb.2017.03.022

Latgé, J. P. (1999). Aspergillus fumigatus and aspergillosis. *Clinical Microbiology Reviews, 12*(2), 310–350. https://doi.org/10.1128/CMR.12.2.310

Lat-Lat, H., Sani, R., Hassan, L., Sheikh-Omar, A., Jeyabalan, S., Hishammfariz, M., Rohani, K., Azlan, E., & Ramli, P. (2010). Lungworm of cattle in Malaysia. *Tropical Biomedicine, 27*, 236–240.

Levot, G. (2000). Resistance and the control of lice on humans and production animals. *International Journal for Parasitology, 30*(3), 291–297. https://doi.org/10.1016/s0020-7519(99)00203-9

Light, J. E., Smith, V. S., Allen, J. M., Durden, L. A., & Reed, D. L. (2010). Evolutionary history of mammalian sucking lice (Phthiraptera: Anoplura). *BMC Evolutionary Biology, 10*, 292. https://doi.org/10.1186/1471-2148-10-292

Lightowlers, M. W. (1996). Vaccination against cestode parasites. *International Journal for Parasitology, 26*(8–9), 819–824. https://doi.org/10.1016/s0020-7519(96)80048-8

Liu, L., Xie, M., & Wei, D. (2022). Biological detoxification of mycotoxins: current status and future advances. *International Journal of Molecular Sciences, 23*(3), 1064. https://doi.org/10.3390/ijms23031064

Liver Fluke and Rumen Fluke. (n.d.). *Cattle parasites*. Retrieved June 21, 2024, from https://www.cattleparasites.org.uk/liver-fluke-and-rumen-fluke/

López, A., & Martinson, S. A. (2017). Chapter 9 - respiratory system, mediastinum, and Pleurae1. In J. F. Zachary (Ed.), *Pathologic basis of veterinary disease* (6th ed., pp. 471–560.e1). Mosby. https://doi.org/10.1016/B978-0-323-35775-3.00009-6. isbn:978-0-323-35775-3.

Machtinger, E. T., Leppla, N. C., & Hogsette, J. A. (2016). House and stable fly seasonal abundance, larval development substrates, and natural parasitism on small equine farms in Florida. *Neotropical Entomology, 45*(4), 433–440. https://doi.org/10.1007/s13744-016-0376-z. Epub 2016 Feb 22.

Mahlobo, S. I., & Zishiri, O. T. (2021). A descriptive study of parasites detected in ticks of domestic animals in Lesotho. *Veterinary Parasitology: Regional Studies and Reports, 25*, 100611. https://doi.org/10.1016/j.vprsr.2021.100611

Mans, B. J., Venzal, J. M., & Muñoz-Leal, S. (2022). Editorial: Soft ticks as parasites and vectors. *Frontiers in Veterinary Science, 9*, 1037492. https://doi.org/10.3389/fvets.2022.1037492. PMCID: PMC9662785, PMID: 36387405.

Marsella, R. (2014). Dermatophytosis. In D. C. Sellon & M. T. Long (Eds.), *Equine infectious diseases* (2nd ed., pp. 411–414.e1). W.B. Saunders. https://doi.org/10.1016/B978-1-4557-0891-8.00050-6

Mas-Coma, S., Valero, M. A., & Bargues, M. D. (2009). Chapter 2: Fasciola, lymnaeids and human fascioliasis, with a global overview on disease transmission, epidemiology, evolutionary genetics, molecular epidemiology and control. *Advances in Parasitology*, 41–146. https://doi.org/10.1016/s0065-308x(09)69002-3

Matini, M., Fallah, M., Maghsood, A. H., Saidijam, M., & Fasihi Harandi, M. (2019). *Echinococcus granulosus sensu stricto* in livestock and human in Hamadan, Western Iran. *Iranian Journal of Parasitology, 14*(2), 288–296.

Mayer, F. L., Wilson, D., & Hube, B. (2013). Candida albicans pathogenicity mechanisms. *Virulence, 4*(2), 119–128. https://doi.org/10.4161/viru.22913

McCulloch, J. B., Owen, J. P., Hinkle, N. C., Mullens, B. A., & Busch, J. W. (2020). Genetic structure of northern fowl mite (Mesostigmata: Macronyssidae) populations among layer chicken flocks and local house sparrows (Passeriformes: Passeridae). *Journal of Medical Entomology, 57*(1), 122–130. https://doi.org/10.1093/jme/tjz136

Melo, U. P., Ferreira, C., & Barreto, S. W. M. (2024). Pulmonary aspergillosis in a horse: a case report. *Brazilian Journal of Veterinary Medicine, 46*, e004723. https://doi.org/10.29374/2527-2179.bjvm004723

Mendoza-de Gives, P., López-Arellano, M. E., Olmedo-Juárez, A., Higuera-Pierdrahita, R. I., & von Son-de Fernex, E. (2023). Recent advances in the control of endoparasites in ruminants from a sustainable perspective. *Pathogens (Basel, Switzerland), 12*(9), 1121. https://doi.org/10.3390/pathogens12091121

Moretti, A., & Sarrocco, S. (2016). Fungi. In B. Caballero, P. M. Finglas, & F. Toldrá (Eds.), *Encyclopedia of Food and Health* (pp. 162–168). Academic Press. https://doi.org/10.1016/B978-0-12-384947-2.00341-X

Mostrom, M. (2016). Mycotoxins: Toxicology. In *Encyclopedia of Food and Health* (pp. 43–48). https://doi.org/10.1016/b978-0-12-384947-2.00480-3

Moyo, B., & Masika, P. J. (2009). Tick control methods used by resource-limited farmers and the effect of ticks on cattle in rural areas of the Eastern Cape Province, South Africa. *Tropical Animal Health and Production, 41*(4), 517–523. https://doi.org/10.1007/s11250-008-9216-4. Epub 2008 Aug 14.

MSD Veterinary Manual. (2023). *Aspergillosis in poultry*. Retrieved from https://www.msdvetmanual.com/poultry/aspergillosis/aspergillosis-in-poultr

Mullen, G. R., & Durden, L. A. (Eds.). (2009). *Medical and veterinary entomology*. Academic Press. https://www.sciencedirect.com/book/9780128140437/medical-and-veterinary-entomology

Murillo, A. C., & Mullens, B. A. (2020). Collecting and monitoring for northern fowl mite (Acari: Macronyssidae) and poultry red mite (Acari: Dermanyssidae) in poultry systems. *Journal of Integrated Pest Management, 20*(6), 12. https://doi.org/10.1093/jisesa/ieaa032

Nasirian, H. (2023). Monitoring of hard tick parasitism in domestic ruminants: A scale evidence for policymakers. *Veterinary Parasitology: Regional Studies and Reports, 41*, 100878. https://doi.org/10.1016/j.vprsr.2023.100878. Epub 2023 Apr 18.

National Agricultural Library, et al. (2024). *Parasites, diseases, and control measures*. USDA's Contributions to Veterinary Parasitology, U.S. Department of Agriculture. https://www.nal.usda.gov/exhibits/speccoll/exhibits/show/parasitic-diseases-with-econom/parasitic-diseases-with-econom. Accessed 21 June 2024

National Center for Biotechnology Information. (n.d.). Malassezia Infections. In *StatPearls*. StatPearls Publishing. Retrieved from https://www.ncbi.nlm.nih.gov/books/NBK553091/

Oguntomole, O., Nwaeze, U., & Eremeeva, M. E. (2018). Tick-, flea-, and louse-borne diseases of public health and veterinary significance in Nigeria. *Tropical Medicine and Infectious Disease, 3*(1), 3. https://doi.org/10.3390/tropicalmed3010003

Panuska, C. (2006). Lungworms of ruminants. *The Veterinary Clinics of North America. Food Animal Practice, 22*(3), 583–593. https://doi.org/10.1016/j.cvfa.2006.06.002

Pérez de León, A. A., Mitchell, R. D., 3rd, & Watson, D. W. (2020). Ectoparasites of cattle. *The Veterinary Clinics of North America. Food Animal Practice, 36*(1), 173–185PMID: 32029183. https://doi.org/10.1016/j.cvfa.2019.12.004

Piegari, G., Pepe, P., De Biase, D., d'Aquino, I., Bosco, A., Cringoli, G., Papparella, S., Rinaldi, L., & Paciello, O. (2021). Immunopathological response, histological changes, parasitic burden, and egg output in sheep naturally infected by *Dicrocoelium dendriticum*. *Animals: An Open Access Journal from MDPI, 11*(2), 546. https://doi.org/10.3390/ani11020546

Poglayen, G., Marchesi, B., Dall'Oglio, G., Barlozzari, G., Galuppi, R., & Morandi, B. (2016). Lung parasites of the genus Metastrongylus Molin, 1861 (Nematoda: Metastrongilidae) in wild boar (Sus scrofa L., 1758) in Central-Italy: An eco-epidemiological study. *Veterinary Parasitology*. https://doi.org/10.1016/j.vetpar.2015.12.007

Prabhu, M., Chirukandoth, S., Prema, N., Sakthivel, P. C., Anil, R., & Mathagowder, I. (2023). Occurrence of Dictyocaulus filaria infection in an organized sheep farm of Tamil Nadu, India. *Indian Journal of Veterinary and Animal Sciences Research, 52*, 72–80.

Pradhan, A. K., & Karanth, S. (2023). Chapter 28: Zoonoses from animal meat and milk. In M. E. Knowles, L. E. Anelich, A. R. Boobis, & B. Popping (Eds.), *Present knowledge in food safety* (pp. 394–411). Academic. https://doi.org/10.1016/B978-0-12-819470-6.00029-9

Prathap-Kumar, K., Balakrishna-Nair, M. N., & Punniyamurthy, N. (2023). Inhalation toxicity of mycotoxins in farm animals. *Iranian Journal of Veterinary Research, 24*(2), 94–95. https://doi.org/10.22099/IJVR.2023.47378.6833

Rajput, Z. I., Hu, S., Chen, W., Arijo, A. G., & Xiao, C. (2006). Importance of ticks and their chemical and immunological control in livestock. *Journal of Zhejiang University - Science B, 7*(11), 912–921. https://doi.org/10.1631/jzus.2006.B0912

Riet-Correa, F., Cook, D., Micheloud, J. F., Machado, M., Mendonça, F. S., Schild, A. L., & Lemos, R. A. (2024). A review on mycotoxins and mycotoxicoses in ruminants and Equidae in South America. *Toxicon, 247*, 107827. https://doi.org/10.1016/j.toxicon.2024.107827

Rizwan, H. M., Usman, M., Naeem, M. A., Farid, M. U., Younus, M., Sajid, M. S., Tahir, U. B., Luqman, N., Abbas, H., Ateeq, M. K., Taseer, M. S. A., & Asif, M. (2022). Prevalence of ruminant paramphistomosis and comparative histopathology of the infected rumens in Narowal District, Punjab, Pakistan. *Helminthologia, 59*(4), 377–384. https://doi.org/10.2478/helm-2022-0041

Rosa, C. A. R., Ribeiro, J. M. M., Fraga, M. J., Gatti, M., Cavaglieri, L. R., Magnoli, C. E., Dalcero, A. M., & Lopes, C. W. G. (2006). Mycoflora of poultry feeds and ochratoxin-producing ability of isolated Aspergillus and Penicilliumspecies. *Veterinary Microbiology, 113*(1–2), 89–96. https://doi.org/10.1016/j.vetmic.2005.10.031

Ruiz-Herrera, J., Elorza, M. V., Valentín, E., & Sentandreu, R. (2006). Molecular organization of the cell wall of Candida albicans and its relation to pathogenicity. *FEMS Yeast Research, 6*(1), 14–29. https://doi.org/10.1111/j.1567-1364.2005.00017.x

Sahito, H., et al. (2017). Prevalence of cattle lice; Haematopinus tubercula and ticks; Haemaphysalis bispinosa on cattle at Region Sukkur, Sindh - Pakistan. *International Journal of Research Studies in Biosciences (IJSRB), 5*(12), 1–5. https://doi.org/10.20431/2349-0365.0512001

Scott, D. W. (2018). Skin diseases. In *rebhun's diseases of dairy cattle* (pp. 357–388). https://doi.org/10.1016/b978-0-323-39055-2.00007-3

Seyedmousavi, S., Guillot, J., Arné, P., de Hoog, G. S., Mouton, J. W., Melchers, W. J., & Verweij, P. E. (2015). Aspergillus and aspergilloses in wild and domestic animals: a global health concern with parallels to human disease. *Medical Mycology, 53*(8), 765–797. https://doi.org/10.1093/mmy/myv067

Seyedmousavi, S., Guillot, J., de Hoog, G. S., Verweij, P. E., Vitale, R. G., Martinez, M., … & Mignon, B. (2018). Fungal infections in animals: A patchwork of different situations. *Medical Mycology, 56*(Supplement_1), S165–S187. https://doi.org/10.1093/mmy/myx104

Siles-Lucas, M., & Hemphill, A. (2002). Cestode parasites: Application of in vivo and in vitro models for studies on the host-parasite relationship. *Advances in Parasitology, 51*, 133–230. https://doi.org/10.1016/s0065-308x(02)51005-8

Silva, G. G., Zaldívar, M. F., Oliveira, L. A. R., Mariano, R. M. S., Lair, D. F., Souza, R. A., Galdino, A. S., Chávez-Fumagalli, M. A., Silveira-Lemos, D., Dutra, W. O., et al. (2023). Advances in non-chemical tools to control poultry hematophagous mites. *Veterinary Sciences, 10*(10), 589. https://doi.org/10.3390/vetsci10100589

Simões, D., de Andrade, E., & Sabino, R. (2023). Fungi in a one health perspective. *Encyclopedia, 3*(3), 900–918. https://doi.org/10.3390/encyclopedia3030064

Šlapeta, J. (2022, August). *Face flies of cattle*. In MSD Veterinary Manual. Retrieved from https://www.msdvetmanual.com/integumentary-system/flies/face-flies-of-cattle

Sonenshine, D. E., & Roe, R. M. (Eds.). (2014). *Biology of ticks volume 2* (Vol. 2). Oxford University Press.

Spanamberg, A., Ravazzolo, A. P., Araujo, R., Franceschi, N., & Ferreiro, L. (2023). Bovine ringworm: Detection of Trichophyton verrucosum by SYBR-Green real-time PCR. *Medical Mycology Case Reports, 39*, 34–37. https://doi.org/10.1016/j.mmcr.2023.01.002

Sparagano, O. A. E., & Ho, J. (2020). Parasitic mite fauna in Asian poultry farming systems. *Frontiers in Veterinary Science, 7*, 400. https://doi.org/10.3389/fvets.2020.00400

Spieler, N., & Schnyder, M. (2021). Lungworms (*Metastrongylus* spp.) and intestinal parasitic stages of two separated Swiss wild boar populations north and south of the Alps: Similar parasite spectrum with regional idiosyncrasies. *International Journal for Parasitology. Parasites and Wildlife, 14*, 202–210. https://doi.org/10.1016/j.ijppaw.2021.03.005

Springer, A., Durden, L. A., Kiene, F., Klein, A., Rakotondravony, R., Ehlers, J., Greiman, S. E., Blanco, M. B., Zohdy, S., Kessler, S. E., Strube, C., & Radespiel, U. (2023). Molecular phylogenetics of the sucking louse genus *Lemurpediculus* (Insecta: Phthiraptera), ectoparasites of lemurs, with descriptions of three new species. *International Journal for Parasitology. Parasites and Wildlife, 20*, 138–152. https://doi.org/10.1016/j.ijppaw.2023.02.002. PMID: 36845223; PMCID: PMC9945782.

Strydom, T., Lavan, R. P., Torres, S., & Heaney, K. (2023). The economic impact of parasitism from nematodes, trematodes and ticks on beef cattle production. *Animals, 13*(10), 1599. https://doi.org/10.3390/ani13101599

Sykes, J. E., Nagle, T. M., & White, S. D. (2014). Malassezia infections. In *Canine and Feline Infectious Diseases* (pp. 570–573). https://doi.org/10.1016/b978-1-4377-0795-3.00059-4

Szewc, M., De Waal, T., & Zintl, A. (2021). Biological methods for the control of gastrointestinal nematodes. *Veterinary Journal (London, England: 1997), 268*, 105602. https://doi.org/10.1016/j.tvjl.2020.105602

Tachack, E. B., Oviedo-Socarrás, T., Pastrana, M. O., Pérez-Cogollo, L. C., Benavides, Y. H., Pinto, C. R., & Garay, O. V. (2022). Status of gastrointestinal nematode infections and associated epidemiological factors in sheep from Córdoba, Colombia. *Tropical Animal Health and Production, 54*(3), 171. https://doi.org/10.1007/s11250-022-03170-2

Tam, T. T., Lan, N. T. K., & Doanh, P. N. (2020). Morphological differences and molecular phylogenetic relationship of two tapeworm species, Moniezia expansa and Moniezia benedeni, collected from domestic ruminants in northern Vietnam. *Parasitology International, 74*, 101998. https://doi.org/10.1016/j.parint.2019.101998

Tell, L. A. (2005). Aspergillosis in mammals and birds: impact on veterinary medicine. *Medical Mycology, 43*, Issue Supplement_1, pp. S71–S73. https://doi.org/10.1080/13693780400020089

Tessema, W., Getachew, M., & Tora, E. (2024). Prevalence and risk factors of lungworm infection in small ruminants in selected districts of Wolaita Zone, Southern Ethiopia. *Journal of Parasitology Research, 2024*, 6303598. https://doi.org/10.1155/2024/6303598

Tolan, R. W. (2011). Fascioliasis due to *Fasciola hepatica* and *Fasciola gigantica* infection: An update on this 'neglected' neglected tropical disease. *Laboratory Medicine, 42*(2), 107–116. https://doi.org/10.1309/LMLFBB8PW4SA0YJI

Uilenberg, G. (1995). International collaborative research: Significance of tick-borne hemoparasitic diseases to world animal health. *Veterinary Parasitology, 57*(1-3), 19–41. https://doi.org/10.1016/0304-4017(94)03107-8

Verocai, G. G., et al. (2020). Diagnostic methods for detecting internal parasites of livestock. *Veterinary Clinics of North America: Food Animal Practice, 36*(1), 125–143. https://doi.org/10.1016/j.cvfa.2019.12.003

Villalba, J. J., Miller, J., Ungar, E. D., Landau, S. Y., & Glendinning, J. (2014). Ruminant self-medication against gastrointestinal nematodes: Evidence, mechanism, and origins. *Parasite (Paris, France), 21*, 31. https://doi.org/10.1051/parasite/2014032

Wakenell, P. (2016). Management and medicine of backyard poultry. In B. L. Speer (Ed.), *Current therapy in avian medicine and surgery* (pp. 550–565). W.B. Saunders. https://doi.org/10.1016/B978-1-4557-4671-2.00024-0

Wall, R., & Shearer, D. (2008). *Veterinary ectoparasites: Biology, pathology and control*. Wiley. https://books.google.com/books/about/Veterinary_Ectoparasites.html?id=xJrbivF54u0C

Yadav, A., Panadero, R., Katoch, R., Godara, R., & Cabanelas, E. (2017). Myiasis of domestic and wild ruminants caused by Hypodermatinae in the Mediterranean and Indian subcontinent. *Veterinary Parasitology, 243*, 208–218. https://doi.org/10.1016/j.vetpar.2017.07.007. Epub 2017 Jul 10.

Yang, H., Song, G., & Lim, W. (2020). Effects of mycotoxin-contaminated feed on farm animals. *Journal of Hazardous Materials, 389*, 122087. https://doi.org/10.1016/j.jhazmat.2020.122087

Yasin, M. G., Alim, M. A., Anisuzzaman, Ahasan, S. A., Munsi, M. N., Chowdhury, E. H., Hatta, T., Tsuji, N., & Mondal, M. M. H. (2018). Trematode infections in farm animals and their vector snails in Saint Martin's Island, the southeastern offshore area of Bangladesh in the Bay of Bengal. *The Journal of Veterinary Medical Science, 80*(4), 684–688. https://doi.org/10.1292/jvms.17-0308

Zafari, S., Mohtasebi, S., Sazmand, A., Bahari, A., Sargison, N. D., & Verocai, G. G. (2022). The prevalence and control of lungworms of pastoral ruminants in Iran. *Pathogens (Basel, Switzerland), 11*(12), 1392. https://doi.org/10.3390/pathogens11121392

Zeece, M. (2020). *Food additives. In Introduction to the chemistry of food* (pp. 251–311). Academic Press. https://doi.org/10.1016/B978-0-12-809434-1.00007-4

Fungus as Feed Additives in Animal Nutrition and Health

Nishu Sharma, Rahul Khanna, Shiwani Guleria Sharma, and Gurvinder Singh Kocher

Abstract Traditional animal feed does not fulfill the specifications of balanced food necessary for livestock which are ultimately crucial for their proper growth and productivity performance. These additives, derived from various fungal species enhance digestibility, complex feed breakdown, nutrient absorption and improve growth rate in livestock. Fungal enzymes, such as cellulase, amylase, and protease, are known to break down complex constituents of animal feed. Additionally, conventional feed does not provide the benefits of probiotics and prebiotics which is very essential for maintaining the healthy gut microbiome of the animal. Therefore, there is an urgent need to scrutinize the supplements that not only provide essential nourishment to the host but also encourage optimal growth, and development along with the improvement of the productivity of the livestock. Furthermore, fungal feed additives are gaining momentum in animal nutrition because of their astounding ability to amplify growth and boost health as they act as substitutes for antibiotics, acting as a sustainable source to combat antibiotic resistance in animals. Cell walls, polysaccharides, ergosterol, adenosine, and triterpenes of fungi feed additives are known to provide antimicrobial, antiphlogistic, anti-fatigue, antioxidant, and immunity-boosting effects. The present chapter explains the comprehensive discussion on feed additives, their classification, and role along with the effect of supplementing fungal feed additives in livestock diets.

Keywords Additives · Fungi · Mushrooms · Polysaccharides · Probiotics

---

N. Sharma · R. Khanna · S. G. Sharma (✉) · G. S. Kocher
Department of Microbiology, College of Basic Sciences and Humanities, Punjab Agricultural University, Ludhiana, Punjab, India
e-mail: shiwani@pau.edu

A. Gupta (ed.), *Prospects of Fungal Biotechnologies for Livestock Volume 1*,
Fungal Biology, https://doi.org/10.1007/978-3-031-90994-8_2

1 Introduction

Fungus, the eukaryotic organisms, include yeast, must, rust, mushrooms, and toadstools. They have a prodigious history of being exploited for food by humans and animals. They are decomposers, free-living, parasites, and symbiotic (Wijayawardene et al., 2022). Tending to grow from single hyphae, they can degrade all forms of hydrocarbons present on Earth. They decompose organic materials, releasing essential elements such as carbon, oxygen, nitrogen, and phosphorus back into the environment. They have taken the process of degradation to the next level because of their efficiency in facilitating enzymatic processes through the production of a range of hydrolytic enzymes (Shang et al., 2020). They are classified into the following five groups based on their life cycle and morphology: Ascomycetes, Basidiomycetes, Zygomycetes, Oomycetes, and Deuteromycetes (Lücking et al., 2021).

Fungi have a long-standing history of utilization in the production of a diverse array of foods and beverages by fermentation. Conventional fermentation processes that incorporate the use of fungi and yeasts yield fermented dairy products, soya sauce, alcoholic beverages such as wine, low and high-alcoholic wine, and probiotic drinks (Chai et al., 2022). Additionally, mushrooms, which represent the fruiting bodies of macrofungi, serve as significant food sources, offering considerable nutritional benefits due to lesser caloric content and enhanced proteins, vitamins, and antioxidant content (Chugh et al., 2022). Animal feed alone does not have the potential to provide essential nutrients and required functional compositions, necessitating the requirement of feed additives necessary for encouraging superior health and productivity in livestock.

Basic and standard feed formulations are rich in carbohydrates but typically do not contain adequate levels of vitamins, bioactive compounds, anti-oxidants, and essential minerals such as calcium and phosphorus, which are vital for metabolic functions, bone growth, and immune response (Dufossé et al., 2021). Moreover, natural feed sources like hay, straw, and silage lack the digestive hydrolyzing enzymes needed for the effective breakdown of complex polysaccharides, proteins, and fibers, resulting in poor nutrient absorption because of the presence of anti-nutritional compounds (Wise, 2024). Additionally, conventional and basic feeds often do not include probiotics and prebiotics, which are vital for maintaining a balanced gut microbiome and enhancing digestion and immune health (Lipiński et al., 2021).

Antioxidants have an inevitable contribution in preventing cells from oxidative damage, as well as specific fatty acids, which are important for immune and reproductive functions, are frequently lacking in the feed given to animals (Calder, 2021). Furthermore, plant-based feeds often do not supply sufficient quantities of essential amino acids like lysine, methionine, and threonine which are critical for protein synthesis and growth (Kumar et al., 2023). Consequently, the addition of feed additives is essential to rectify these deficiencies, ensuring a well-rounded diet that meets the physiological requirements of animals and promotes their growth and productivity.

Fungal food additives are becoming increasingly popular in animal nutrition due to their potential to enhance growth performance, improve gut health, and bolster the immune response. These additives, including enzymes, probiotics, and fungal metabolites, are derived from various fungi and have shown robust outcomes in improving the overall productivity of livestock (Bilal et al., 2023). *Saccharomyces cerevisiae*, *Lentinulaedodes* spp., *Aspergillus oryzae, Amylomyces rouxii, Pleurotus* spp., *Antrodia cinnamomea*, *Neurospora oncomensis,* and *Cordyceps militaris* are among the fungal species that have lured scientists for their bioactive compounds and beneficial health effects. These fungi are known to be the powerhouse of protein, polysaccharides, antioxidants, phenolic compounds, glucans, ergosterol, and laccases concerning good health and increased production of milk, meat, and protein in livestock (Chuang et al., 2021).

This chapter delves into recent investigations to apprehend the effect of fungal food additives on animal health and performance. Recent studies emphasize the efficacy of hydrolyzing enzymes produced by fungus that can convert xylanase, cellulose, and phytase in improving growth performance across different animal species. For instance, *Aspalathus linearis* extracts, rich in fungal enzymes, are reported to increase growth ability and feed conversion ratios in broilers (Shang et al., 2020). Similarly, *Trichoderma reesei* in swine diets produces xylanase and β-glucanase and has led to significant improvements in average daily gain and feed efficiency. Apart from promoting growth, Fungal probiotics, including Saccharomyces cerevisiae and *Aspergillus oryzae* are well-documented for their positive impact on gastrointestinal health. These probiotics contribute to the maintenance of a balanced gut microbiota, not allowing pathogenic bacteria to proliferate, and improve nutrient absorption efficiency.

The introduction of yeast in poultry has shown an increase in nutrient absorption to a significant level by enhancing villus height and crypt depth (Bilal et al., 2023). Correspondingly, the introduction of *Aspergillus oryzae* in pig diet, reduced incidents of diarrhoea. Fungal feed additives act as immunomodulators and enhance immune response in animals. Zhang et al. (2020) confirmed that adding -glucans derived from *Saccharomyces cerevisiae* in aquaculture, boosted the immune system of fish and enhanced its resistance against pathogen bacteria and viruses. Other than health benefits, fungal food additives also confer a significant increase in economic growth. There is a reduction in feed cost and increase in production yield, because of the enhanced feed efficiency and growth rates.

The cost of the addition of antibiotics, inorganic minerals, and chemicals for treatment and growth development decreases with the help of fungal components in feed because of their antimicrobial and phytase-producing ability (Aranda-Aguirre et al., 2021). Wang et al. (2020) observed an improvement in the yield and composition of milk with the addition of extracts of *Aspergillus oryzae* after fermentation, increasing the profitability of dairy farms. These benefits highlight the economic viability of integrating fungal additives into animal feed formulations. Being economical benefactor, fungal components in feed are well-documented for their safety regulation which enhances their chances for global adoption.

For the precise monitoring of mycotoxins, quality, and consistency of fungal products. Recent regulatory frameworks have established guidelines for the approval and use of these additives, ensuring their safety for animal health and consumer safety (EFSA, 2023). Feed additives can also contribute to environmental sustainability by reducing the environmental footprint of animal production. Phytase, for example, helps in the reduction of phosphorus excretion, minimizing environmental pollution (Zhao et al., 2023). Mycotoxin binders reduce the risk of mycotoxin contamination in feed, thus preventing potential health issues and reducing the need for veterinary interventions. According to recent government surveys and market evaluations, feed additives incorporating fungi have gained significant traction in the Indian animal feed industry. Approximately 25–30% of poultry, 15–20% of dairy cattle, and 10–15% of swine feed are combined with fungal additives (AICRP-PB, 2020; Sharma et al., 2021). This trend reflects the growing recognition of the benefits of fungal feed supplements in promoting health and productivity in animals. The fate of feed additives resides with the continued development of innovative products and technologies that enhance their efficacy and safety so that they can eternally be introduced in animal feed, being easily accessible to low-income farmers and the public.

2 Feed Additives and Their Role

Animal feed additives are utilized globally across various livestock sectors, including poultry, for multiple purposes. These additives serve to supply recommended daily need nutrients, appetizing and savory feed, improve body growth, and optimize efficacious consumption. The demand from the dairy and meat industries has surged rapidly due to rising population pressures (Nofal et al., 2022). Achieving food security, which is the second Sustainable Development Goal (SDG) focused on "zero hunger," along with the 12th SDG emphasizing "responsible consumption and production," is increasingly prioritized within the animal feed industry. These goals are being pursued more vigorously to address the challenges of sustainable food production. The U.S. is a major consumer of feed additives, with the widespread use of nutritional and zootechnical additives to enhance overall growth in poultry, swine, and cattle (NRC, 2012). Furthermore, In India, the employment of fungal feed additives such as enzymes and probiotics is mounting, obsessed with the need to improve nutrient digestibility and boost animal performance in the dairy and poultry sectors (Kumar et al., 2020).

The Official Journal of the European Union (2003) states feed additives as ingredients, microorganisms, or preparations that are intentionally incorporated into animal feed or water, distinct from feed materials and premixtures, to achieve specific functions. These functions include enhancing the characteristics of feed. They should:

- Improve the animal product quality.
- Influence the attractiveness and intensity of color in ornamental animals.

- Meet the dietary requirements of animals.
- Mitigating the environmental impact of animal production.
- Positively affects the performance of animals.
- modify gastrointestinal flora or improve the digestibility of feed,
- Exhibit coccidiostats, histomonostatic, and antimicrobial properties.

In this regard, feed additives should not:

- Be interfering with human health, animal well-being, and ecology.
- Contains wrong ingredient and counterfeit user.
- Interfere with the genetic makeup of the animal.

Antibiotics, except coccidiostats and histomonostats, are not permitted as feed additives. However, antibiotics play a crucial role as essential medications for treating bacterial infections (Guo et al., 2020). The growing issue of antibiotic resistance in human medicine has raised noteworthy health concerns regarding the employment of antibiotics in animals. This situation necessitates scrutiny of antibiotic use in livestock to mitigate potential risks associated with resistance development (Jindal et al., 2021). The role of various types of feed additives is presented in Table 1.

2.1 Roles of Feed Additives

2.1.1 Additives Promoting Amount of Feed Intake

Antioxidants are crucial components in animal feed, as they prevent the oxidative rancidity of polyunsaturated fats, a process that can severely compromise all the attributes of the feed. This rancidity not only diminishes the energy content but also leads to the degradation of vital vitamins, including A, D, E, and several B complex vitamins (Rafiei & Khajali, 2021). The addition of synthetic antioxidants such as hydroxytoluene (BHT) and ethoxyquin has been demonstrated to effectively counter these adverse effects. Research by Hernandez et al. (2020) has shown that incorporating antioxidants into animal feed can enhance feed intake and growth. Similarly, another study on poultry found that adding ethoxyquin at 150 ppm resulted in a 5% increase in feed intake and a 3% improvement in weight gain, which directly increased meat production. Likewise, the use of BHT at 200 ppm in dairy cow diets is known to boost milk outputs and overall health, as they are reported to help preserve feed quality and vitamin durability (Alhussien et al., 2021). Furthermore, antioxidants are vital for extending the shelf life and safety of animal feed.

Table 1 Different feed additives used globally and their functions

S. No.	Feed additives	Function	References
1.	Anhydrous ammonia	Source of non-protein nitrogen	Reis et al. (2020)
2.	Anionic salts	Makes salty and increases blood calcium level	Shahzad et al. (2018)
3.	*Aspergillus oryzae*	Digest fiber and maintain pH	Kim and Kim (2021)
4.	Beta carotene	Improve reproductive health and immune response	Baldi et al. (2020)
5.	Choline	Prevent fatty liver	Zanton et al. (2018)
6.	Decoquinate	Prevent coccidiosis	Trotta et al. (2023)
7.	Lasalocid	Ionophore	Potter and Arnold (2021)
8.	Magnesium oxide	Alkalizer	Lopes et al. (2020)
9.	Methionine hydroxy analogue	Minimize fatty liver	Patton and Huber (2019)
10.	Monensin	Ionophore and prevent coccidiosis	Duffield and Lissemore (2020)
11.	Niacin	Coenzyme	Zhao et al. (2019)
12.	Probiotics	Prevent the growth of pathogens and detoxify harmful metabolites	Jami and Mizrahi (2018)
13.	Propionic acid	Mold inhibitor	Wieland and Liebhard (2019)
14.	Sodium bentonite	Clay minerals used as a binder	Mohammadi and Rouzbehan (2019)
15.	Urea	Source of non-protein nitrogen	Marini et al. (2021)
16.	Sodium bicarbonate	Maintain pH	Erdman and Cunningham (2019)
17.	*Saccharomyces cerevisiae*	Digest fiber, produce antioxidants and polyphenols	Kiarie and Mills (2019)
18.	Zinc methionine	Immunomodulatory effect	Spears and Kegley (2020)

2.1.2 Additives Increasing Feed Acceptability

The feed should be appealing to the animals for overall acceptability. Therefore, additives should enhance aroma, flavor, taste, and palpability. In poultry nutrition, additives play crucial roles in enhancing feed quality and optimizing production outcomes. Xanthophylls, including lutein and zeaxanthin, are added to poultry feed to intensify the yellow coloration of broiler skin and egg yolks, a trait that significantly influences consumer preferences (Mozin et al., 2023). Other compounds that are used are Anticaking agents, they are essential for maintaining the flowability of feed powders and preventing clumping caused by moisture. Common agents include salt or long-chain fatty acids, which inhibit feed particles from sticking together through hydrophobic interactions (Nielsen et al., 2020). Calcium phosphate, which improves feed quality while also serving as a calcium source, and potassium and sodium ferrocyanide, which prevent moisture-induced clumping, are also widely used. Similarly, firming and crisping agents, which maintain the texture and turgidity of vegetable tissues by preserving internal water pressure, are not commonly used in poultry feeds but are important in processed food products (Patra et al., 2022). Furthermore, Sequestrants like EDTA (ethylenediaminetetraacetic acid) are used to immobilize metal ions, which can act as pro-oxidants and degrade feed quality (Zhang et al., 2020; Guo et al., 2021). To intensify the taste, Sweeteners, such as sugar and saccharin, are occasionally added to poultry feed to enhance palatability and encourage consumption. While sugar adds caloric content, saccharin is a non-caloric sweetener used for its intense sweetness and minimal digestibility (Kawabata & Tabata, 2022).

2.1.3 Increasing Digestion and Absorption

Fungal additives in feed produce hydrolytic enzymes that catalyze biochemical reactions and are crucial for optimizing poultry feed. It has been reported by Kim and Kim (2021) that beta-glucanase produced by fungi is employed to enhance the digestibility of plant-based feed components. Cereal-based feeds, containing complex polymers like beta-glucan, can hinder starch digestion and lead to energy loss. For example, beta-glucanase supplementation in barley-based diets has been shown to increase energy availability by up to 15% and eliminate issues such as sticky droppings (Xia et al., 2021). This enzyme breakdown allows for more efficient utilization of energy, improving feed efficiency and supporting poultry health, especially under stressful conditions (Peng et al., 2020). Other additives that increase the bio-availability of nutritional compounds and absorption are *Saccharomyces cerevisiae*-based. These microbes confer probiotic effects in the gut of animals. A few other examples of probiotics, commonly used as feed additives, include strains such as *Lactobacillus*, *Bifidobacterium*, *Enterococcus faecium*, and spore-forming *Bacillus* species (Ouwehand & Von Wright, 2021). These microorganisms are incorporated into animal feeds to enhance gut health and improve overall animal performance. They can produce phytase enzymes to degrade anti-nutritional compounds like phytic acid (Gänzle, 2021). For instance, *Lactobacillus* strains have been detected to enhance feed conversion rate by up to 8% in poultry (Shakouri et al., 2020), while *Bifidobacterium* can enhance immune responses and decrease the incidence of gastrointestinal ailments (Nava et al., 2021).

2.1.4 Additives in Preventing Infections

Antimicrobials are not recommended for use as feed additives, as they increase the chances of antibiotic resistance, leading to an increased burden on veterinary treatments. Therefore, some additives are suggested to control infections in livestock (WHO, 2017). Antifungal additives denote a crucial category of feed supplements aimed at preventing pathogenic mold growth in feeds susceptible to fungal contamination, particularly from species such as *Aspergillus flavus* and *Penicillium cyclopium*. Common antifungal agents include propionic acid, acetic acid, and sodium propionate, which are typically supplemented with high-moisture grains. Additionally, antifungal compounds like nystatin and copper sulfate are employed in concentrate feeds to inhibit mold development (Oladele et al., 2022). On the other hand, anticoccidials are extensively used to suppress the growth of coccidia, protozoa that reside in the intestinal lining of livestock, thereby decreasing the prevalence of coccidiosis (Chapman et al., 2020). Moreover, anthelmintics are employed in practical feeding scenarios to address parasitic infections by targeting and reducing helminth populations in livestock, ultimately enhancing their overall health and productivity (Matté et al., 2023).

3 Classifications of Feed Additives

Classification of feed is done based on the following approaches: (1) Based on the regulations of the European Commission; (2) Holistic approach; (3) Origin and function (Pandey et al., 2019).

3.1 Based on the Regulations of the European Commission

3.1.1 Sensory Additives

The group of additives that ameliorates the diet palatability by boosting appetite primarily affects diet flavor or color. For instance, taste enhancers or sweeteners. Piglets are encouraged to consume a ration by adding vanilla extracts (Rychen & Toussaint, 2018).

3.1.2 Nutritional Additives

They supply nutrients such as vitamins, amino acids, or trace minerals for the optimal growth of the animal. Mostly, these nutrients are added in a concentrated form as natural ingredients in the animal ration (Rychen & Toussaint, 2018).

3.1.3 Zootechnical Additives

They enhance the nutritional profile and livestock production. They not only provide special nutrients but also assist in more efficacious use of them. The inclusion of an enzyme or a microbial by-product, improves the gut condition, thus making effective extraction of nutrients. Therefore, they are known as pro-nutrients, i.e., substances that upgrade the diet's nutrient content without necessarily providing nutrients directly (Rychen & Toussaint, 2018).

3.1.4 Coccidiostats and Histomonostats

They directly affect poultry health and are usually used to enhance poultry's intestinal health. They have a direct effect on intestinal parasitic organisms (Feed Additive Classifications 2018).

3.1.5 Technological Additives

This group of additives greatly impacts the practical characteristics of the ration. They have no impact on the nutritional quality of the feed directly but they may boost handling or hygiene features indirectly. An organic acid included for food preservation provides the best example of them (Rychen & Toussaint, 2018).

3.2 Based on the Holistic Approach

Indian Council of Agricultural Research, Ministry of Agriculture, Government of India provided a holistic approach for the classification of feed. They are classified as given below (Faraz, 2018):

3.2.1 Additives That Influence Feed Stability, Manufacturing, and Its Characteristics

They are further subdivided into the following categories: (a) antifungal, (b) antioxidant, and (c) pellet binders.

3.2.2 Additives That Modify Animal Growth, Feed Efficiency, Metabolism, and Performance

They are further classified as: (a) Feed flavors; (b) Digestion modifiers: They have subcategorized include enzymes, prebiotics, buffers, acidifiers, ionophores, antibloat substances, isoacids, salivation inducers, probiotics; (c) Metabolism modifiers: Hormones and beta-adrenergic compounds; (d) Growth promotants: Antibiotics, chemotherapeutic, prebiotics, and probiotics agents.

3.2.3 Additives That Modify Animal Health

They include: (a) Drugs and (b) Immunomodulators.

3.2.4 Additives That Modify Consumer Acceptance

Xanthophylls are the subclass under this category of additives that modify feed for acceptance by consumers.

3.3 *Based on the Origin and Function*

These are further subdivided into categories mentioned below:

3.3.1 Additives That Enhance Feed Intake

Antioxidants

They are defined as substances that fend off the oxidative damage to lipids. This oxidative damage called rancidity causes further destruction of the vitamins A, D, E, and B complex. Products resulting from their breakdown impact the protein content of the feed by reacting with lysine residues. A prominent example of an antioxidant is ethoxyquin or BHT (butylated hydroxytoluene).

Flavoring Substances

They enhance the palatability of feed. Flavoring agents are the substances that are added to the ration to enhance its intake by the animal under situations when highly unpalatable medicines are being added to the feed.

3.3.2 Additives That Enhance the Quality and Acceptability of the Feed

Anticaking Agents

They are the substances that take up moisture from the feed by surface adhesion. They are mixed with the dry feed and prevent the clumping of particles for free-flowing of the product. Fatty acids, sodium ferrocyanide calcium phosphate, oxides of potassium and magnesium, salts of Al, Mg, Ca, silicic acid, sodium aluminum silicate, sodium calcium aluminum silicate, and calcium aluminum silicate are some examples.

Humectants

They keep products moist such as bread and cakes.

Firming and Crisping Agents

The compounds work by maintaining the pressure of water inside the tissues and making them turgid.

Sequestrants

Elements such as Cu and Fe present in feed act as prooxidant catalysts and therefore they are required to be immobilized which can be achieved by the use of sequestrants. Sequestrants capture the metal ions and prevent them from being involved in oxidative reactions. Ethylenediaminetetraacetic acid is a good example of a sequestrant.

Sweeteners

They are the common components of food but some forms of sweeteners are not easily digestible and cause stomach problems. Saccharin is a substance without any calorific value.

3.3.3 Additives That Facilitate Digestion and Absorption

Grit

Grit is added to the fibrous feed fed to poultry to increase surface area for digestion and absorption. Some examples of grit used as additives in the feed are coquina shells, limestone, and oyster shells.

Buffers and Neutralizers

To meet the growing requirement of energy and requirement, high doses of concentrated feeds are given which lowers the pH of the rumen that cannot be tolerated by the microflora of the rumen. This further favours the growth of acidophiles resulting in acidosis and eventually disturbing digestion. This can be alleviated by the inclusion of buffers and neutralizers in the feed. For example, carbonates, bicarbonates, hydroxides, oxides, phosphate salts, ammonium chloride, and sodium sulfates.

Chelates

Chelates are cyclic compounds that control the absorption of several minerals. A chelated element may be released as an ion at the wall of the intestine whereby they may be absorbed readily. Chelates can be natural substances (chlorophyll, cytochromes, hemoglobin, and vitamin B12) or they may be synthetic in nature such as ethylenediaminetetraacetic acid. Chelates also play an important role as feed additives. Three subtypes of chelates act as feed additives.

Enzymes

Enzymes are the proteinaceous substances existing in plants, animals, and microorganisms that can catalyze a variety of biochemical reactions. In recent years, microbial enzymes have been generated at an industrial scale for feed production. Poultry feed is a composite mixture of vegetable and plant matter that is degraded and modified enzymatically. The enzymes ameliorate poultry feeding by changing the efficiency of its utilization, reforming poorly digested cereals or feed substances, and supplying additional digestive enzymes to tolerate stressful conditions such as hot and humid climates. Polymers such as glucose, arabinose, and xylose present in cereal grains are not easily digestible. This can lead to energy loss due to the hindrance of starch digestion by enzymes present in the intestine by layering starch granules and due to the incapability of the animal's enzyme to degrade polymers. The addition of the enzymes in the system enhances the polymer degradation along with the availability of the energy value (Pirgozliev et al., 2019).

For this purpose, the choice of the enzyme should be made correctly based on the criterion described below:

- At the pH range of the animal's digestive system, it must be agile and able to persist during haulage through the gastrointestinal passage.
- It must be available in a physical form for safe and easy mixing with different forms of the feed.
- It should maintain its high standardized activity before and even after inclusion into the feed.
- It should be able to survive normal pelleting conditions.

Non-starch polysaccharides (NSPases) are the widely used industrial enzymes that split the non-starchy polysaccharides present in wheat, barley, and triticale. Phytases of microbial origin are the enzymes that cleave the phytate complexes in plant-derived substances (Pirgozliev & Bedford, 2013).

3.3.4 Additives That Promote Growth and Production

Antibiotics

Antibiotics are compounds furnished by living organisms in small concentrations and are known to have both bacteriostatic and bactericidal attributes. Originally, they were produced for the control of pathogens and have a very crucial role in the medical and veterinary sectors. Penicillin, oxytetracycline, flavomycin, erythromycin, bacitracin, streptomycin, tyrothricin, gramicidin, chlortetracycline and are some antibiotics that are reported to have growth-promoting effects. They should only be used for the development and expanding pigs, chicks, and turkey to annihilate pork, bacon, or poultry, respectively. They should not be inculcated in the feed of ruminants, breeding pigs, and poultry stock. The recommended dose of the antibiotics should be evenly mixed with the feed.

Probiotic and Prebiotic

The gut microflora is the prime factor in Vigor and illness progression. Probiotic is the term used for the health condition that is due to the existence of favorable bacteria or owing to the supply of nutrients that boost the growth of the advantageous microbes growing endogenously. Classically, it is a dietary supplement containing microbes that have a serviceable impact on the host species and ameliorate their intestinal microbial balance (Sanchez & Rivas-Estilla, 2006). *Lactobacilli*, *Bifidobacterium, Enterococcus faecium*, spore-forming *Bacillus* spp., and Saccharomyces are the important genera of bacteria and yeast that are the most frequently used as prebiotic agents. Probiotics are strain-dependent specified formulations that bestow different strains the property to be beneficial for different diseases, with some strains being detrimental while others are beneficial to individuals.

Benefits of probiotics include:

- Direct antagonistic effect on harmful organisms by releasing some antibacterial compounds or by reducing their rivalry for nutrients.
- Changing the microbial metabolic patterns in the gastrointestinal system.
- Immunity booster.
- Toxin nullification.

A prebiotic is a food constituent that is not digestible and boosts the growth of a certain specific number of colon bacteria, thus exerting beneficial effects on the animal. Gibson and Roberfroid (1995) reported that prebiotics change the microflora composition of the colon with the dominance of especially *Lactobacilli* and *Bifidobacterium*. They provide nutrients to beneficial indigenous microflora present in the host species and administered probiotic strains. Two main classes of prebiotics exist, i.e., (a) Fructans [fructooligosaccharides (FOS) and inulin] and (b) Galactans (galactooligosaccharides or GOS) (Patterson & Burkholder, 2003).

3.3.5 Additives That Alter Metabolism

Hormones

Hormones are the chemical substances produced by the ductless glands. They are then transferred to another location within the body whereby they act by triggering certain physiological responses. Estrogens, androgens, progestogens, hormones, and thyroxine are some examples of hormones that are used for enhancing the growth and fattening of animals producing meat.

Implant

These are the hormones or hormone-like substances that slowly and constantly release the active chemicals that are eventually absorbed in the blood and implanted in the ear. For example, diethylstilbestrol (DES).

3.3.6 Additives That Affect the Health of the Livestock

Antibloating Substances

Substances poloxalene and several other products are used as preventive agents for pasture bloating.

Antifungal Compounds

Feed that is susceptible to the attack of fungi such as *Aspergillus flavus*, *Penicillium cyclopium*, etc. is supplemented with mold inhibitors. Before adding, the feed is dried well (<10% moisture). Some examples include propionic, acetic acid, and sodium which are added to the feed with high moisture content. Nystatin and copper sulfate are some antifungals added for concentration of feed and mold prevention.

Anticoccidials

Anticoccidial substances avert the growth of protozoa and other organisms that thrive inside the intestinal lining of livestock.

Anthelmintics

Addition of anthelmintics compounds for reducing parasitic infections.

Phytogenics

These are the plant-derived secondary metabolites, phytochemicals, or phytobiotics used for the upgrading of feed quality, performance, and livestock health. These substances have gained wide momentum among the farmer community. They include herbs, spices, essential oils, and oleoresins that are known as growth promoters and have therapeutic properties (Bravo et al., 2014; Pirgozliev et al., 2015). The chemical structures of plant extracts vary widely. The effects of phytogenic are greatly dependent on the chemistry of the compounds but their mode of action remains unknown (Acamovic & Brooker, 2005).

4 Fungi as a Food Source

Fungi are increasingly recognized as a feasible and valuable resource of nutrients for animal feed, offering numerous advantages such as improved digestion, enhanced immune function, and a lower environmental footprint. Recently, the incorporation of fungi into animal feed has garnered considerable interest due to their rich nutritional composition, which encompasses essential amino acids, vitamins, minerals, and bioactive compounds (Strong et al., 2022). *Saccharomyces cerevisiae* has the potential to produce bioactive compounds in fermentation and can be fed as a single cell protein. Second, Quorn has a history of being used as a protein source that is derived from *Fusarium venenatum*. The third strain that is incorporated into the diet widely is *Yarrowia lipolytica*, which has been approved under the Novel Food Regulation. Another notable fungal SCP product is PEKILO, which is derived from *Paecilomyces variotii* that grows on the waste of the paper industry and is a very popular feed for poultry and fish. Some other examples can be inculcated into the feed (Dunuweera et al., 2021). Further, fungi such as *Aspergillus oryzae* and *Trichoderma viride* are used to ferment crop residues, improving their digestibility and nutrient availability. This process increases the crude protein content and decreases anti-nutritional factors, making the feed more suitable for ruminants (Zhang et al., 2020). Khan et al. (2024) concluded that mycoprotein, derived from *Fusarium venenatum*, is a high-protein, low-fat fungal product that has been extensively studied and utilized as a meat substitute for animals and humans. A study in swine and poultry evaluated that there is a reduction of greenhouse gas emissions associated with traditional livestock farming when they are fed with fungal-origin protein (Khan et al., 2024).

4.1 Nutritional Value of Fungi as Food

Fungi are enriched in proteins, fibers, vitamins, and minerals, making them an outstanding addition to animal feed. They have a total protein count of 20–50% of dry weight, comparable to conventional protein sources such as soy and fish meal (Zhang et al., 2021). Additionally, fungi provide all essential amino acids, including lysine and methionine, which are often lacking in plant-based feeds. The fiber content found in fungi is in the form of beta-glucans and chitin, contributes to improved gut health, and boosts the immune systems of animals. Beta-glucans are recognized for their immunomodulatory properties, which can reduce infection rates and enhance overall health (Kumar et al., 2020). Furthermore, fungi are an appreciated source of vitamins including B-complex, vitamin D, and vitamin K, along with essential minerals such as zinc, selenium, and copper, all of which play critical roles in various metabolic processes (Rai et al., 2021).

Fungi-based food products are recognized for their high fiber content, which contains chitin and branched β-glucans (Şengül & Ufuk, 2022). This naturally produced fibrous matrix is characterized by low water solubility, contributing to its dietary benefits. Fungal biomass typically consists of about 25% dietary fiber (Derbyshire & Delange, 2021) which can be metabolized into short-chain fatty acids, like acetate and butyrate. Additionally, microbial oil from fungi offers several advantages, such as ease of scaling production, less extensive labor, and environmental benefits.

4.2 Environmental Impact

The usage of fungi in animal feed proposes several environmental benefits. Fungi can be cultivated on agricultural waste and by-products, thus reducing waste and contributing to a circular economy. This practice not only lowers the environmental footprint of animal production but also provides an economical feed source. Additionally, the production of fungal biomass requires less land and water compared to traditional feed crops, making it a more sustainable option (Ravi et al., 2021). The cultivation of fungal biomass on agricultural residues can reduce waste generation by as much as 30%, promoting more sustainable agricultural practices. Additionally, fungal biomass production is highly resource-efficient, requiring significantly less land and water compared to traditional feed crops.

A study by Sharif et al. (2021) concluded that producing fungal protein utilizes only 10% of the land and 15% of the water needed for soybean cultivation, a common protein source in animal feeds. This reduction in resource consumption leads to a lower overall environmental impact, aligning with global sustainability objectives. A survey of animal feed producers in the European Union found that 85% of respondents considered fungal proteins essential for minimizing the environmental footprint of their operations (Ravi et al., 2021). In conclusion, using fungi-based feeds can decrease greenhouse gas emissions from livestock production, highlighting their environmental advantages.

4.3 Economic Relevance

Economic analyses have shown that incorporating fungal products into animal feed can be cost-effective, especially when considering the long-term benefits such as improved animal health, reduced veterinary costs, and enhanced productivity. The scalability of fungal cultivation and fermentation processes further supports their viability as a commercial feed additive (Smith et al., 2020). It was researched by Sethi et al., (2023) observed that broilers that were fed with diets augmented with fungal proteins had a 10% higher growth rate compared to those fed with traditional feeds, leading to reduced feed costs and increased profitability for poultry producers. Moreover, the scalability of fungal cultivation and fermentation processes supports their viability as a commercial feed additive. Modern biotechnological advancements have enabled the mass production of fungal biomass, which can be accomplished at a reduced price as compared to conventional feed ingredients. This scalability is crucial for meeting the growing demand for sustainable and cost-effective animal feeds. Miller and Johnson (2021) highlighted that the cost of producing fungal protein at scale can be 20–30% lower than producing equivalent amounts of soybean meal, making it a competitive alternative in the market.

The long-term economic benefits of using fungal products in animal feed are also substantial. Improved animal health translates to fewer incidences of disease and lower mortality rates, which can significantly reduce veterinary expenses. Additionally, healthier animals tend to have better productivity in terms of milk yield, egg production, and meat quality, contributing to higher overall revenues for farmers (Thakur & Kumar, 2021).

5 Functional Elements of Fungi

Fungi employ several enzymes for the digestion of the culture media in which it is cultivated (Filiatrault-Chastel et al., 2021). Several secondary metabolites and other products are generated during this process. These functional elements include cell walls, polysaccharides, ergosterol, adenosine, and triterpenes. They are antimicrobial, antiphlogistic, anti-fatigue, antioxidant, and boost immunity (Strong et al., 2022).

5.1 Triterpenes

Mushrooms contain a functional sterol called triterpene. These substances can impede the enzyme glucosidase further lowering the blood sugar concentration in animals. Triterpenes in *Ganoderma lucidum* reduce liver damage caused by LPS/D--Galactosamine-induced by lowering the expression of both tumor necrosis factor-alpha (TNF-α) and interleukin-6 (IL-6) expression (Hu et al., 2020). Anostane in *G. lucidum* increased the expression of heme oxygenase-1 and reduced the inflammatory reaction in RAW264.7 cells. Similarly, other workers also reported the antibacterial effects and antioxidant and antiviral functions of *P. mutilis* and mushrooms, respectively.

5.2 Polyphenols and Flavanoids

Plants and fungi contain polyphenols. Flavonoids present the most prominent example of polyphenols existing in nature. Mohanta (2020) reported that fungi have large numbers of genes and enzymes for flavonoid synthesis. Animal feed additives contained mainly contained agriculture byproducts or ferments. Fermented products are bestowed with high concentrations of plant-based flavonoids (Chuang et al., 2021). Phenols increase the expression of Nrf-2 and glutamate-cysteine ligase catalytic (GCLC) which are antioxidant systems. Hence, they ameliorate the metabolism of lipids and the quality of meat (Agbana et al., 2023).

5.3 Ergosterol

Fungi contain an anti-inflammatory compound called ergosterol that was first found in yeast (Chuang et al., 2021). It hinders the expression of lipopolysaccharide-induced RAW264.7 NF–κB and diminishes the inflammatory response of TNF-α and other substances (IC50 = 24.5 μg/mL). It can inhibit the enzyme cyclooxygenase-2 (COX-2) but does not affect nitric oxide production. After UV exposure, it is changed into vitamin D. In animals, vitamin D intake enhances serum levels and eventually increases antioxidant and lipid metabolism properties (Al-Rabia et al., 2021).

5.4 Adenosine

Adenosine is produced by ATP lysis and commonly prevails in fungi and animals (Iamtham et al., 2022). It functions in the regulation of energy utilization and diminishes the inflammatory response in animal cells. For the regulation of immune capacity, adenosine binds to G proteins, i.e., A1, A2A, A2B, and A3. This further inhibits the generation of deleterious enzymes and proteins. It combines with G protein receptors to improve the growth of the macrophage (A1) and represses the inflammatory system. Cordycepin present in *Cordyceps militaris* upregulates the pro-apoptosis genes such as P53, BCL2-associated X protein, caspase-3, and caspase-9. It downregulates the gene expression of antioxidants (Chuang et al., 2021).

5.5 Fungal Cell Wall and Polysaccharides

β-Glucan, mannoprotein, and chitin are the chief components of the fungal cell wall. β-Glucan, mannoprotein along chitin modulate immunity, enhance the antioxidant system, and diminish the inflammatory reaction (Liu et al., 2021). Yeast cell walls are a common example of fungal polysaccharides and predominantly consist of mannan-oligosaccharides and β- glucans, with a lesser amount of chitin. Mannan-oligosaccharides and β -glucans diminish the harmful effects of ochratoxin A on broilers and regulate their immunity while another function of β-glucans is the decrease in the inflammatory reaction and the prevention of the harmful effect of LPS (Peng et al., 2020).

Polysaccharides of *Cordyceps militaris* increase the phagocytosis of immune cells and ameliorate humoral and cell-mediated immunity. They are also known to enhance macrophage, monocyte, and lymphocyte activity and decrease oxidative stress by increasing SOD and Gpx activity (Das et al., 2021). Mushroom polysaccharides not only act as a type of prebiotic but also have antioxidant and anti-inflammatory abilities. Owing to the antioxidant and anti-inflammatory abilities of mushroom carbohydrates, they can be employed as animal food additives (Junior et al., 2020).

6 Fungi as Feed Additives

Several fungal species have been investigated such as *S. cerevisiae*, *A. oryzae*, *A. cinnamomea*, *Trichoderma pseudokoningii*, *Cordyceps militaris*, *Pleurotus ostreatus*, *Pleurotus eryngii*, and *Aureobasidium pullulans* to elucidate their role as feed additives. Feed additives are either added directly or a portion of a diet constituent is changed with either a probiotic, fermented material, or mushroom waste compost. They may increase the ability to produce and the health of the animal (Chuang et al., 2021).

The brewing industry employs fungal species, i.e., *S. cerevisiae* and *A. oryzae* which are the traditional probiotics. It has been earlier reported that co-fermentation of 0.1% of *S. cerevisiae* or *A. oryzae* with the enzyme phytase could decrease the inflammatory reaction in broilers. This was due to a reduction in the *Clostridium perfringens* count and suppression of the expression of the mRNA related to the inflammation (Kulkarni et al., 2022). Several enzymes were found in *A. oryzae* such as cellulase and hemicellulase that play important roles in feeding fiber and enhancing nutrient adsorption (de Oliveira Simas et al., 2024).

Trichoderma pseudokoningii was highly efficacious as a feed additive for poultry. However, its use as feed is not legal (Sanguine et al., 2022). It was previously found that adding *T. pseudokoningii* powder at concentrations of 0.1%, 0.2%, or 0.4% enhanced the animal growth and feed conversion ratio, and levels of lactate and lactic acid bacteria in cecum.

Traditional Chinese medicines involve the use of *C. militaris* and *A. cinnamomea* as they possess antioxidative, anti-inflammatory, and antitumor attributes (Hsieh et al., 2020). Mushrooms act as a potential feed additive as their waste residue is enriched with mycelium and other bioactive components. It was earlier reported that the waste residue of *C. militaris* contained higher polysaccharide content as compared to *Pleurotus eryngii*, *Pleurotus sajor-caju*, or *Fammulina velutipes*. They further found a significant enhancement in the antioxidant capacity via activation of the expression of the Keap-1 and Nrf-2 mRNA (Hsieh et al., 2020).

Another sought-after food additive is mushroom waste compost. It is enriched in mycelium and wood fiber that can enhance antioxidants and adipolysis in animals. However, its effect on adipose tissue metabolism is not clear. The enhancement of the expression of adipolysis-related mRNA enhances the expression may be due to the expression of KCTD15 and adiponectin (Chuang et al., 2021). Anti-obesity effect of vitamin D has been gaining momentum among scientists (Papoutsis et al., 2020). Vitamin D addition could increase antioxidant activity, and reduce cholesterol and triglyceride content in serum. They increase adipolysis by increasing the expression of the peroxisome proliferator-activated receptor (PPAR) and perilipin-2.

Hong et al. (2024) reported that the PPARγ and CCAAT enhancer-binding protein α (CEBP α) could be enhanced by *Pleurotus ostreatus* and its bioactive like anthraquinone could be amplified and eventually lead to the decreased accumulation of adipose accumulation. Several workers investigated and found that increasing the expression of Nrf-2 and GCLC mRNA, in mushroom waste compost could increase the antioxidant activity and reduce MDA content in poultry (Chuang et al., 2021).

In a nutshell, whether they are employed in Chinese medicine or as edible species, fungi act as a beneficial and potent candidate for animal health and protect stressful environments (Chuang et al., 2021).

7 Conclusion

In modern animal husbandry, feed additives have become indispensable as they offer a plethora of probabilities for enhancing the health, growth, and productivity of livestock. These include vitamins, minerals, enzymes, probiotics, and antioxidants. They play crucial roles in addressing nutritional inadequacy, increasing feed efficiency, and mitigating the ill effects of diseases. The classification of feed additives constitutes a wide array of compounds each having specific action tailored to different aspects of animal nutrition and health. The use of probiotics and prebiotics which have a positive effect on gut health and the functioning of the immune system has been emerging significantly and replacing conventional antibiotic therapy due to the development of antimicrobial resistance. The employment of enzymes originating from fungi species is known to hydrolyze complex feed components, thereby improving nutrient availability and digestion. Additionally, antioxidants protect animals from oxidative stress, contributing to better health and longevity. Traditionally, Fungi were not linked with animal feed but with advanced studies and recent findings, they have transpired as a remarkable food source owing to their nutrient-rich profile and sustainable production methods.

European Union has approved several fungal species such as *Saccharomyces cerevisiae, Fusarium venenatum,* and *Yarrowia lipolytica*, as they are potent protein sources. These fungi can grow several types of wastes in agriculture, paper industry, and food processing, and contribute towards the development of a circular economy and reduction of environmental pollution. Furthermore, fungal biomass production needs smaller footprints of land, water, and resources in comparison to traditional feed crops, making it a more sustainable and viable option. The astounding potential of fungi to produce various bioactives such as antioxidants, polyphenols, and tannins further underscores their importance as feed additives. Fungal enzymes can degrade plant polymers in feed, enhancing nutrient availability and feed efficiency. Moreover, fungi can synthesize antimicrobial peptides and other metabolites that promote gut health and protect against pathogens. The integration of these functional elements into animal feed can lead to better health and productivity and lesser dependence on synthetic additives.

Inculcating fungi into animal feed as additives offers multifaced benefits. It has been reported previously in many studies the inclusion of fungal enzymes such as cellulase, amylase, and protease assist efficient degradation of complicated polymers such as carbohydrates, proteins, and fibers present as feed components, ensuring more efficacious use of nutrients in poultry. This can improve weight gain and high-protein meat. Another advantage of fungi-based food additives is that they significantly reduce the cost by decreasing the requirement for additional feed supplements. They increase the absorption of nutrients, thus lesser quantity of feed is needed to accomplish the optimal growth as compared to the traditional additives which contributes towards the lower feed costs. These bioactive components of fungal feed give protection to the animals from detrimental diseases by subsiding the bacterial, fungal, and viral growth in the gastrointestinal tract. The antimicrobial attributes of fungal bioactives reduce the probability of the incidence of diseases. These factors further minimize the treatment cost and contribute to improved overall animal health. The environmental benefits of using fungi-based feed additives for animals bestow additional advantages as cultivation of fungal biomass usually requires very less resources compared to traditional feed ingredients. Moreover, some strains of fungi can be cultivated even on agricultural wastes such as crop residues, which are otherwise an environmental menace. This contributes to the circular economy, reducing waste and making feed production more sustainable. Some specific strains of fungi synthesize some products that can modify the rumen microflora of animals like cows and sheep. They thus encourage the growth of microorganisms that produce lesser amounts of methane as a byproduct of digestion.

Fungal feeds are gaining extensive popularity day by day, but the augmentation of cultivation processes remains the major constraint associated with it. More economical production can be accomplished by the evolution of efficacious downstream. Several studies are being conducted worldwide for the development of an economical and sustainable fungi-based production process. However, while developing the process, vigilance regarding the production of mycotoxin-free products should be taken into account. Further, this can be accomplished by genetic alteration, leading to research initiatives like EUROFUNG, which focus on strain improvement programs utilizing advanced technologies. Moreover, studies on the selection of valuable enzymes produced by fungi and procedures of downstream them need to be standardized to raise the scope of fungal food additives in animals.

Acknowledgments The authors are thankful to Punjab Agricultural University, Ludhian, Punjab.

References

Acamovic, T., & Brooker, J. (2005). Biochemistry of plant secondary metabolites and their effects in animals. *The Proceedings of the Nutrition Society, 64*(3), 403–412.

Agbana, J. A., Emmanuel, O. S., Aliyu, A. K., Ete, O. M. I., et al. (2023). Effect of dietary supplementation of *Pleurotus ostreatus* polyphenol extract on antioxidant enzymes profile and antioxidant gene expressions in weaned rabbits. *Nigerian Journal of Animal Science and Technology, 6*(1), 91–99.

AICRP-PB (All India Coordinated Research Project on Poultry Breeding). (2020). *Annual report*. ICAR-Directorate of Poultry Research.

Alhussien, M. N., Tiwari, S., Panda, B. S. K., Pandey, Y., Lathwal, S. S., Dang, A. K., et al. (2021). Supplementation of antioxidant micronutrients reduces stress and improves immune function/response in periparturient dairy cows and their calves. *Journal of Trace Elements in Medicine and Biology, 65*, 126718.

Al-Rabia, M. W., Mohamed, G. A., Ibrahim, S. R. M., Asfour, H. Z., et al. (2021). Anti-inflammatory ergosterol derivatives from the endophytic fungus *Fusarium chlamydosporum*. *Natural Product Research, 35*(23), 5011–5020.

Aranda-Aguirre, E., Robles-Jimenez, L. E., Osorio-Avalos, J., Vargas-Bello-Pérez, E., Gonzalez-Ronquillo, M., et al. (2021). A systematic review on the role of exogenous enzymes on the productive performance at weaning, growing and finishing in pigs. *Veterinary and Animal Science, 14*, 100195.

Baldi, A., Cheli, F., Savoini, G., et al. (2020). Role of beta-carotene in the health and performance of dairy cows: A review. *Italian Journal of Animal Science, 19*(1), 324–336.

Bilal, R.M., Tahir, M.A., Shahzad, A., Farag, M.R., Siddiq, A., El-Shall, N.A., Dhama, K., Elnesr, S.S., Alagawany, M. (2023). Yeast and derived products: their uses in preventing mycotoxins in poultry feeds. *World's Poult Sci J, 79*(2), 351–378.

Bravo, D., Pirgozliev, V., Rose, S. P., et al. (2014). A mixture of carvacrol, cinnamaldehyde, and capsicum oleoresin improves energy utilization and growth performance of broiler chickens fed maize-based diet. *Journal of Animal Science, 92*, 1531–1536.

Calder, P. C. (2021). Health benefits of omega-3 fatty acids. In *Omega-3 delivery systems* (pp. 25–53). Academic.

Chai, K. F., Ng, K. R., Samarasiri, M., Chen, W. N., et al. (2022). Precision fermentation to advance fungal food fermentations. *Current Opinion in Food Science, 47*, 100881.

Chapman, H. D., Barta, J. R., Blake, D., Gruber, A., Jenkins, M., Smith, N. C., Suo, X., Tomley, F. M., et al. (2020). A selective review of advances in coccidiosis research. *Advances in Parasitology, 80*, 93–171.

Chuang, W. Y., Lin, L. J., Shih, H. D., Shy, Y. M., Chang, S. C., Lee, T. T., et al. (2021). Intestinal microbiota, anti-inflammatory, and anti-oxidative status of broiler chickens fed diets containing mushroom waste compost by-products. *Animals, 11*(9), 2550.

Chugh, R. M., Mittal, P., Mp, N., Arora, T., Bhattacharya, T., Chopra, H., Cavalu, S., Gautam, R. K., et al. (2022). Fungal mushrooms: A natural compound with therapeutic applications. *Frontiers in Pharmacology, 13*, 925387.

Das, G., Shin, H. S., Leyva-Gómez, G., Prado-Audelo, M. L. D., Cortes, H., Singh, Y. D., Patra, J. K., et al. (2021). *Cordyceps* spp.: A review on its immune-stimulatory and other biological potentials. *Frontiers in Pharmacology, 11*, 602364.

de Oliveira Simas, A. L., de Alencar Guimarães, N. C., Glienke, N. N., Galeano, R. M. S., de Sá Teles, J. S., Kiefer Giannesi, G. C., et al. (2024). Production of phytase, protease and xylanase by *Aspergillus niveus* with rice husk as a carbon source and application of the enzymes in animal feed. *Waste and Biomass Valorization*, 1–13.

Derbyshire, E. J., & Delange, J. (2021). Fungal protein—What is it and what is the health evidence? A systematic review focusing on mycoprotein. *Frontiers in Sustainable Food Systems, 5*(18), 1–14.

Duffield, T. F., & Lissemore, K. D. (2020). Effects of monensin on health and production of dairy cows: A systematic review and meta-analysis. *Journal of Dairy Science, 103*(1), 571–584.

Dufossé, L., Fouillaud, M., Caro, Y., et al. (2021). Fungi and fungal metabolites for the improvement of human and animal nutrition and health. *Journal of Fungi, 7*(4), 274.

Dunuweera, A. N., Nikagolla, D. N., Ranganathan, K., et al. (2021). Fruit waste substrates to produce single-cell proteins as alternative human food supplements and animal feeds using baker's yeast (Saccharomyces cerevisiae). *Journal of Food Quality*, 9932762.

EFSA Panel on Biological Hazards (BIOHAZ), Koutsoumanis K, Allende A, Álvarez-Ordóñez A, Bolton D, Bover-Cid S, Herman L (2023) Update of the list of qualified presumption of safety (QPS) recommended microorganisms intentionally added to food or feed as notified to EFSA. *EFSA J 21*(1): e07747.

Erdman, R. A., & Cunningham, K. D. (2019). Effects of sodium bicarbonate on ruminal pH and lactation performance in dairy cows. *Journal of Dairy Science, 102*(5), 4661–4671.

Faraz, A. (2018). Feed additives and their use in livestock feed, *3*. Farmer Reformer 3, 10.

Feed additive classifications based on Europe regulation and guideline. (2018). Bevenovo Co Limited.

Filiatrault-Chastel, C., Heiss-Blanquet, S., Margeot, A., Berrin, J. G., et al. (2021). From fungal secretomes to enzyme cocktails: The path forward to bioeconomy. *Biotechnology Advances, 52*, 107833.

Gänzle, M. G. (2021). Lactic acid bacteria and their bacteriocins: The role of probiotics in gut health. *Frontiers in Microbiology, 12*, 691682.

Gibson, G.R., & Roberfroid, M.B. (1995). Dietary modulation of the human colonic microbiota: introducing the concept of prebiotics. *J Nutr, 125*(6), 1401–1412.

Guo, H., Gu, J., Wang, X., Nasir, M., Yu, J., Lei, L., Wang, Q., et al. (2020). Elucidating the effect of microbial inoculum and ferric chloride as additives on the removal of antibiotic resistance genes from chicken manure during aerobic composting. *Bioresource Technology, 309*, 122802.

Guo, J., Ma, Q., & Zhang, S. (2021). Calcium salts of EDTA as sequestrants in feed: Applications and effectiveness. *Journal of Agricultural and Food Chemistry, 69*(4), 1204–1212.

Hernandez, F., Madrid, J., Garcia, V., Orengo, J., Megias, M. D., et al. (2020). Influence of two plant extracts on broilers performance, digestibility, and digestive organ size. *Poultry Science, 93*(3), 539–546.

Hong, S., Park, S., Lee, J., Park, S., Park, J., Lee, Y., et al. (2024). Anti-obesity effects of *Pleurotus ferulae* water extract on 3T3-L1 adipocytes and high-fat-diet-induced obese mice. *Nutrients, 16*(23), 4139.

Hsieh, Y. C., Lin, W. C., Chuang, W. Y., Chen, M. H., Chang, S. C., Lee, T. T., et al. (2020). Effects of mushroom waster medium and stalk residues on the growth performance and oxidative status in broilers. *Animal Science, 34*(2), 265–275.

Hu, Z., Du, R., Xiu, L., Bian, Z., Ma, C., Sato, N., Hattori, M., Zhang, H., Liang, Y., Yu, S., Wang, S., et al. (2020). Protective effect of triterpenes of *Ganoderma lucidum* on lipopolysaccharide induced inflammatory responses and acute liver injury. *Cytokine, 127*, 154917.

Iamtham, S., Kaewkam, A., Chanprame, S., Pan-utai, W., et al. (2022). Effect of *Spirulina* biomass residue on yield and cordycepin and adenosine production of *Cordyceps militaris* culture. *Bioresource Technology Reports, 17*, 100893.

ICAR-Directorate of Poultry Research, Rajendranagar, Hyderabad, Telangana, India. in Food: Practical Applications 149–173.Jami, E., & Mizrahi, I. (2018). The effect of probiotics on the health and performance of livestock: A review. *Animal Microbiome, 1*(1), 2–15.

Jindal, P., Bedi, J., Singh, R., Aulakh, R., Gill, J., et al. (2021). Phenotypic and genotypic antimicrobial resistance patterns of Escherichia coli and Klebsiella isolated from dairy farm milk, farm slurry and water in Punjab, India. *Environmental Science and Pollution Research, 28*, 28556–28570.

Junior, J. A., Vieira, Y. A., Cruz, I. A., Vilar, D. S., Aguiar, M. M., Torres, N. H., Bharagava, R. N., Lima, Á. S., deSouza, R. L., Ferreira, L. F. R., et al. (2020). Sequential degradation of raw vinasse by a laccase enzyme producing fungus *Pleurotus sajor-caju* and its ATPS purification. *Biotechnology Reports, 25*, e00411.

Kawabata, F., & Tabata, S. (2022). Bitter taste perception in chickens. *The Journal of Poultry Science, 59*(1), 8–15.

Khan, R., Brishti, F. H., Arulrajah, B., Goh, Y. M., Abd Rahim, M. H., Karim, R., Saari, N., et al. (2024). Mycoprotein as a meat substitute: Production, functional properties, and current challenges-a review. *International Journal of Food Science and Technology, 59*(1), 522–544.

Kiarie, E., & Mills, A. (2019). The use of *Saccharomyces cerevisiae* in animal nutrition to improve performance and health. *Journal of Animal Science and Biotechnology, 10*(1), 3–10. https://doi.org/10.1186/s40104-019-0324-9

Kim, M. H., & Kim, D. (2021). Effects of *Aspergillus oryzae* fermentation extract on feed intake, milk production, and rumen fermentation parameters in lactating dairy cows. *Animal Feed Science and Technology, 274*, 114850.

Kulkarni, R. R., Gaghan, C., Gorrell, K., Sharif, S., Taha-Abdelaziz, K., et al. (2022). Probiotics as alternatives to antibiotics for the prevention and control of necrotic enteritis in chickens. *Pathogens, 11*(6), 692.

Kumar, P., Mishra, S., Singh, R., et al. (2020). Beta-glucan as a feed additive for improving gut health and immune response in animals. *Animal Feed Science and Technology, 261*, 114394.

Kumar, K., Sinha, R. R. K., Singh, P. K., Kishor, A., Kumar, P., Chauhan, S. S., Kumar, R., et al. (2023). Effects of organic selenium and probiotic supplementation on the performance of indigenous and crossbred dairy cows during summer in the tropics. *Ruminants, 3*(2), 158–171.

Lipiński, K., Mazur-Kuśnirek, M., Antoszkiewicz, Z., Makowski, Z., Śliżewska, K., Siwicki, A., Gesek, M., et al. (2021). The effect of synbiotics and probiotics on the growth performance, gastrointestinal function and health status of turkeys. *Archives of Animal Nutrition, 75*(5), 376–388.

Liu, Y., Wu, Q., Wu, X., Algharib, S. A., Gong, F., Hu, J., Wang, Y., et al. (2021). Structure, preparation, modification, and bioactivities of β-glucan and mannan from yeast cell wall: A review. *International Journal of Biological Macromolecules, 173*, 445–456.

Lopes, F., Silva, L. F. P., Dill-Espe, M. C., et al. (2020). Effects of magnesium oxide supplementation on ruminal fermentation and milk production in dairy cows. *Journal of Dairy Science, 103*(9), 8620–8629.

Lücking, R., Aime, M. C., Robbertse, B., Miller, A. N., Aoki, T., Ariyawansa, H. A., Schoch, C. L., et al. (2021). Fungal taxonomy and sequence-based nomenclature. *Nature Microbiology, 6*(5), 540–548.

Matté, E. H., Luciano, F. B., Evangelista, A. G., et al. (2023). Essential oils and essential oil compounds in animal production as antimicrobials and anthelmintics: An updated review. *Animal Health Research Reviews, 24*(1), 1–11.

Miller, R. M., & Johnson, H. R. (2021). Advances in fungal feed additives for enhancing gut health in livestock. *Livestock Science, 244*, 104372.

Mohammadi, R., & Rouzbehan, Y. (2019). Effects of sodium bentonite supplementation on nutrient digestibility and performance of Holstein steers. *Animal Feed Science and Technology, 253*, 14–23.

Mohanta, T. K. (2020). Fungi contain genes associated with flavonoid biosynthesis pathway. *Journal of Functional Foods, 68*, 103910.

Mozin, S., Hatta, U., Sarjuni, S., Adjis, A., et al. (2023). The effect of lutein on egg component. In *IOP conference series: Earth and environmental science* (Vol. 1253, No. 1, p. 012047). IOP Publishing.

Nava, G. M., Kolkman, J., Ramer-Tait, A. E., et al. (2021). Probiotics and gut health in livestock: A review of recent developments. *Animal Feed Science and Technology, 273*, 114793.

Nielsen, P., Bohr, S. S., Madsen, M., et al. (2020). The efficacy of anticaking agents in poultry feeds. *Feed Science and Technology, 269*, 49–56.

Nofal, A., Azzazy, M., Ayyad, S., Abdelsalm, E., Abousekken, M. S., Tammam, O., et al. (2022). Evaluation of the brown alga, *Sargassum muticum* extract as an antimicrobial and feeding additive. *Brazilian Journal of Biology, 84*, e259721.

NRC (National Research Council). (2012). *Nutrient requirements of swine* (11th ed.). National Academies Press.

Official Journal of the European Union (2003) REGULATION (EC) No 1831/2003 of The European Parliament and of the Council of 22 September 2003 on additives for use in animal nutrition.

Oladele, O. O., Ameji, N. O., Odey, M., Gurumyen, G. Y., Adanu, A. W., Bitrus, A. A., Unanam, E. S., et al. (2022). Management of acute aspergillosis in newly hatched chicks: A case report. *Nigerian Veterinary Journal, 43*(4), 64–73.

Ouwehand, A., & Von Wright, A. (2021). *The lactic acid bacteria: Volume 1: The genera Lactobacillus, Lactococcus, Leuconostoc, Pediococcus, and Carnobacterium*. Springer.

Pandey, A. K., Kumar, P., Saxena, M. J., et al. (2019). Feed additives in animal health. Springer International Publishing, Cham. In *Nutraceuticals in veterinary medicine* (pp. 523–543).

Papoutsis, K., Grasso, S., Menon, A., Brunton, N. P., Lyng, J. G., Jacquier, J. C., Bhuyan, D. J., et al. (2020). Recovery of ergosterol and vitamin D2 from mushroom waste - potential valorization by food and pharmaceutical industries. *Trends in Food Science and Technology, 99*, 351–366.

Patra, A., Prasath, V. A., Sutar, P. P., Pandian, N. K. S., Pandiselvam, R., et al. (2022). Evaluation of effect of vacuum frying on textural properties of food products. *Food Research International, 162*, 112074.

Patterson, J. A., & Burkholder, K. M. (2003). Application of prebiotics and probiotics in poultry production. *Poultry Science, 82*(4), 627–631.

Patton, R. S., & Huber, J. T. (2019). Efficacy of methionine hydroxy analog in improving lactation performance in dairy cows. *Animal Feed Science and Technology, 255*, 114233.

Peng, Q. H., Cheng, L., Kang, K., Tian, G., Mohammad, A. M., Xue, B., Wang, L. Z., Zou, H. W., Mathew, G. G., Wang, Z. S., et al. (2020). Effects of yeast and yeast cell wall polysaccharides supplementation on beef cattle growth performance, rumen microbial populations and lipopolysaccharides production. *Journal of Integrative Agriculture, 19*, 810–819.

Pirgozliev, V., & Bedford, M. R. (2013). Energy utilisation and growth performance of chicken fed diets containing graded levels of supplementary bacterial phytase. *The British Journal of Nutrition, 109*, 248–253.
Pirgozliev, V., Bravo, D., Mirza, M. W., Rose, S. P., et al. (2015). Growth performance and endogenous losses of broilers fed wheat-based diets with and without essential oils and xylanase supplementation. *Poultry Science, 94*, 1227–1232.
Pirgozliev, V., Rose, S. P., Ivanova, S., et al. (2019). Feed additives in poultry nutrition. *Bulgarian Journal of Agricultural Science, 25*(Suppl 1), 8–11.
Potter, E. L., & Arnold, D. H. (2021). Lasalocid: Mode of action and effects on cattle performance. *Journal of Animal Science, 99*(8), skab215.
Rafiei, F., & Khajali, F. (2021). Flavonoid antioxidants in chicken meat production: Potential application and future trends. *World's Poultry Science Journal, 77*(2), 347–361.
Rai, M., Pandit, R., Gaikwad, S., Kövics, G., et al. (2021). Nutritional properties of edible mushrooms and their significance as a functional food. *Innovative Food Science and Emerging Technologies, 37*, 246–257.
Ravi, R., Ganesh, S., Bhat, R., et al. (2021). Fungal biotechnology in food and feed processing: Industrial applications and future perspectives. *Current Opinion in Biotechnology, 38*, 62–71.
Reis, R. A., Costa, H. G., Vieira, R. A. M., et al. (2020). Effects of anhydrous ammonia on the nutritional value of tropical forages. *Journal of Animal Science and Biotechnology, 11*(1), 8–15.
Rychen, G., & Toussaint, H. (2018). Evaluation of enzymes for animal nutrition by the EFSA in the European Union. Academic Press, New York City. In *Enzymes in human and animal nutrition* (pp. 481–486).
Sanchez, A. R. R., & Rivas-Estilla, A. M. (2006). Role of probiotics in hepatic ischemia-reperfusion injury. *Journal of Gastroenterology and Hepatology, 21*, 647–656.
Sanguine, I. S., Cavalheiro, G. F., Garcia, N. F. L., dos Santos, M. V., Gandra, J. R., Buschinelli, R. H. D., Leite, R. S. R., et al. (2022). Xylanases of *Trichoderma koningii* and *Trichoderma pseudokoningii*: Production, characterization and application as additives in the digestibility of forage for cattle. *Biocatalysis and Agricultural Biotechnology, 44*, 102482.
Şengül, M., & Ufuk, S. (2022). Therapeutic and functional properties of beta-glucan, and its effects on health. *The Eurasian Journal of Food Science and Technology, 6*(1), 29–41.
Sethi, M., Prasher, I.B., Kapila, S. (2023). Sustainable utilization of fungal resources: applications in plant and animal health. In: Fungal resources for sustainable economy: current status and future perspectives. *Springer Nature Singapore*, pp 39–68.
Shahzad, M. A., Rehman, Z. U., Hussain, J., Khan, R. U., Ahmad, S., et al. (2018). Effect of feeding anionic salts on blood mineral status and performance of transition dairy cows. *Asian-Australasian Journal of Animal Sciences, 31*(6), 924–929.
Shakouri, M. D., Hossaini, S., Rafiee, M., et al. (2020). Impact of *Lactobacillus*-based probiotic supplementation on growth performance and feed conversion efficiency in broiler chickens. *Poultry Science, 99*(6), 3108–3115.
Shang, Y., Zhang, H., Yu, Y., et al. (2020). Effects of dietary yeast β-glucan on growth performance, immune responses, and gut microbiota in broiler chickens. *JASB, 11*(1), 1–10.
Sharif, M., Zafar, M. H., Aqib, A. I., Saeed, M., Farag, M. R., Alagawany, M., et al. (2021). Single cell protein: Sources, mechanism of production, nutritional value and its uses in aquaculture nutrition. *Aquaculture, 531*, 735885.
Sharma, V., Pathak, N. N., Yadav, R. N., et al. (2021). Impact of fungal phytase on growth performance and nutrient digestibility in broilers. *World's Poultry Science Journal, 9*(2), 123–130.
Smith, A. M., Thomas, S. M., Garcia, H. M., et al. (2020). Economic feasibility of fungal-based feed additives in animal production systems. *Agricultural Economics, 51*(2), 221–231.
Spears, J. W., & Kegley, E. B. (2020). Effects of zinc methionine on growth performance and immune function in beef cattle. *Journal of Animal Science, 98*(2), skz398.
Strong, P. J., Self, R., Allikian, K., Szewczyk, E., Speight, R., O'Hara, I., Harrison, M. D., et al. (2022). Filamentous fungi for future functional food and feed. *Current Opinion in Biotechnology, 76*, 102729.

Thakur, M., & Kumar, P. (2021). Impact of fungal-based feed additives on the growth performance and immune response in ruminants. *J Dairy Sci, 104*(7), 7912–7924.

Trotta. R. J., Kreikemeier, K. K., Foote, S., McLeod, K. R., Harmon, D. L. (2023). Influence of anti-coccidial compounds and phytogenic saponin extracts on in vitro and in vivo ruminal fermentation and methane production of cattle. *Animals, 13*(14), 2308.

Wang, H., Long, W., Chadwick, D., Velthof, G.L., Oenema, O., Ma, W., Wang, J., Qin, W., Hou, Y., Zhang, F. (2020). Can dietary manipulations improve the productivity of pigs with lower environmental and economic cost? A global metaanalysis. *Agric Ecosyst Environ, 289*, 106748.

Wieland, M., & Liebhard, P. (2019). The use of propionic acid as a feed preservative for dairy cows. *Journal of Animal Science, 97*(12), 5015–5024.

Wijayawardene, N. N., Hyde, K. D., Dai, D. Q., Sánchez-García, M., Goto, B. T., Magurno, F., et al. (2022). *Outline of Fungi and fungus-like taxa.*

Wise, T. (2024). *Effects of exogenous enzymes on dietary nutrient digestibility for broiler chickens and growing pigs* (Doctoral dissertation, Purdue University).

World Health Organization. (2017). *WHO guidelines on use of medically important antimicrobials in food-producing animals.*

Xia, W. H., Wang, L., Niu, X. D., Wang, J. H., Wang, Y. M., Li, Q. L., Wang, Z. Y., et al. (2021). Supplementation with beta-1, 3-glucan improves productivity, immunity and antioxidative status in transition Holstein cows. *Research in Veterinary Science, 134*, 120–126.

Zanton, G. I., Bowman, G. R., Clark, J. H., et al. (2018). Effects of rumen-protected choline supplementation on lactation performance and metabolism of periparturient dairy cows. *Journal of Dairy Science, 101*(8), 7079–7093.

Zhang, Q., Li, X., Yang, M., et al. (2020). Sequestrants in animal feeds: Function and efficacy. *Animal Feed Science and Technology, 264*, 114467.

Zhang, Y., Wang, D., Chen, Y., Liu, T., Zhang, S., Fan, H., Li, Y., et al. (2021). Healthy function and high valued utilization of edible fungi. *Food Science and Human Wellness, 10*(4), 408–420.

Zhao, R., Hu, Q., Ma, G., Su, A., Xie, M., Li, X., Chen, G., Zhao, L. (2019). Effects of Flammulina velutipes polysaccharide on immune response and intestinal microbiota in mice. *J Funct Foods, 56*, 255–264.

Zhao, S., Zhang, H., Zhang, Y., Wang, L., et al. (2023). Phytase supplementation in animal feed: Impact on phosphorus utilization and environmental sustainability. *Journal of Animal Science and Biotechnology, 14*, 25.

The Effects of Fungal Feed Additives in Animals

Bhavya Sharma, Lavanya Singh, Gurumurthy Dummi Mahadevan, and Nahid Masood Siddiqui

Abstract Fungal probiotics boost animal health. There is potential for using fungus probiotics as feed supplements. Fungal feed additives, as we know from earlier research, increase antioxidant capacity and reduce inflammatory responses, which further improves animal health and growth performance. Fungi such as *Saccharomyces cerevisiae, Antrodia cinnamomea, Aspergillus oryzae*, etc. are rich in many functional groups like polysaccharides, adenosine, and laccases, which are important for the immune system, antioxidants, and anti-inflammatory properties. It also increases body weight and improves egg production. Any residues in feeds have an impact on the amount of feed used, how nutrients are used, and ultimately the health and productivity of the animals. Fungal feed additives are very valuable, therefore, it is important to understand how to use and protect them (as shown in Fig. 1).

Keywords Fungal feed additives · Animal nutrition · Digestibility · Mycotoxins · Gut health · Probiotics and prebiotics · Immunomodulatory properties

1 Introduction

The fungal additive is a biodegradable and natural substitute for chemical additives and antibiotics. For this very reason, the role of fungus in livestock feed additives is significant in the present livestock industry. The main composition of the fungal feed additive consists of fungi or products resulting from fungi (Abdelhameed & Metwally, 2022). Better digestion, nutrient intake, and higher immunity are some of the primary functions of fungi feed additives. The presence of fungal probiotic enzymes and metabolites further enhances an animal's health development and overall performance. They usually do this by breaking down carbohydrates, proteins, and fibers which would have otherwise not been digested into simple

B. Sharma · L. Singh · G. D. Mahadevan · N. M. Siddiqui (✉)
Amity Institute of Biotechnology, Amity University, Noida, Noida, Uttar Pradesh, India
e-mail: nsiddiqui@amity.edu

A. Gupta (ed.), *Prospects of Fungal Biotechnologies for Livestock Volume 1*,
Fungal Biology, https://doi.org/10.1007/978-3-031-90994-8_3

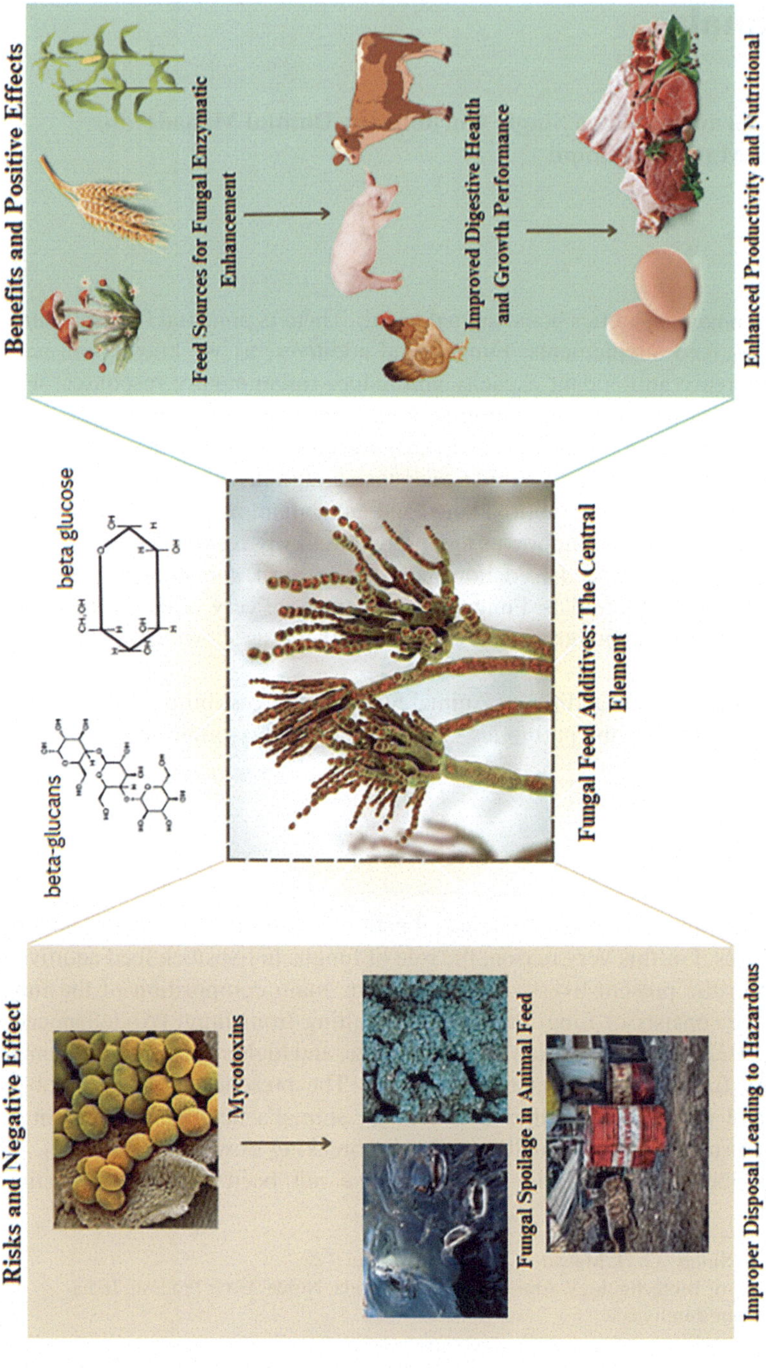

Fig. 1 Abstract image

components the body part of an animal can easily absorb (Amara & El-Baky, 2023). They are said to contribute to promoting efficiency in the growth of animals through the establishment of gut health, lowered pathogenic bacteria, and establishing a well-balanced microbiome (Amara & El-Baky, 2023). Advances in genetics and biotechnology have further developed the nutritional feed additives made from fungi more recently. Yet, feeding additives is not a recent discovery for animal nutrition. Many old or traditional practices involved fermentation by fungi on feed, but scientists did not understand the actual role fungi play in the quality of feed until later (Mapook et al., 2022). Around the mid-twentieth century, research into fungal probiotics and feed additives experienced some development. Experts were becoming acquainted with the prospects of applications with fungi in particular if this can pass through as probiotics, thus assisting in the enhancement of the animal's health (An et al., 2020). Some specific species of fungal have been reported to be useful for gut flora efficiency in the digestive system of animals. They are, among others, *Saccharomyces cerevisiae, Aspergillus oryzae, Trichoderma*, etc. (Table 1) (Ejiofor et al., 2021).

One particularly interesting area to explore with regard to the effects of fungus feed additives is how they can help enhance the quality of life for our animal friends. Fungal feed additives are important for keeping the overall health and performance of animals due to their role in maintaining gut health, enhancing absorption of nutrition, and consequently protecting against harmful chemicals such as mycotoxins (Ayalew et al., 2022). These additives mainly contribute to animal nutrition because they can enhance digestion, improve well-being, and increase productivity by adding them to the food of an animal (Bahram & Netherway, 2021). Fungus feed additions have a big impact on the health and well-being of animals. These additions involve mycotoxin binders, yeast, etc., which are used to help advance nutrient absorption, enhance gastrointestinal health, and protect animals from harmful chemicals such as mycotoxins (Ejiofor et al., 2021). Adding fungal feed supplementations to their rations would enable farmers and other livestock producers to fatten up their performance, enhance the digestion of their cattle, and generally feel better. The advantages provided by the fungus-based feed supplements underpin the significance of fungal feeds in animal feeding processes. All research into fungal feed

Table 1 Examples of fungal feed additive

Fungi type	Examples	Roles in feed additives
Aspergillus spp.	Aspergillus *niger*, A. *orzyae*	Produces enzymes (e.g., amylase, cellulase) to improve digestion
Saccharomyces spp.	Saccharomyces *cerevisiae*	Enhances gut microbiota and nutrient uptake
Trichoderma spp.	Trichoderma *reesei*	Produces cellulase and hemicellulose for fiber breakdown
Penicillium spp.	Penicillium *roqueforti*	Produces secondary metabolites with potential health benefits

is completely focused on the development of safe, functional, and more environmentally friendly alternatives to improve the health performance of livestock (Chuang et al., 2020).

Most studies regarding the use of fungal feed additives focus on the enhancement of growth as well as feed efficiency of animals through improved digestibility and nutrient utilization. The fungi break down the complex feed components through the production of enzymes, thus making them more available for absorption by the animal. Better growth results from increased nutrient assimilation (Abdala-Díaz et al., 2024). Quite importantly, another aspect in which the fungal additives play a role is the improvement of gut health and immunity. Fungal probiotics have a positive impact on the balance of gut microbiomes, decreased growth of pathogenic bacteria, enhanced immunity, and promotion of health in the gastrointestinal tract (Bampidis et al., 2019). Fungal feed additives, particularly for ruminants, offer environmental benefits through a reduction in methane levels. This therefore further promotes more environmentally friendly systems of production in the livestock sector. Further, these additives are targeted as a natural alternative to chemical additives and antibiotics, helpful in catering to the global alarm about antibiotic resistance and serving a developing consumer preference for natural and safe animal products (Calla-Quispe et al., 2020).

2 Functional Properties of Fungi

2.1 Enzymatic Functions of Fungi

Fungi produce several enzymes. The production of cellulases is probably the most important enzymatic function of fungi. The main function of cellulases is to break down cellulose which is a major component of cell walls (Carr et al., 2023). This process is essential for herbivore animals, especially ruminants. The indigestible fibers from the waste of plants are broken into simple sugars like glucose by the process, thereby raising feed digestibility and releasing energy (Cheng et al., 2022). Except for cellulase, proteases are also produced by fungi. Proteases break proteins into smaller peptides or directly to amino acids and help in their digestion, thereby releasing proteins for absorption and utilization in growth, muscle building, and health (Cheng et al., 2021). Through hydrolysis of triglycerides into free fatty acids and glycerol, the action of fungal lipases further contributes to fat metabolism, thus aiding better digestion and energy production from fats (Kholif et al., 2024). Enhancement of feed utilization efficiency is by united actions of enzymes, which leads to higher growth rates, With the integration of the modern formulation of animal feed the promotion of overall animal health has taken place.

Cellulases for fiber digestion: Fungi secrete cellulases which are hydrolytic enzymes, cellulose is the major component of cell walls and cellulases degrade it to glucose and other simple sugars (Cochereau et al., 2022). The process is very important for ruminants. Animals cannot digest cellulose alone, fungal cellulases are one of the key factors that should be used in improving feed digestibility (Corbu et al., 2023). It allows animals to fetch a higher amount of nutrients and energy from their feed for better growth and productivity. Also, allowing animals to consume a greater number of fibers in their feed for the generation of usable energy for cellulases improves overall feed efficiency (Floudas et al., 2022). This is key towards maximum feed value, majorly for diets well represented by roughages. In animal nutrition, for better digestion, nutrition absorption and health application of fungal cellulases is essential.

Proteases for protein breakdown: For easier assimilation of proteins for animals, fungal proteases break protein into smaller peptides and amino acids. For effective utilization of proteins in animal diets during muscle building, tissue repair, and all other metabolic processes the process is very essential (Ghosh et al., 2023). The ability to enhance the digestibility of plant proteins otherwise often so resistant to full utilization by animals is one of the main advantages of fungal proteases (Guan et al., 2023). Proteins are broken up by proteases into their very raw components so that amino acids can be made available for optimal growth, for effective functioning of the immune system, and for the good health of animals (Hess et al., 2020). Enzymic activity results in the promotion of feed efficiency and minimizes intake of expensive protein supplements, thus making nutrition cheap. For example, certain fungal proteases are of special importance to young and monogastric animals, such as pigs and poultry, as they produce only a limited number of proteases on their own, thereby ensuring better digestibility of nutrients at an early stage of production (Hu et al., 2023).

Lipases and their role in fat: The enzymes that are responsible for catalyzing the hydrolysis of triglycerides are lipases, large proportion of fat in the diets of animals constitutes them (Hüttner et al., 2020). Free fatty acids and glycerol are their main products. This is the main process for fat digestion and absorption among animals, as fats are an excellent source of energy (Uwineza et al., 2024) and important biological functions such as hormone production and structural cell membrane integrity are provided by them (Wang et al., 2024). The degradation of dietary fats through activities of lipases from fungi facilitates improved fat metabolism in animals, conversion of fats into available energy occurs smoothly. Growth, reproduction, and performance to continue unabated are enabled by improved fat utilization (Salazar-Alekseyeva et al., 2024). Lipases can be used to increase the efficient utilization of energy by formulating animal feed (Stoev, 2024). One may then need lower fats and high-fat feed additives in the feeds, which results in increased feed conversion rates.

2.2 Fungal Polysaccharides

Complex carbohydrates present within the fungal cell walls are fungal polysaccharides, which when utilized in animal nutrition impart significant health benefits. Some of the key polysaccharides, such as chitin and beta-glucans (as shown in Fig. 2), have been made to make animals healthy (Lu & Liu, 2024).

The structural component of fungal cell walls is chitin, which carries antimicrobial and immune-enhancing properties. The activation of the immune system by stimulating important immune cells such as macrophages is done by beta-glucans, enhancing a stronger immune response and improving gut health through the enhancement of beneficial gut bacteria. Gut barrier function is also done by these compounds, enhancing digestion and assimilation of nutrients (Mirończuk-Chodakowska et al., 2021); outstanding antioxidant properties that neutralize oxidizing free radicals and help reduce oxidative stress are also properties of fungal polysaccharides. Protection of animals from the effects of stress, disease, and aging is also done by them, thereby contributing to better health, growth, and productivity in livestock.

Chitin and beta-glucans are two important polysaccharides found in the cell walls of fungi that provide numerous health benefits in animal nutrition. Chitin, a fibrous structural component, is often broken down into chitosan (Mirończuk-Chodakowska et al., 2021), which exhibits antimicrobial properties and helps to enhance the immune system. This makes animals more resilient to infections and supports overall health. Beta-glucans, on the other hand, are well-known for their

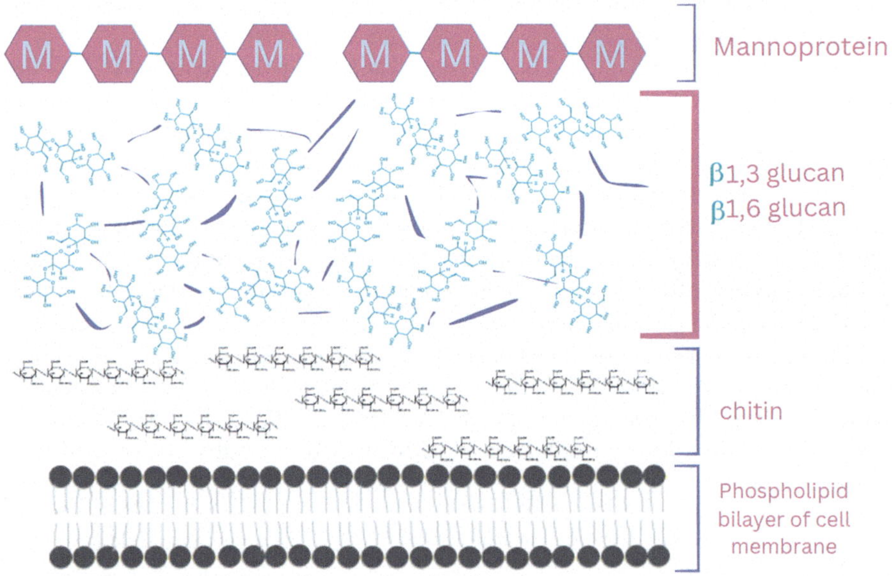

Fig. 2 Schematic overview of fungi cell wall composition

immunomodulatory effects (Zhu et al., 2023). They stimulate immune cells like macrophages, boosting the animal's immune response and helping to protect against diseases (Weiland et al., 2023). In addition to immune support, beta-glucans promote gut health by encouraging the growth of beneficial gut bacteria and improving gut barrier function, which leads to better digestion and nutrient absorption. Together, chitin and beta-glucans offer valuable benefits, enhancing immunity, gut health, and overall animal well-being.

Chitin and beta-glucans are two of the largest polysaccharides present in the walls of fungal cells, both are responsible for greatly enhancing animal nutrition with their benefits on health (Wan Mohtar et al., 2022). Chitin functions as a fibrous structural component, chitosan is a decomposed form of chitin that has antimicrobial properties and enhances the immunity of the body, thereby making animals more resistant to infections, making them healthier and better for well-being (Müller et al., 2023). Beta-glucans are highly immunomodulatory. Macrophages are active animals' immune cells; they endow an animal's immunity system with enhanced protection against diseases. Besides enhancing immunity, better gut health through increased production of healthy intestinal flora besides fortifying the gut barrier and allowing improved digestion and absorption of nutrients is promoted by beta-glucan (Wan Mohtar et al., 2022). Combining chitin and beta-glucans provides the most beneficial useful effects that enhance immunity, gut health, and general well-being of animals.

Fungal polysaccharides, mainly beta-glucans, possess antioxidant effects that maintain health in animals. Reduction of damage brought by oxidative processes by neutralizing the harmful free radicals is the role of these polysaccharides (Sun & Shahrajabian, 2023). Factors causing this form of stress include diseases, dietary deficiency, or other environmental stresses, all of which disrupt normal growth and effective immune response in an animal. Fungal polysaccharides minimize inflammation and protect tissue against damage as they act as antioxidants (Pawar et al., 2023), leading to better immune response and recovery from stress or illness. Positive acceleration is healthier aging in livestock and enhances productivity, in this respect, fungal polysaccharides are considered an essential factor that could be inculcated into animal nutrition to keep the animal healthy and well-conditioned over long periods (Rutkowski et al., 2023).

2.3 Essential Vitamins and Minerals from Fungi

Fungi are storehouses of several other vitamins and minerals that significantly impact animal nutrition. Riboflavin, niacin, and folate from fungal species are among the available B-vitamins that animals employ in energy metabolism, development of red blood cells, and positive contribution to optimal growth and reproductive health, and overall vigor is a function of the general nervous system. Apart from vitamins, the bioavailability of some critical minerals like zinc, iron, and magnesium that are readily absorbed by animals is improved by fungi, in some of the physiological processes within animals' bodies, such as immune system functioning, enzyme activation, and bone development Such minerals play a critical role (Desamero et al., 2021; Wang et al., 2022). Therefore, through the provision of such critical vitamins and minerals, fungi have a fundamental role in improving the nutritional profile of animal feed, therefore resulting in better health, growth, and productivity levels in livestock.

For the metabolic and physiological functions of animals, fungal-derived B-vitamins are necessary. Riboflavin, niacin, and folate are the most widely utilized B-vitamins, which are associated with energy metabolism or the conversion of feeds into usable energy by animals for growth, maintenance, or reproduction. Among others, the use of riboflavin is an absolute necessity, while niacin was initially identified as contributing significantly to the metabolism of fats, carbohydrates, and proteins (Wang et al., 2022). In DNA synthesis as well as red blood cell production, folate plays a crucial role, thus plays a crucial role in healthy growth and immune function. The nervous system functions are also enhanced by fungal-derived B-vitamins, therefore ensuring proper neurological health in animals. Fungi enhance the nutritional quality of animal feed by supplying such essential nutrients, thus contributing to better health, performance, and productivity in livestock.

The bioavailability of critical minerals used in animal nutrition is vastly enhanced by fungal additives, hence, absorption of the body is made simple by them and the body utilizes those nutrients (Miyauchi et al., 2020b). Some of the minerals like zinc, iron, and magnesium, among others, which influence its digestion efficiency, are found in organic forms within fungi (Tanaka et al., 2022), for a healthy immune function and work as cofactors for enzymes. Zinc and iron are important while magnesium is vital to bone health and nerve function. Improvement in the bioavailability of these minerals as well as many more means that animals will gain superior growth (Mirończuk-Chodakowska & Witkowska, 2020), immunologically robust functions, and improved overall health and productivity from their nourishment. Overload supplementation needs to be reduced with increased mineral uptake, which means that these mycological additives are extremely effective and economical ingredients in animal feeds (Vitale et al., 2020).

2.4 *Fungal Secondary Metabolites and Their Effects*

These fungi molecules are bioactive and, as such, confer some health and productivity benefits to the animals. They differ from the primary metabolites since they carry out specific functions that are highly responsible for defense mechanisms together with signaling (Nagy et al., 2023). Connected with their use in animal nutrition, they belong to health-promoting functions; being growth promoters, antimicrobials, and anti-inflammatory drugs, but natural.

Growth Promoters: Some microbial secondary metabolites possess the effect of promoting animal growth and increased feed efficiency by the uptake of more nutrient up-take. They help in metabolic intensification, resulting in animals gaining in weight and total production (Richter et al., 2022). Many of the fungal metabolites, including organic acids and alkaloids, protect animals from opportunistic pathogenic organisms through their antimicrobial activity. Since these metabolites inhibit the proliferation of bacteria, fungi, and other microbes, gut health can improve, thereby reducing disease incidence (Wang et al., 2022). Antimicrobial activity may be a promising alternative to antibiotics in feeds, and its use may replace or even reverse the high level of resistance that has been developed to antibiotics currently used in feeds. Secondary metabolites of fungi have excellent anti-inflammatory and antioxidant effects as they reduce the potential of inflammation, thus protecting cells from oxidative stress. This leads to supporting the immune system and faster recovery from disease or injury (Tanaka et al., 2022). Besides, lowering the level of inflammation can benefit the general well-being of animals—a factor that should result in better growth and production.

3 Functional Uses of Fungi

Fungi enhance the optimized assimilation of feed in animal systems and optimize the degrading process for feed components so that efficient complex feed degradations are maximally improved, which largely enhances feed efficiencies and health. Bioactive compounds and enzymatic powers enhance better digestion, nutrient absorption, and living livestock farming with less adverse effects on the environment.

3.1 Role of Fungi in Aiding Digestion

Fungi simplify the process of digestion in rather simple ways. For example, some complex or otherwise indigestible parts of animal feed are broken down by specialized enzymes in fungi. Better digestion occurs because better nutrient extraction improves feed utilization and promotes a healthful gut environment. This talking points out one of the important roles of the fungus in digestion, which is their contribution toward breaking down cellulose and hemicellulose (Lange et al., 2019), two major constituents of the plant cell wall. The fungi degrade these fibrous materials into the more readily absorbable sugars for the animal. This is invaluable for ruminants whose diet consists of high-fibrous feedings such as grass and hay (Li et al., 2020a). Such difficult structures, therefore, are broken down by these fungi that use energy and nutrients found in the sources available for animals to allow comprehensive digestion and uptake of nutrients.

Feed Efficiency Enhancement and Waste Reduction: Besides improving digestibility, these fungi play an equal role in feed efficiency. Simply stated, it means the same size feed intake will ingest most of the nutrients and change them into better growth, health, and productivity with fewer inputs of feed (Miyauchi et al., 2020a). Except for maximizing nutrient absorption, fungal additives do not throw much of the feed out as waste without being digested. It has lower feed costs and impacts on the environment through livestock farming, in as much as the undigested material, especially in ruminants, reduces manure production and hence their greenhouse gas emissions (Wang et al., 2019). In simple words, mycology does play an important role in the breakdown and decomposition of complex plant materials towards digestive improvement and enhancing feed usage. Enhanced Nutrient Absorption, Minimized Feed Loss, and Ecological-Friendly Animal Husbandry from the Fungi Application in Animal Nutrition (Li et al., 2020b).

3.2 Probiotics and Prebiotics

Fungi play a crucial role in controlling the beneficial gut microbiota, which encourages a healthy environment in the gastrointestinal tract of animals, enzymes and various bioactive compounds can be produced by fungi, they help in feeding the beneficial bacteria while inhibiting the hostile pathogens (Meyer et al., 2020). Triggering of healthy digestion and enhancement in nutrient absorption is achieved in the obliging gut microbiome environment, thereby enhancing the immune system. Fungal probiotics, such as some species of *Aspergillus* and *Saccharomyces,* growth of beneficial bacteria is induced by them, which include *Lactobacillus* and *Bifidobacterium,* in the breakdown of complex feed components (Kemboi et al. 2020). As discussed, a balanced gut microbiota, therefore, improved digestion and reduced gastrointestinal infections, and inflammation are their contributions, with a declined chance of getting ill. Betterment in health, growth, and productivity of livestock by fostering a healthier gut environment are contributions of fungi. One of the most promising sources of prebiotics, a significant number of bioactive compounds are contained by them, including polysaccharides such as β-glucans, mannans, and chitin (Mirończuk-Chodakowska et al., 2021). These compounds are not cleaved in the human intestine, and they reach the colon intact, where the growth and activity of useful gut flora are selectively stimulated by them, such as Bifidobacteria and Lactobacilli (Kemboi et al. 2020) prebiotics, improving digestive health because they modulate the balance of gut microbiota are the functions of polysaccharides as fungi. Furthermore, fungi bring various metabolic benefits, such as antioxidant, anti-inflammatory, and antimicrobial attributes (EFSA Panel on Biological Hazards (BIOHAZ) et al., 2019), which makes them an excellent source of any contribution in the line of natural improvement for gut health.

3.3 Immunomodulatory Properties of Fungi

Fungi are famous for powerful immunomodulatory activities and, therefore, general immunity. A vast majority of bioactive compounds isolated from fungi can modulate the immune system, either making it stronger in disease and inflammation-fighting or even developing resistance to stress and diseases (Jia et al., 2023). Most of those chemicals that modulate take part in crucial steps of both natural and specific host immune responses and thus have specific therapeutic potential in the treatment of conditions connected with autoimmunity, hypersensitivity disorders, and other immune diseases (Swift et al., 2021).

The wide range of bioactive molecules produced by fungi encompasses β-glucans and polysaccharides, terpenoids, and lectins that are well-known to interact with the immune system (Roik et al., 2022). Of those, β-glucans of the cell wall of most fungi have been studied in considerable detail as activators of macrophages, dendritic cells, and natural killer cells capable of enhancing the capacity of the host to recognize and eliminate pathogenic and tumor cells (Langer et al., 2019). At the same time, fungal polysaccharides activate GALT which amplifies activation of immune cells and mucosal immunity. Medicinal mushrooms represent some of the best-represented examples of immunomodulators for therapies oriented to the treatment of Reishi (Ganoderma lucidum), Cordyceps sinensis, and Shiitake (Lentinula edodes).

Stress Reduction and Disease Resistance: Especially the adaptogenic strains of fungi such as Cordyceps and Reishi have proved useful in helping the body cope with stress by regulating hormones associated with stress, such as cortisol (Desamero et al., 2021). This regulation, meant to give control over physiological expressions of stress, enhances resistance to diseases and results in proper immunity because chronic stress is immunosuppressive and makes the body susceptible to infections and chronic diseases (Swift et al., 2021). With the help of fungi to provide more resilience to stress, the person would improve his well-being besides boosting the cell-mediated immune system fighting diseases from the body. It has been proved that the fungal compounds that can reduce inflammation and oxidative stress are both critical factors of most chronic diseases thus promoting long-term health with resistance to diseases (Zheng et al., 2020).

4 Potential Uses of Fungal Feed Additives

4.1 Fungi as Growth Promoters in Livestock

Livestock production imposes one of its heaviest financial burdens-the feed cost. Strong, healthy animals on minimal feeds are the effective efficiency optimization behind feed efficiency. Fungal feed additives, such as yeast supplements, have proven a good route for that equitative balance (Lu & Liu, 2024). Saccharomyces cerevisiae is amongst the most significantly characterized yeasts and their capacity to enhance nutrient digestion, and absorption has won over the field. Therefore, for farmers, this means that it is actually possible to get greater energy from the very same intake of feed which enables faster growth and accelerated weight gain-factors determining profitability (Wan Mohtar et al., 2022). That is, many of the experiments confirmed increases in the FCR of the supplemented animals with yeast probiotics. The feed conversion ratio is an important indicator in animal husbandry, referring to the amount of feed required to be fed for gaining a given increase in body weight. The lower the FCR, the greater the feed efficiency; thus, there is a more economical and environmentally friendly answer. Indeed, broilers would take much less time to market weight if yeast is incorporated in poultry feeds (Pawar et al., 2023). Furthermore, with ruminants such as cattle, yeast favors breaking down and better nutrient uptake on feed materials, which contain a lot of fiber. Important: higher growth rates and healthier animals—thanks to the intelligent exploitation of fungus feed additives (Tanaka et al., 2022). This differentiation between species underlines good potential for feeding additives based on fungi in modern agriculture. Fungal feed additives have the flexibility that qualifies them to be useful in various livestock species. These are critical in growth trends and performance enhancement in poultry, pigs, and ruminants (Mirończuk-Chodakowska & Witkowska, 2020). For poultry, high-growth rates such as broilers often put significant stress on digestive systems and more often lead to suboptimal nutrient absorption and greater susceptibility to disorders in the gastrointestinal tract (Stoev, 2024). Subsequently, probiotics yeast thus guarantees a friendly gut and is established such that harmful bacteria are subsequently minimized as nutrient uptake becomes better in leading towards earlier marketing.

In pigs, the gastrointestinal tracts are more complex, and yeasts improve gut flora balance. This is improved digestion with enhancement of the immune system (Roik et al., 2022). More predictable growth rates will be possible, together with better health in litter, therefore, better control of variability in production. The benefits of mycotic feed supplements are particularly developed among ruminants, such as cattle and sheep, given the complexity of fermentive processes undergone in the rumen (Swift et al., 2021). Yeast facilitates higher efficiency in fiber digestion to absorb nutrients for such benefits as increases in the milk yield of dairy cows and the development of muscles in beef cattle. In principle, fungi represent an important method of promoting animal health and productivity within an extensive range of species.

4.2 Antifungals Substitutions to Antibiotics in Fungi

In the last 2 years, animal agriculture has been shaken into realizing the alarming growth in antibiotic resistance. From the early use of antibiotics to deal with animal health management, the overuse today is complicated and poses a threat to both animal and human health through the transmission of resistant pathogens (Li et al., 2020b). Fungal feed additives like yeast-based probiotics are promising natural alternatives to antibiotics since these will most probably stimulate better immunity in animals and reduce the outer application of antibiotics (Kemboi et al. 2020). Saccharomyces cerevisiae plays an extremely key role because it promotes a healthier environment in the gut area where a considerable percentage of the immune system is located. It accomplishes this with its feature of promoting an improved increase in the beneficial bacteria balance and the reduction in pathogenic harmful microorganisms, including Escherichia coli and Salmonella, which would otherwise require antibiotics to maintain the gut (Naranjo-Ortiz & Gabaldón, 2020). Thereby, this competition for available nutrients and attachment sites on the gut mucosa creates an unfavorable environment for potentially pathogenic bacteria and protects against infections. This immune-boosting effect gives feed additives based on fungi a strategic role in reducing dependency on antibiotics (Zheng et al., 2020).

Synergism between fungi and antimicrobial chemicals fungal feed additives replace antibiotics and co-work synergistically in antipathogens with other such agents, organic acids, and plant extracts with respect to improving the health of livestock (Nagy et al., 2023; Vitale et al., 2020). Yeast, for example, has been seen to synergize the action of organic acids, thus lowering the pH of the gut, an environment unfavorable for pathogenic bacteria survival inside. Such a synergistic relation is fortifying the gastropod mucosa against pathogenic microbes. Not only can these be done in tandem to prevent disease while improving digestion and nutrition assimilation, leading to healthier livestock that requires fewer interventions and, therefore, in the long run, reducing antibiotics (Zhu et al., 2023). While the world contemplates the eventual use of zero antibiotics in agriculture, the feed additives of fungi are the holistic and sustainable answer that should meet both animal welfare and public health concerns because they minimize exposure to the spread of antibiotic resistance.

5 Positive Effects of Fungal Feed Additives

5.1 Enhancement of Digestion and Nutrient Absorption

Fungal feed additives, especially those enriched with enzyme-producing strains, received much attention lately because most of them tend to upgrade digestion in livestock (Mougin et al., 2023). Most of the contents of plant-based feeds cannot be digested by various animals, including poultry, such as cellulose and hemicellulose, because most of these animals are unable to break down such complex compounds using their digestive enzymes. Fungal enzymes such as cellulases, xylanases, and β-glucanases break the cell walls of fibrous plant parts filling the gap for enzymes. This will enhance nutrient accessibility for absorption. For instance, Aspergillus niger and Trichoderma reesei are some of the recorded fungi that produce enzymes, which have activity especially on NSPs in cereals, especially on wheat and barley. NSPs increase gut content viscosity, which slows down the process of nutrient absorption (Lattimer & Haub, 2010). Reduced viscosity levels made the otherwise efficient process of extraction and absorption more feasible. Primarily, it enabled the proper application of the necessary nutrients such as carbohydrates and proteins to be absorbed through their system. This enzymatic activation ensures that animals uptake vital nutrients such as glucose, amino acids, and fatty acids more effectively, which is critical for energy generation and growth (Newsholme, 2021). Mycovestigates also regulate the activity of ANFs in the feed of both legumes and oilseeds. Phytic acid and tannins are two compounds that reduce the uptake of essential nutrients through the binding process (Samtiya et al., 2020). Phytase, a notable enzyme derived from fungi, is renowned for its remarkable capacity to break down intricate compounds like phytic acid and tannins, thereby releasing vital minerals such as phosphorus, calcium, and magnesium, which are essential for effective absorption (Rizwanuddin et al., 2023). The above innovation not only adds nutritional value to the feed but also reduces the requirement for phosphorus supplementation, which results in significant environmental benefits as it reduces the surplus of phosphorus present in agricultural runoff (Lian et al., 2024). Besides, fungal enzymes increase the retention time of feed in the gut. The increased retention time leads to increased exposure to digestive enzymes, which improves the breakdown and absorption of nutrients. Practical experiments have demonstrated that when supplemented with fungal enzymes, livestock appear to have feed conversion ratios greater than others. This has the implication that the supplemented animals may require less feed to be equal in weight to the non-supplemented ones (Bedford, 2018). Additionally, the breakdown of proteins into smaller peptides and amino acids and fats to free fatty acids and glycerol implies that both protein and lipid digestion are enhanced. This improves the bioavailability of lipids and fat-soluble vitamins, including A, D, E, and K, whose increased absorption benefits animal health (Andrès et al., 2024).

5.2 Improved Animal Growth and Productivity

The application of feed additives from fungi has provided great improvements in growth and productivity for animals. Particularly, it enhances animal productivity by maximizing feed efficiency while minimizing waste (Chuang et al., 2020). The efficiency with which animals convert feed into measurable outputs, such as muscle growth, milk, or eggs, directly impacts growth rates and overall productivity (Patience et al., 2015). Fungal additives enhance digestive processes and nutrient assimilation, ensuring optimal utilization of dietary nutrients by animals, leading to accelerated growth and improved performance (Ajomiwe et al., 2024). Fungal enzymes, particularly cellulases, and xylanases are important for breaking down non-starch polysaccharides and indigestible fibers, which are made mainly of high-fiber diets (Kant et al., 2024). In the absence of such enzymes, digestion and absorption decrease, and hampered growth follows. As these problems are corrected by the fungal additives, there is an increase in the growth rate, and feed conversion efficiency is improved (Chuang et al., 2020). It seems to suggest that lower feed intakes into livestock would be enough to reach the set benchmark growth levels and thereby have lower feeding costs while improving the overall economic efficiency of farming. Moreover, it is further likely that feed additives can also be derived from fungi thereby increasing the production of products whose origin is animals such as meat, milk, and eggs. This can be attributed to the potency of fungal enzymes in increasing the efficiency of digestion in dairy cattle of fibrous feeds and subsequently leading to increased milk production (Amara & El-Baky, 2023). The ability of nutrients to be more readily absorbed through the action of enzymes used on chickens translates to increased egg production rates and increased growth rates (Ravindran & Abdollahi, 2021). Reports are mainly indicative of the fact that poultry and pig industries grow at an excellent rate and produce better quality products when the fungal additives form part of their feeding system. Apart from stimulating growth and productivity, fungal feed additives reduce major stress factors in livestock. The stress factors may include poor digestion, diseases, and adverse environmental conditions limiting growth and productivity (Chuang et al., 2020). The use of fungal additives has enhanced stress management by promoting gut health and immune strength. In this manner, the growth and productivity of animals are protected regardless of adverse environmental conditions (Zhang, 2022).

5.3 Fungal Feed Additives as Immuno-Enhancers

Fungal feed additives have been proven to be necessary as a means of utilizing the natural bioactive compounds available in the cell walls of fungi for the immunoenhancement of animals. Interestingly, this context has underlined the immunostimulatory potential of beta-glucans and mannans. Beta-glucans, which are a subject of extensive research for their immune-enhancing properties, interact with the receptors on key immune cells, including macrophages, neutrophils, and dendritic cells (EFSA Panel on Additives and Products or Substances used in Animal Feed (FEEDAP) et al., 2024). It would induce a chain of immuno-mechanism, resulting in an amount of cytokine acting like a signaling molecule for the body's immune mechanisms against pathogens (EFSA Panel on Contaminants in the Food Chain (CONTAM) et al., 2023). In this manner, the immunity of the livestock will get stronger from the feed additives of fungi and make the resistance of the animal more potent towards the pathogen invasions. Fungal feed additives encourage the growth and stimulation of lymphocytes, which form the heart of the adaptive immune system. The lymphocytes play a very key role in the formation of antibodies and long-term immunity. Therefore, the immunity power of the livestock to fight off infections is significantly boosted (Salem et al., 2023). The dual activation of both macrophages and lymphocytes not only ensures a more rapid clearance of infections but also promotes tissue repair, reducing inflammation. This strengthening of the immune system is particularly important in intensive farming systems where animals are usually exposed to high levels of pathogens due to their closeness and other environmental stressors (Uwineza et al., 2023). Mannans, together with beta-glucans, are essential in enhancing immunity; they bind to pathogenic bacteria such as Salmonella and Escherichia coli, thereby preventing these pathogens from attaching to the intestinal lining. This mechanism also enables the immune system to concentrate on general well-being instead of constantly trying to seek out pathogens (Urgessa et al., 2024). Furthermore, a reduced need for antibiotics exists in the lower gastrointestinal pathogens, and this is one of the crucial calls for worldwide initiatives that would help mitigate antimicrobial resistance. Such mycological feed supplements are highly beneficial for grower animals whose immune mechanisms are more prone to infections (Huo et al., 2024). As studies continue, it clearly shows that animals fed on high-fungi-supplemented diets possess increased immunity indices of antibody titers and infection resistance (Shurson et al., 2023). In the final analysis, this results in lower diseases, fewer antimicrobial treatments, and leads to a more sustainable livestock operation.

5.4 *Gut Health and Microbiome Regulation Through Fungal Feed Additives*

Gut health has long been acknowledged as an important aspect of the overall productivity and health of animals. Fungal feed additives have, therefore, come to play a very central role in this regard. The additives encourage the richness of microorganisms in the gut microbiome, which plays a significant role in the digestion, immune response, and prevention of diseases. This promotes a healthy, balanced, and diverse microbiome, and such a microbiome leads to better nutrient absorption. It also enhances protection against diseases in livestock. Among these fungal enzymes are β-glucanases and xylanases that break up complex carbohydrates and some indigestible fibers in feed. These end products become short-chain fatty acids. SCFAs are prebiotics that will feed beneficial gut bacteria like Lactobacillus, Bifidobacterium, and Enterococcus (Kumar et al., 2023). These bacteria, through their metabolic activities, release various compounds that strengthen the layer of the gut, which will enhance the integrity of the gut barrier. An intensive gut barrier is stronger at resisting the entry of harmful invaders like Escherichia coli and Clostridium perfringens, known to cause digestive diseases in animals (EFSA Panel on Biological Hazards (BIOHAZ) et al., 2023). Fungal feed additives increase the probiotic bacteria population but decrease the pathogenic bacteria population. This equilibrium maintains a healthy balance in the microbiome. Thus, it enables maximum efficiency in digestion without causing disorders like diarrhea, enteritis, or dysbiosis. They stimulate mucin secretion which are glycoproteins that coat the intestines with a protective layer of mucous (Lin & Xu, 2022). This mucous coat protects from systemic inflammation, thereby ensuring the general welfare of an animal by preventing viruses and toxins from penetrating into the bloodstream. Another significant benefit of fungal feed additives is that they can improve feed conversion efficiency; they nourish a healthy and balanced microbiome within the alimentary canal. As a result, animals can use their feed better, convert nutrients into energy and body mass with higher efficiency, and gain more benefits (Naz et al., 2023).

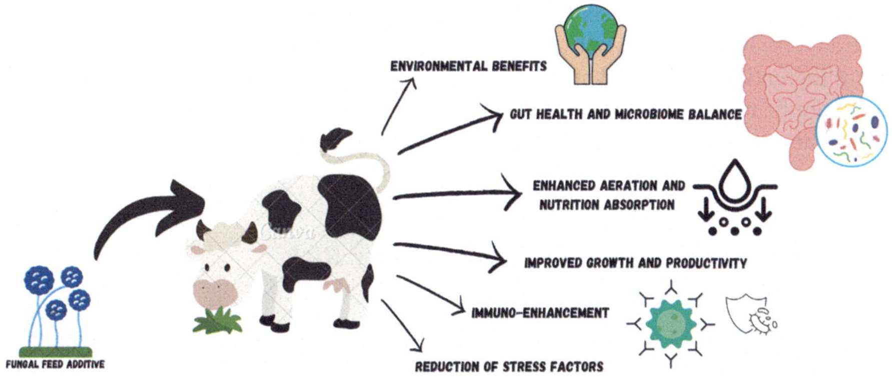

Fig. 3 Positive effects of feed additive

Therefore, the growth rate increases, and productivity levels enhance drastically in the enterprises that have specialized in producing meat, milk, or eggs. Nutrients assimilated by animals reduce irritant gut contents of the additives developed from fungi. Such products smoothen the inflammatory as well as the oxidative reaction produced by deleterious pathogens while at the same time increasing the development of good flora in the gut. Chronic inflammation limits nutrient uptake and is also associated with stunted growth. This makes the animals more prone to infections than they would have been otherwise (Olmo et al., 2022). However, the overall welfare of the animals improves, and the incidence of gastrointestinal diseases decreases through the gut health environment created by fungal additives (Fig. 3), thereby decreasing the medical treatments that happen. Feed fungal additives also contribute significantly to having a healthier environment in a farm, given that it will effectively diminish the production of harmful microbial by-products that include ammonia and hydrogen sulfide. Not only did these substances compromise gut health, but also pollutant compounds in confined animal settings, according to EFSA Panel on Biological Hazards (BIOHAZ) et al. (2022). The reduction of these toxic byproducts has been shown to significantly improve animal welfare and increase productivity in farming operations while dramatically lowering levels of stress factors (Fig. 3).

6 Negative Effects of Fungal Feed Additives

Fungal feed additives are attracting a lot of attention since they have many benefits such as improvement in digestive systems, increase in immunity, and better animal health in general (Chuang et al., 2020). True, though, that there are negative consequences related to mistreatment or over-supplementation. Over-supplementation and complicated interactions with other feed components and medications introduce adverse health effects in livestock (Nutrition: Micronutrient Intake, Imbalances, and Interventions, 2024). This chapter spotlights significant concerns with respect to the application of fungal feed additives in regard to the risks of toxicity, allergic and inflammatory effects, and interaction with other components in the diet.

6.1 Over-Supplementation and Toxicity Risk

Fungal feed additives are potent agents of feeding efficacy and animal productivity in a controlled level of addition. However, over-supplementation involves a significant risk in relation to it, primarily through toxicity (Bryden, 2012). Mycotoxins are secondary metabolites that produce severe complications in animal feed. Some of the dangerous mycotoxins include aflatoxins, ochratoxins, and fumonisins. At very low levels, these toxins can contaminate feed due to poor storage or processing if maintained under stringent, controlled conditions (Awuchi et al., 2021). It is widely known that mycotoxins are harmful to animals' health. Long-term exposure to these toxins, even in low concentrations, can cause liver and kidney damage, weaken the immune system, and hinder growth (Zain, 2011). For example, aflatoxins damage the liver and lead to chronic immunosuppression in livestock, making them more susceptible to infections. This endangers both the health of individual animals and the overall productivity and economic viability of livestock farming operations (Godde et al., 2021). The threat is not limited to animal health. Aflatoxins can enter the food chain and reach humans through contaminated animal products: milk, meat, and eggs. Aflatoxins are carcinogens; their presence in food items intended for human consumption results in significant public health risks (Mahato et al., 2019). In addition to the danger that mycotoxins pose, overuse of fungal enzymes like xylanases, glucanases, and cellulases may interfere with nutrient uptake processes (Awuchi et al., 2021). These enzymes are essential in the breakdown of complex carbohydrates into simpler, more digestible forms, which improves feed efficiency (Ravindran, 2013). However, overuse of these enzymes could enhance the breakdown of nutrients into abnormal patterns of nutrient uptake (Ojha et al., 2018). This may cause a high rate of metabolism of essential nutrients, such as proteins, vitamins, and minerals, leading to deficiencies that affect growth, productivity, and health (Munteanu & Schwartz, 2022). Such nutritional imbalances prove particularly detrimental in young, growing animals or those in high-demand physiological states, such as lactating females, where the need for balanced nutrition is critical. Given these risks, the handling of such fungus-based feed additives requires proper regulation and monitoring. Their storage, processing, and administration through feed additives must be carried out under controlled conditions and frequently monitored to avoid contamination or overactivity of the enzymes (Fumagalli et al., 2021).

6.2 Allergic or Inflammatory Reactions in Animals

Though the majority of the bioactive compounds present in fungal feed additives are nontoxic and functional, some of them have been reported to induce allergic or inflammatory reactions in sensitive animals. The most obvious of these are beta-glucans due to their immunomodulatory effects. While beta-glucans have, on the whole, been utilized in potentiating immune function, overstimulation of the immune can occur through enhancement or augmentation of inflammatory events in some instances. Beta-glucans can cause adverse reactions in patients with underlying health conditions or genetic susceptibility to immune hypersensitivity (Kortei et al., 2022). These reactions can include respiratory issues, skin irritation, and gastrointestinal problems such as diarrhea. Sometimes, these may lead to chronic inflammation, which may significantly affect the welfare and production of the animals. Chronic inflammation weakens the intestinal ability of the host animal to absorb nutrients while suppressing the appetite, thus causing feed intake reduction and stunted growth (Abd El-Hack et al., 2022). The animals become more susceptible to diseases and subsequent infections caused by maintained inflammatory reactions. This chronic stimulation decreases their ability to build a proper immune system that will combat infections efficiently. Consequently, this causes a vicious cycle of deteriorating health that makes animals increasingly susceptible to pathogens, which in turn worsens their inflammatory state (Cui et al., 2021). Such pathogenic effects occur mainly in young animals whose immune mechanisms are still under development. Gross exposure of such young animals to slim exposure of fungal feed additives can grossly trigger uneven immune responses, leading to retardation of growth and predisposition to developmental abnormalities (Lübeck & Lübeck, 2022). Therefore, such risks should be addressed by proper scrutiny of the application of fungal feed additives, particularly in sensitive or vulnerable populations. Veterinary supervision should thus be accompanied by careful regulation of the dose applied so that it does not compromise the animal's health while utilizing it effectively (Ochieng et al., 2021).

6.3 Interaction with Other Feed Components

The use of fungal feed additives can be beneficial by themselves but may function unpredictably with other dietary constituents in an animal, including supplements, minerals, vitamins, or drugs. Among the major issues is that fungal enzymes may change the metabolism and absorption of other nutrients in a different way (Fumagalli et al., 2021). These enzymes, which are intended to break down complex carbohydrates, can overprocess feed components when they are present in excess (Li et al., 2021). This overprocessing can lead to the underutilization of key nutrients, causing imbalances in the diet of the animal. Mineral absorption is one area that is easily affected by fungal feed additives. The minerals calcium, phosphorus, and magnesium are fundamental to the health of bones and metabolism and livestock reproduction processes (Romero et al., 2021). However, the reduction of these minerals in the system by fungal enzymes can affect the system's ability to absorb them, leading to deficiency and poor growth, and overall health. The "young ones" which include young animals and lactating females are very vulnerable to mineral deficiencies since they have high mineral contents. Calcium and phosphorus deficiency leads to rickets in young animals and reduces milk production in lactating females. These are some of the serious negative impacts on productivity. In addition to interfering with the nutrients, fungal feed additives interfere with the effectiveness of drugs that are administered to animals. These include antibiotics. The enzymatic activity in mycotoxin additives may break down antibiotics too fast and make it impossible for them to be effective against infections (Li et al., 2020c). This is a significant challenge to large-scale farming where the role of these antibiotics in preventing and controlling outbreaks is well understood. The partial loss of antibiotic efficacy affects the health of animals while concurrently stimulating the growth of antibiotic-resistant bacteria, a cause of great concern to both animal and human health (McGenity et al., 2020). Preservatives often used in animal feed can interact with supplements containing fungus-based feed. Some preservatives may impede the enzymatic processes of fungal supplements that break down feed components. This can result in the rapid growth of harmful molds or bacteria (Lopes et al., 2019). This dual hazard compromises the benefits of fungal adjuvants while promoting the survival of spoilage microorganisms, thereby increasing the risk of contamination and harm to animals (Peng et al., 2018). Thus, an integrated feed management system is required, incorporating fungal feed additives into a balanced diet. Regular monitoring of nutrient absorption, drug efficacy, and feed quality prevents adverse interactions and optimizes livestock health and productivity (B. Chen et al., 2020).

7 Impact of Mycotoxins on Animals

Mycotoxins, toxic secondary metabolites that different species of fungi produce, pose a great threat to modern agricultural practices, particularly to the health and safety of animals. These harmful compounds have more often contaminated crops and animal feed, resulting in a wide and extensive range of negative effects on livestock productivity and welfare. Moreover, the penetration of mycotoxins in animal products into the food chain is a serious threat since it suggests a formidable risk factor for human health. To that effect, adequate knowledge of the different types of mycotoxins, routes of contamination, and their ensuing health impacts should be in place to develop targeted policies and measures for risk management to ensure both animal and human health (Kemboi et al. 2020).

7.1 Common Mycotoxins and Health Impacts on Animals

These species of fungi, which include Aspergillus, Fusarium, and Penicillium, produce various mycotoxins. Each of these mycotoxins can have adverse effects on animal health. Of these, aflatoxins, fumonisins, ochratoxins, zearalenone, and deoxynivalenol (DON) are considered the most prevalent and hazardous mycotoxin-producing fungi. Aflatoxins, mainly produced by Aspergillus flavus and Aspergillus parasiticus, are highly toxic and can cause hepatotoxicity and immunosuppression, and potentially contribute to carcinogenesis (Stoev, 2024). These toxins are commonly associated with crops like maize and peanuts when improper storage conditions lead to moisture accumulation. Fumonisins are another very important class of mycotoxins primarily produced by Fusarium verticillioides and Fusarium proliferatum, and maize is highly contaminated (Li et al., 2024). These toxins impact lipid metabolism, causing equine neurologic disease and porcine respiratory problems. Ochratoxins are nephrotoxic mycotoxins of the Aspergillus and Penicillium species, causing kidney damage mainly in poultry and pig production. An example of deoxynivalenol is one of the many toxins primarily associated with cereals like wheat, barley, and maize, within the family of trichothecene. They are recognized to suppress protein synthesis since their administration in animals causes low growth rates, feed conversion, and gastrointestinal problems. Zearalenone is a product of Fusarium species, which is an estrogen responsible for a range of reproductive disorders in animals, especially pigs, due to infertility, miscarriages, and other forms of estrogenic manifestations.

7.2 Mechanisms of Mycotoxin Contamination in Animal Feed

The three key areas involved in the cycle of ripening, harvest, and storage are those where the feedstocks to animals get contaminated by mycotoxins. Among the three environmental factors involved—temperature, humidity, and rainfall—most of the infection arises from fungi. Aspergillus, Fusarium, and Penicillium, which account for the first three genera in relation to mycotoxin production, are mainly attributed to warm wet weather. The crops are more prone at the time of experiencing conditions of drought or even pest attacks, providing room for fungi to establish themselves (Zhen et al., 2021). Drying post-harvest by leaving the materials with poor ventilation, encourages mycotoxin production besides enhancing fungal growth. Fungi may remain latent under adverse conditions but reactivate when favorable conditions are restored; this raises the possibility of mycotoxin production long after the harvest period has expired. Since mycotoxins are chemically stable, they cannot be easily eliminated during most feed manufacturing processes such as drying and pelleting, which means that it is impossible to eliminate them once they have contaminated the feed (Ellatif et al., 2022). Cross-contamination is one of the biggest issues in feed industries; even a tiny quantity of mycotoxin contamination in a feed ingredient leads to the introduction of mycotoxins into feeds that otherwise would have been free of contamination because of processing or storage. Given that even slight exposure to mycotoxins can have harmful effects on livestock, it is crucial that strict monitoring and testing procedures be established to prevent contaminated feed from entering animals (An et al., 2023).

7.3 Health Effects of Mycotoxins on Animals

The risk of health damage due to mycotoxins to animals depends on the type of mycotoxin, the concentration inhaled or ingested, and the particular species exposed. If an animal is exposed to an acute level of high concentrations of mycotoxins, it may lead to immediate and acute health problems, including failure of organs, internal bleeding, and even death. For example, aflatoxins have been known to harm the liver and induce jaundice in cattle and poultry. Fumonisins have been known to cause equine leukoencephalomalacia, a deadly neurological disease in horses, which sometimes is referred to as "crazy horse disease." Mycotoxins at lethal doses cause a gradual deterioration in the health of the animal; hence, there is a significant reduction in its productivity over time. Another well-known effect of mycotoxin exposure is immunosuppression, which leaves animals susceptible to infections (Hosseini et al., 2023). This presents a considerable challenge in livestock production systems where animals are kept in close proximity to one another and diseases can easily spread. For instance, aflatoxins depress the function of the immune system because they reduce the lymphocyte count; therefore, animals become more susceptible to pathogenic infections, which leads to increased mortality and rocketing veterinary costs. In addition, DON and fumonisins among other mycotoxins induce gastrointestinal problems, which have adverse effects on the process of nutrient absorption in their organisms (Chen et al., 2023). Deoxynivalenol, also called "vomitoxin," elicits the whole spectrum of symptoms in animals from vomiting, diarrhea, and reduced feed intake to final growth performance and weight loss. Chronic exposure to these mycotoxins also lowers feed conversion efficiency, where more feed is needed to get the same amount of weight gain, which affects farm profitability negatively. Other mycotoxins aside from those that have direct implications for the gastrointestinal and immune systems are related to the malfunction of reproduction in animals. For instance, consider Zearalenone, a mycotoxin with estrogenic activity that leads to disruption of hormonal balance, mainly among pigs, which often leads to infertility and even stillbirths resulting in great losses for the farmers. Yet another nephrotoxic mycotoxin is ochratoxin causing renal damage as well as chronic kidney disease. Ochratoxin is postulated to cause very serious issues within poultry and pig farming. However, the aftereffects of mycotoxins do not merely relate to health disorders associated with animals but also cause extreme consequences for human beings through food. For instance, aflatoxins are trans-bound in dairy products as aflatoxin M1 which is toxic to human beings. That is a greater risk in areas with no proper control and monitoring system for mycotoxin contamination. Mycotoxins constitute a complex threat to animal health, which affects growth and production, reproductive performance, and immune functions in the animal, while affecting safety, food safety, and economic implications of agriculture.

8 Risk of Using Mycotoxin-Derived Feed Additives

Livestock production that uses feed additives derived from fungi is promising to improve the efficiency and productivity of that sector. However, feeding livestock with fungi-based additives necessitates a holistic risk assessment relating to long-term safety, impact on the environment, and regulatory oversight. Therefore, just as people in the agricultural industry adopt this mode of additive, it calls for critical evaluation of such factors to ensure the safety and sustainability of the process.

8.1 Long-Term Safety of Mycotoxin-Producing Feed Additives

The long-term impact of the use of fungal feed additives on animal health is not well understood. These feed supplements, which include bioactive compounds such as polysaccharides, beta-glucans, and enzymes, have been shown to have excellent short-term benefits; however, the long-term effects of their use are still not well understood (Dhaver et al., 2023). Fungal feed additives enhance the digestion of fibers and carbohydrates; therefore, the feed efficiency improves. However, it has also raised questions regarding animals relying on such exogenous enzymes. Higher dependency makes natural digestive functions weaken in animals, which, at last, breaks the metabolic activities (Sharma et al., 2021). This is particularly important with multi-generation livestock as chronic cumulative effects may undermine their adaptability to digestive capabilities. The risks of mycotoxin exposure should by no means be overlooked; low-quality or poorly stored fungal feed additives serve as a conduct to mycotoxins: dangerous metabolites that pose serious threats to the health of humans and animals alike. Mycotoxins have been associated with immunosuppression, liver damage, and many more chronic effects (Kemboi et al. 2020). Control measures should be implemented through quality assurance procedures and monitoring to minimize risks of contamination in animal feed systems. There is also a significant need for research on the interaction between fungal feed additives and antibiotics. This longer-term exposure to fungal-origin additives and antibiotics is further escalating a worrying concern, related to antimicrobial resistance (Shan et al., 2019). This puts people at a severe risk concerning public health as it gives rise to the generation of antibiotic-resistant bacterial strains simultaneously with decreasing the ability of antibiotics to control infections in livestock.

8.2 Environmental Impact of Fungal Additives

While the fungal feed additives enhance nutrient uptake and minimize feed loss, environmental impacts call for more attention. Among these is nutrient loss by run-off; large amounts of nitrogen and phosphorus, which are abundant in nutrient-rich feeds supplemented with fungal additives, are lost (Peles et al., 2019). The excess nutrients that are leached out can eventually end up in aquatic environments, thereby causing eutrophication. This promotes the growth of harmful algal blooms that consume dissolved oxygen in water bodies, thus disrupting the balance of these water ecosystems (Hartinger & Zebeli, 2021). The effects on marine biodiversity are devastating. Furthermore, fungal feed additives prevent the diminishment of methane produced by ruminant animals, like cows. Methane is a notoriously potent greenhouse gas that is produced through digestive processes. While introducing fungal amendments can help improve the efficiency of digestion, methane reduction may have limited potential; therefore, a limitation in their benefit to the environment in this scope of climate change mitigation becomes quite evident (Venkatesh & Keller, 2019). The entire cycle, from producing, processing, and distributing mycotic feed additives, will have an effect on the overall environmental impact. The direct energy inputs due to such processes especially linked with the imported additives will have to be weighed based on the sustainable dimensions. Fungi growth during storage contributes to environmental waste and calls for strong measures regarding its storage, handling, and other methods which should not put its users under exposure from adverse situations, especially toxic effects (Bampidis et al., 2021).

8.3 Regulatory and Safety Issues

The regulatory framework controlling the use of fungal feed additives is complex and dynamic. For example, for such a feed additive to be permitted for agricultural use, the European Food Safety Authority has to demonstrate that it satisfies pre-established standards concerning safety and efficacy. However, the pace of advancement in fungal biotechnology, especially in the discovery of new bioactive compounds and transgenic organisms is sometimes faster than what the current regulatory systems are designed to address, thereby casting doubt over whether adequate regulation is in place (Le Cocq et al., 2020). The feed additive regulatory environment varies between regions and makes international trade and compliance rather complex. For instance, mycotoxin contamination regulations vary from one country to the other, which makes importing and exporting animal feed products a rather complicated task. This lack of harmonized international standards has increased the danger of unsafe products penetrating the food chain, thus sounding an alarm call for uniform regulation of products globally for the protection of both animal and human health (Novak et al., 2019). The other regulatory issue comes in terms of the interaction of fungal feed additives with antibiotics. The interactions might bring about certain complications that would influence treatment efficacy, particularly on the effects of antibiotic resistance (Chen et al., 2023). The risks involve significant animal welfare and food safety concerns that call for strict regulations and appropriate guidelines to combat such threats. Another aspect that may also reflect regulation is consumer perception. With an increased outcry of the public demanding organic agricultural practices and sustainability, the control of synthetic additives use is also expected to increase GMO oversight from regulatory agencies (Stoev, 2024). For instance, organic certifying bodies demand that a farmer prove a minimum lack of dependence on such chemicals, which puts them in a sensitive balance between such demands by regulatory bodies and the need to feed their livestock. As awareness of consumers grows and concerns about such additives mount, we are sure to see tighter labeling regulations and more stringent regulations being placed on feed additives.

9 Conclusion

Fungal feed additives are a new means to improve livestock health and productivity in a sustainable manner. Fungal feed additives are full of enzymes, polysaccharides, and bioactive compounds, responsible for beneficial roles such as increased digestion, uptake of nutrients, and immunity. The breakdown of the ingredients of feed, including cellulose, proteins, and fats, by fungal enzymes improves nutrient extraction while leading to healthier growth, feed efficiency, and performance of animals.

In addition to these, fungal probiotics save the use of antibiotics, which is increasingly an important issue because the problem of antibiotic resistance increases in sight of animal agriculture. Fungal feed additives also reduce methane production in ruminants by providing greater environmental sustainability and offering lower feed costs associated with improved nutrient efficiency.

However, there are risks associated with its over-supplementation and mycotoxin feed contamination. As much as the benefits of the fungal-based additives are fairly well documented, there is a need for increased research on their long-term effects for safe application and efficiency in modern farming for livestock. Conclusion: Fungal feed additives are a valuable and eco-friendly alternative in livestock nutrition.

References

Abd El-Hack, M. E., El-Saadony, M. T., Salem, H. M., El-Tahan, A. M., Soliman, M. M., Youssef, G. B. A., Taha, A. E., Soliman, S. M., Ahmed, A. E., El-kott, A. F., Al Syaad, K. M., & Swelum, A. A. (2022). Alternatives to antibiotics for organic poultry production: Types, modes of action and impacts on bird's health and production. *Poultry Science, 101*(4). https://doi.org/10.1016/j.psj.2022.101696

Abdala-Díaz, R. T., Casas-Arrojo, V., Castro-Varela, P., Riquelme, C., Carrillo, P., Medina, M. Á., Cárdenas, C., Becerra, J., & Manríquez, C. P. (2024). Immunomodulatory, antioxidant, and potential anticancer activity of the polysaccharides of the fungus Fomitiporia chilensis. *Molecules, 29*(15), 3628. https://doi.org/10.3390/molecules29153628

Abdelhameed, R. E., & Metwally, R. A. (2022). Assessment of beneficial fungal microorganism's bio-efficacy in stimulating morphological and physiological parameters of Allium cepa plants grown in soil amended with fish wastes. *BMC Plant Biology, 22*(1). https://doi.org/10.1186/s12870-022-03965-3

Ajomiwe, N., Boland, M., Phongthai, S., Bagiyal, M., Singh, J., & Kaur, L. (2024). Protein nutrition: Understanding structure, digestibility, and bioavailability for optimal health. *Food, 13*(11), 1771. https://doi.org/10.3390/foods13111771

Amara, A. A., & El-Baky, N. A. (2023). Fungi as a source of edible proteins and animal feed. *Journal of Fungi, 9*(1), 73. https://doi.org/10.3390/jof9010073

An, C., Ma, S., Shi, X., Xue, W., Liu, C., & Ding, H. (2020). Diversity and antimicrobial activity of endophytic fungi isolated from Chloranthus japonicus Sieb in Qinling Mountains, China. *International Journal of Molecular Sciences, 21*(17), 5958. https://doi.org/10.3390/ijms21175958

An, Y., Wang, H., Zong, Z., Gao, Z., Shi, C., Li, S., & Khas-Erdene. (2023). Effects of adding Sophora alopecuroides to high concentrate diet on rumen fermentation parameters and microbial diversity of sheep. *Frontiers in Veterinary Science, 10*. https://doi.org/10.3389/FVETS.2023.1200272

Andrès, E., Lorenzo-Villalba, N., Terrade, J., & Méndez-Bailon, M. (2024). Fat-soluble vitamins A, D, E, and K: Review of the literature and points of interest for the clinician. *Journal of Clinical Medicine, 13*(13), 3641. https://doi.org/10.3390/jcm13133641

Awuchi, C. G., Ondari, E. N., Ogbonna, C. U., Upadhyay, A. K., Baran, K., Okpala, C. O. R., Korzeniowska, M., & Guiné, R. P. F. (2021). Mycotoxins affecting animals, foods, humans, and plants: Types, occurrence, toxicities, action mechanisms, prevention, and detoxification strategies—A revisit. *Food, 10*(6), 1279. https://doi.org/10.3390/foods10061279

Ayalew, H., Zhang, H., Wang, J., Wu, S., Qiu, K., Qi, G., Tekeste, A., Wassie, T., & Chanie, D. (2022). Potential feed additives as antibiotic alternatives in broiler production. *Frontiers in Veterinary Science, 9*. https://doi.org/10.3389/fvets.2022.916473

Bahram, M., & Netherway, T. (2021). Fungi as mediators linking organisms and ecosystems. *FEMS Microbiology Reviews, 46*(2). https://doi.org/10.1093/femsre/fuab058

Bampidis, V., Bastos, M., Christensen, H., Dusemund, B., Kouba, M., Durjava, M. K., López-Alonso, M., Puente, S. L., Marcon, F., Mayo, B., Pechová, A., Petkova, M., Ramos, F., Sanz, Y., Villa, R. E., Woutersen, R., Brock, T., De Knecht, J., Kolar, B., et al. (2019). Guidance on the assessment of the safety of feed additives for the environment. *EFSA Journal, 17*(4). https://doi.org/10.2903/j.efsa.2019.5648

Bampidis, V., Azimonti, G., Bastos, M. d. L., Christensen, H., Dusemund, B., Fašmon Durjava, M., Kouba, M., López-Alonso, M., López Puente, S., Marcon, F., Mayo, B., Pechová, A., Petkova, M., Ramos, F., Sanz, Y., Villa, R. E., Woutersen, R., Cocconcelli, P. S., Glandorf, B., et al. (2021). Safety and efficacy of a feed additive consisting of endo-1,4-β-xylanase (ECONASE® XT) produced by Trichoderma reesei CBS 140027 as a feed additive for piglets (weaned), pigs for fattening, chickens for fattening, chickens reared for laying, laying hens, turkeys for fattening, turkeys reared for breeding and minor poultry species (Roal Oy). *EFSA Journal, 19*(4). https://doi.org/10.2903/J.EFSA.2021.6536

Bedford, M. R. (2018). The evolution and application of enzymes in the animal feed industry: The role of data interpretation. *British Poultry Science, 59*(5), 486–493. https://doi.org/10.1080/00071668.2018.1484074

Bryden, W. L. (2012). Mycotoxin contamination of the feed supply chain: Implications for animal productivity and feed security. *Animal Feed Science and Technology, 173*(1–2), 134–158. https://doi.org/10.1016/j.anifeedsci.2011.12.014

Calla-Quispe, E., Fuentes-Rivera, H. L., Ramírez, P., Martel, C., & Ibañez, A. J. (2020). Mass spectrometry: A Rosetta stone to learn how fungi interact and talk. *Life, 10*(6), 89. https://doi.org/10.3390/life10060089

Carr, E. C., Barton, Q., Grambo, S., Sullivan, M., Renfro, C. M., Kuo, A., Pangilinan, J., Lipzen, A., Keymanesh, K., Savage, E., Barry, K., Grigoriev, I. V., Riekhof, W. R., & Harris, S. D. (2023). Characterization of a novel polyextremotolerant fungus, Exophiala viscosa, with insights into its melanin regulation and ecological niche. *G3 Genes Genomes Genetics, 13*(8). https://doi.org/10.1093/g3journal/jkad110

Chen, B., Sun, Y., Luo, F., & Wang, C. (2020). Bioactive metabolites and potential mycotoxins produced by cordyceps fungi: A review of safety. *Toxins, 12*(6). https://doi.org/10.3390/toxins12060410

Chen, G., Zhang, K., Tang, W., Li, Y., Pang, J., Yuan, X., Song, X., Jiang, L., Yu, X., Zhu, H., Wang, J., Zhang, J., & Zhang, X. (2023). Feed nutritional composition affects the intestinal microbiota and digestive enzyme activity of black soldier fly larvae. *Frontiers in Microbiology, 14*. https://doi.org/10.3389/FMICB.2023.1184139

Cheng, S., Zou, Y., Kuča, K., Hashem, A., Abd_Allah, E. F., & Wu, Q. (2021). Elucidating the mechanisms underlying enhanced drought tolerance in plants mediated by arbuscular mycorrhizal fungi. *Frontiers in Microbiology, 12*. https://doi.org/10.3389/fmicb.2021.809473

Cheng, M., Wijayawardene, N. N., Promputtha, I., De Vries, R. P., Lan, Y., Luo, G., Wang, M., Li, Q., Guo, X., Wang, F., Liu, Y., & Kang, Y. (2022). Potential fungi isolated from anti-biodegradable Chinese medicine residue to degrade lignocellulose. *Frontiers in Microbiology, 13*. https://doi.org/10.3389/fmicb.2022.877884

Chuang, W. Y., Hsieh, Y. C., & Lee, T. (2020). The effects of fungal feed additives in animals: A review. *Animals, 10*(5), 805. https://doi.org/10.3390/ani10050805

Cochereau, B., Meslet-Cladière, L., Pouchus, Y. F., Grovel, O., & Roullier, C. (2022). Halogenation in fungi: What do we know and what remains to be discovered? *Molecules, 27*(10), 3157. https://doi.org/10.3390/molecules27103157

Corbu, V. M., Gheorghe-Barbu, I., Dumbravă, A. Ş., Vrâncianu, C. O., & Şesan, T. E. (2023). Current insights in fungal importance—A comprehensive review. *Microorganisms, 11*(6), 1384. https://doi.org/10.3390/microorganisms11061384

Cui, Y., Li, J., Deng, D., Lu, H., Tian, Z., Liu, Z., & Ma, X. (2021). Solid-state fermentation by Aspergillus niger and Trichoderma koningii improves the quality of tea dregs for use as feed additives. *PLoS One, 16*(11 November). https://doi.org/10.1371/journal.pone.0260045

Desamero, M. J. M., Chung, S. H., & Kakuta, S. (2021). Insights on the functional role of beta-glucans in fungal immunity using receptor-deficient mouse models. *International Journal of Molecular Sciences, 22*(9). https://doi.org/10.3390/ijms22094778

Dhaver, P., Pletschke, B., Sithole, B., & Govinden, R. (2023). Optimization of xylooligosaccharides production by native and recombinant xylanase hydrolysis of chicken feed substrates. *International Journal of Molecular Sciences, 24*(23). https://doi.org/10.3390/IJMS242317110

EFSA Panel on Additives and Products or Substances used in Animal Feed (FEEDAP), Bampidis, V., Azimonti, G., Bastos, M. d. L., Christensen, H., Durjava, M., Dusemund, B., Kouba, M., López-Alonso, M., López Puente, S., Marcon, F., Mayo, B., Pechová, A., Petkova, M., Ramos, F., Villa, R. E., Woutersen, R., Dierick, N., Gropp, J., et al. (2024). Guidance on the assessment of the efficacy of feed additives. *EFSA Journal. European Food Safety Authority, 22*(7), e8856. https://doi.org/10.2903/j.efsa.2024.8856

EFSA Panel on Biological Hazards (BIOHAZ), Koutsoumanis, K., Allende, A., Alvarez-Ordóñez, A., Bolton, D., Bover-Cid, S., Chemaly, M., Davies, R., De Cesare, A., Hilbert, F., Lindqvist, R., Nauta, M., Peixe, L., Ru, G., Simmons, M., Skandamis, P., Suffredini, E., Cocconcelli, P. S., Fernández Escámez, P. S., et al. (2019). Update of the list of QPS-recommended biological agents intentionally added to food or feed as notified to EFSA 10: Suitability of taxonomic units notified to EFSA until March 2019. *EFSA Journal. European Food Safety Authority, 17*(7), e05753. https://doi.org/10.2903/j.efsa.2019.5753

EFSA Panel on Biological Hazards (BIOHAZ), Koutsoumanis, K., Allende, A., Alvarez-Ordóñez, A., Bolton, D., Bover-Cid, S., Chemaly, M., Davies, R., De Cesare, A., Hilbert, F., Lindqvist, R., Nauta, M., Peixe, L., Ru, G., Simmons, M., Skandamis, P., Suffredini, E., Cocconcelli, P. S., Fernández Escámez, P. S., et al. (2022). Update of the list of QPS-recommended microbiological agents intentionally added to food or feed as notified to EFSA 16: Suitability of taxonomic units notified to EFSA until March 2022. *EFSA Journal. European Food Safety Authority, 20*(7), e07408. https://doi.org/10.2903/j.efsa.2022.7408

EFSA Panel on Biological Hazards (BIOHAZ), Koutsoumanis, K., Allende, A., Alvarez-Ordóñez, A., Bolton, D., Bover-Cid, S., Chemaly, M., De Cesare, A., Hilbert, F., Lindqvist, R., Nauta, M., Peixe, L., Ru, G., Simmons, M., Skandamis, P., Suffredini, E., Cocconcelli, P. S., Escámez, P. S. F., Maradona, M. P., et al. (2023). Update of the list of qualified presumption of safety (QPS) recommended microbiological agents intentionally added to food or feed as notified to EFSA 17: Suitability of taxonomic units notified to EFSA until September 2022. *EFSA Journal. European Food Safety Authority, 21*(1), e07746. https://doi.org/10.2903/j.efsa.2023.7746

EFSA Panel on Contaminants in the Food Chain (CONTAM), Schrenk, D., Bignami, M., Bodin, L., Chipman, J. K., Del Mazo, J., Grasl-Kraupp, B., Hogstrand, C., Hoogenboom, L. R., Leblanc, J.-C., Nielsen, E., Ntzani, E., Sand, S., Schwerdtle, T., Vleminckx, C., Wallace, H., Gropp, J., Antonissen, G., Rychen, G., et al. (2023). Risks for animal health related to the presence of ochratoxin A (OTA) in feed. *EFSA Journal. European Food Safety Authority, 21*(11), e08375. https://doi.org/10.2903/j.efsa.2023.8375

Ejiofor, T., Mgbeahuruike, A. C., Ojiako, C., Ushie, A. M., Nwoko, E. I., Onoja, I. R., Dada, T., Mwanza, M., & Karlsson, M. (2021). Saccharomyces cerevisiae, bentonite, and kaolin as adsorbents for reducing the adverse impacts of mycotoxin contaminated feed on broiler histopathology and hemato-biochemical changes. *Veterinary World, 14*(1), 23–32. https://doi.org/10.14202/vetworld.2021.23-32

Ellatif, S. A., Abdel Razik, E. S., Al-Surhanee, A. A., Al-Sarraj, F., Daigham, G. E., & Mahfouz, A. Y. (2022). Enhanced production, cloning, and expression of a xylanase gene from endophytic fungal strain Trichoderma harzianum kj831197.1: Unveiling the in vitro anti-fungal activity against phytopathogenic fungi. *Journal of Fungi, 8*(5). https://doi.org/10.3390/JOF8050447

Floudas, D., Gentile, L., Andersson, E., Kanellopoulos, S. G., Tunlid, A., Persson, P., & Olsson, U. (2022). X-ray scattering reveals two mechanisms of cellulose microfibril degradation by filamentous fungi. *Applied and Environmental Microbiology, 88*(17). https://doi.org/10.1128/aem.00995-22

Fumagalli, F., Ottoboni, M., Pinotti, L., & Cheli, F. (2021). Integrated mycotoxin management system in the feed supply chain: Innovative approaches. *Toxins, 13*(8), 572. https://doi.org/10.3390/toxins13080572

Ghosh, S., Rusyn, I., Dmytruk, O. V., Dmytruk, K. V., Onyeaka, H., Gryzenhout, M., & Gafforov, Y. (2023). Filamentous fungi for sustainable remediation of pharmaceutical compounds, heavy metal and oil hydrocarbons. *Frontiers in Bioengineering and Biotechnology, 11*. https://doi.org/10.3389/fbioe.2023.1106973

Godde, C., Mason-D'Croz, D., Mayberry, D., Thornton, P., & Herrero, M. (2021). Impacts of climate change on the livestock food supply chain; a review of the evidence. *Global Food Security, 28*, 100488. https://doi.org/10.1016/j.gfs.2020.100488

Guan, Y., Lv, H., Wu, G., Chen, J., Wang, M., Zhang, M., Pang, H., Duan, Y., Wang, L., & Tan, Z. (2023). Effects of lactic acid bacteria reducing the content of harmful fungi and mycotoxins on the quality of mixed fermented feed. *Toxins, 15*(3), 226. https://doi.org/10.3390/toxins15030226

Hartinger, T., & Zebeli, Q. (2021). The present role and new potentials of anaerobic fungi in ruminant nutrition. *Journal of Fungi, 7*(3). https://doi.org/10.3390/JOF7030200

Hess, M., Paul, S. S., Puniya, A. K., Van Der Giezen, M., Shaw, C., Edwards, J. E., & Fliegerová, K. (2020). Anaerobic fungi: Past, present, and future. *Frontiers in Microbiology, 11*. https://doi.org/10.3389/fmicb.2020.584893

Hosseini, S., Brenig, B., Winitchakorn, S., Kanmanee, C., Srinual, O., Tapingkae, W., & Gatphayak, K. (2023). Genetic assessment of the effect of red yeast (Sporidiobolus pararoseus) as a feed additive on mycotoxin toxicity in laying hens. *Frontiers in Microbiology, 14*. https://doi.org/10.3389/FMICB.2023.1254569

Hu, J., Chen, J., Hou, Q., Xu, X., Ren, J., Ma, L., & Yan, X. (2023). Core-predominant gut fungus Kazachstania slooffiae promotes intestinal epithelial glycolysis via lysine desuccinylation in pigs. *Microbiome, 11*(1). https://doi.org/10.1186/s40168-023-01468-3

Huo, C., Guo, Y., & Zhao, Y. (2024). Enhancing quality of ruminant feed through fungal treatment: Usage of bamboo shoot residues. *PLoS One, 19*(5 May). https://doi.org/10.1371/journal.pone.0302185

Hüttner, S., Johansson, A., Teixeira, P. G., Achterberg, P., & Nair, R. B. (2020). Recent advances in the intellectual property landscape of filamentous fungi. *Fungal Biology and Biotechnology, 7*(1). https://doi.org/10.1186/s40694-020-00106-z

Jia, M., Gong, X., Fan, M., Liu, H., Zhou, H., Gu, S., Liu, Y., & Dong, J. (2023). Identification and analysis of the secretome of plant pathogenic fungi reveals lifestyle adaptation. *Frontiers in Microbiology, 14*, 1171618. https://doi.org/10.3389/fmicb.2023.1171618

Kant, S., Das, S., Roy, S., & Tripathy, S. (2024). Fungal cellulases: A comprehensive review. *The Nucleus*. https://doi.org/10.1007/s13237-024-00501-6

Kemboi, D. C., Antonissen, G., Ochieng, P. E., Croubels, S., Okoth, S., Kangethe, E. K., Faas, J., Lindahl, J. F., & Gathumbi, J. K. (2020). A review of the impact of mycotoxins on dairy cattle health: Challenges for food safety and dairy production in sub-Saharan Africa. *Toxins, 12*(4). https://doi.org/10.3390/toxins12040222

Kholif, A. E., Anele, A., & Anele, U. Y. (2024). Microbial feed additives in ruminant feeding. *AIMS Microbiology, 10*(3), 542–571. https://doi.org/10.3934/microbiol.2024026

Kortei, N. K., Djaba, B. T., Tettey, C. O., Agyemang, A. O., Aninagyei, E., Essuman, E. K., Boakye, A. A., & Annan, T. (2022). Toxicogenic fungi, aflatoxins, and antimicrobial activities associated with some spices and herbs from three selected markets in Ho Municipality, Ghana. *International Journal of Food Science, 2022*, 7195890. https://doi.org/10.1155/2022/7195890

Kumar, A., Verma, V., Dubey, V. K., Srivastava, A., Garg, S. K., Singh, V. P., & Arora, P. K. (2023). Industrial applications of fungal lipases: A review. *Frontiers in Microbiology, 14*, 1142536. https://doi.org/10.3389/fmicb.2023.1142536

Lange, L., Barrett, K., Pilgaard, B., Gleason, F., & Tsang, A. (2019). Enzymes of early-diverging, zoosporic fungi. *Applied Microbiology and Biotechnology, 103*(17), 6885–6902. https://doi.org/10.1007/s00253-019-09983-w

Langer, S. G., Gabris, C., Einfalt, D., Wemheuer, B., Kazda, M., & Bengelsdorf, F. R. (2019). Different response of bacteria, archaea and fungi to process parameters in nine full-scale anaerobic digesters. *Microbial Biotechnology, 12*(6), 1210–1225. https://doi.org/10.1111/1751-7915.13409

Lattimer, J. M., & Haub, M. D. (2010). Effects of dietary fiber and its components on metabolic health. *Nutrients, 2*(12), 1266–1289. https://doi.org/10.3390/nu2121266

Le Cocq, K., Brown, B., Hodgson, C. J., McFadzean, J., Horrocks, C. A., Lee, M. R. F., & Davies, D. R. (2020). Application of monoclonal antibodies in quantifying fungal growth dynamics during aerobic spoilage of silage. *Microbial Biotechnology, 13*(4), 1054–1065. https://doi.org/10.1111/1751-7915.13552

Li, J., Chen, D., Yu, B., He, J., Huang, Z., Mao, X., Zheng, P., Yu, J., Luo, J., Tian, G., & Luo, Y. (2020a). The fungal community and its interaction with the concentration of short-chain fatty acids in the faeces of Chenghua, Yorkshire and Tibetan pigs. *Microbial Biotechnology, 13*(2), 509–521. https://doi.org/10.1111/1751-7915.13507

Li, X., Zhu, Z., Ye, L., Kang, Z., Zhang, X., Huang, Y., Zhang, B., & Zou, Y. (2020b). Comparison of the partial structure and antioxidant activity of polysaccharides from two species of Chinese truffles. *Molecules, 25*(18). https://doi.org/10.3390/molecules25184345

Li, T. H., Che, P. F., Zhang, C. R., Zhang, B., Ali, A., & Zang, L. S. (2020c). Recycling of spent mushroom substrate: Utilization as feed material for the larvae of the yellow mealworm tenebrio molitor (coleoptera: Tenebrionidae). *PLoS One, 15*(8 August). https://doi.org/10.1371/journal.pone.0237259

Li, J., Wang, W., Chen, S., Shao, T., Tao, X., & Yuan, X. (2021). Effect of lactic acid bacteria on the fermentation quality and mycotoxins concentrations of corn silage infested with mycotoxigenic fungi. *Toxins, 13*(10). https://doi.org/10.3390/toxins13100699

Li, W., Zou, G., Bao, D., & Wu, Y. (2024). Current advances in the functional genes of edible and medicinal fungi: Research techniques, functional analysis, and prospects. *Journal of Fungi, 10*(5). https://doi.org/10.3390/JOF10050311

Lian, X., Shi, M., Liang, Y., Lin, Q., & Zhang, L. (2024). The effects of unconventional feed fermentation on intestinal oxidative stress in animals. *Antioxidants, 13*(3), 305. https://doi.org/10.3390/antiox13030305

Lin, L., & Xu, J. (2022). Production of fungal pigments: Molecular processes and their applications. *Journal of Fungi (Basel, Switzerland), 9*(1). https://doi.org/10.3390/jof9010044

Lopes, D. R. G., La Reau, A. J., Duarte, M. d. S., Detmann, E., Bento, C. B. P., Mercadante, M. E. Z., Bonilha, S. F. M., Suen, G., & Mantovani, H. C. (2019). The bacterial and fungal microbiota of Nelore steers is dynamic across the gastrointestinal tract and its fecal-associated microbiota is correlated to feed efficiency. *Frontiers in Microbiology, 10*(JUN), 1263. https://doi.org/10.3389/fmicb.2019.01263

Lu, Y., & Liu, D. (2024). Optimization of polysaccharide conditions and analysis of antioxidant capacity in the co-culture of Sanghuangporus vaninii and Pleurotus sapidus. *PeerJ, 12*(6), e17571. https://doi.org/10.7717/peerj.17571

Lübeck, M., & Lübeck, P. S. (2022). Fungal cell factories for efficient and sustainable production of proteins and peptides. *Microorganisms, 10*(4). https://doi.org/10.3390/microorganisms10040753

Mahato, D. K., Lee, K. E., Kamle, M., Devi, S., Dewangan, K. N., Kumar, P., & Kang, S. G. (2019). Aflatoxins in food and feed: An overview on prevalence, detection and control strategies. *Frontiers in Microbiology, 10*. https://doi.org/10.3389/fmicb.2019.02266

Mapook, A., Hyde, K., Hassan, K., Kemkuignou, B. M., Čmoková, A., Surup, F., Kuhnert, E., Paomephan, P., Cheng, T., De Hoog, S., Song, Y., Jayawardena, R., Al-Hatmi, A., Mahmoudi, T., Ponts, N., Studt-Reinhold, L., Richard-Forget, F., Chethana, K., Harishchandra, D., et al. (2022). Ten decadal advances in fungal biology leading towards human well-being. *Fungal Diversity, 116*(1), 547–614. https://doi.org/10.1007/s13225-022-00510-3

McGenity, T. J., Gessesse, A., Hallsworth, J. E., Garcia Cela, E., Verheecke-Vaessen, C., Wang, F., Chavarría, M., Haggblom, M. M., Molin, S., Danchin, A., Smid, E. J., Lood, C., Cockell, C. S., Whitby, C., Liu, S. J., Keller, N. P., Stein, L. Y., Bordenstein, S. R., Lal, R., et al. (2020). Visualizing the invisible: Class excursions to ignite children's enthusiasm for microbes. *Microbial Biotechnology, 13*(4), 844–887. https://doi.org/10.1111/1751-7915.13576

Meyer, V., Basenko, E. Y., Benz, J. P., Braus, G. H., Caddick, M. X., Csukai, M., de Vries, R. P., Endy, D., Frisvad, J. C., Gunde-Cimerman, N., Haarmann, T., Hadar, Y., Hansen, K., Johnson, R. I., Keller, N. P., Kraševec, N., Mortensen, U. H., Perez, R., Ram, A. F. J., et al. (2020). Growing a circular economy with fungal biotechnology: A white paper. *Fungal Biology and Biotechnology, 7*(1), 5. https://doi.org/10.1186/s40694-020-00095-z

Mirończuk-Chodakowska, I., & Witkowska, A. M. (2020). Evaluation of polish wild mushrooms as beta-glucan sources. *International Journal of Environmental Research and Public Health, 17*(19), 1–17. https://doi.org/10.3390/ijerph17197299

Mirończuk-Chodakowska, I., Kujawowicz, K., & Witkowska, A. M. (2021). Beta-glucans from fungi: Biological and health-promoting potential in the covid-19 pandemic era. *Nutrients, 13*(11). https://doi.org/10.3390/nu13113960

Miyauchi, S., Hage, H., Drula, E., Lesage-Meessen, L., Berrin, J. G., Navarro, D., Favel, A., Chaduli, D., Grisel, S., Haon, M., Piumi, F., Levasseur, A., Lomascolo, A., Ahrendt, S., Barry, K., LaButti, K. M., Chevret, D., Daum, C., Mariette, J., et al. (2020a). Conservedwhite-rot enzymaticmechanismfor wood decay in the basidiomycota genus pycnoporus. *DNA Research, 27*(2). https://doi.org/10.1093/DNARES/DSAA011

Miyauchi, S., Kiss, E., Kuo, A., Drula, E., Kohler, A., Sánchez-García, M., Morin, E., Andreopoulos, B., Barry, K. W., Bonito, G., Buée, M., Carver, A., Chen, C., Cichocki, N., Clum, A., Culley,

D., Crous, P. W., Fauchery, L., Girlanda, M., et al. (2020b). Large-scale genome sequencing of mycorrhizal fungi provides insights into the early evolution of symbiotic traits. *Nature Communications, 11*(1). https://doi.org/10.1038/s41467-020-18795-w

Mougin, J., Lobanov, V., Danion, M., Roquigny, R., Goardon, L., Grard, T., Morin, T., Labbé, L., & Joyce, A. (2023). Effects of dietary co-exposure to fungal and herbal functional feed additives on immune parameters and microbial intestinal diversity in rainbow trout (Oncorhynchus mykiss). *Fish & Shellfish Immunology, 137*, 108773. https://doi.org/10.1016/j.fsi.2023.108773

Müller, M., Kües, U., Budde, K. B., & Gailing, O. (2023). Applying molecular and genetic methods to trees and their fungal communities. *Applied Microbiology and Biotechnology, 107*(9), 2783–2830. https://doi.org/10.1007/s00253-023-12480-w

Munteanu, C., & Schwartz, B. (2022). The relationship between nutrition and the immune system. *Frontiers in Nutrition, 9*. https://doi.org/10.3389/fnut.2022.1082500

Nagy, L. G., Vonk, P. J., Künzler, M., Földi, C., Virágh, M., Ohm, R. A., Hennicke, F., Bálint, B., Csernetics, Á., Hegedüs, B., Hou, Z., Liu, X. B., Nan, S., Pareek, M., Sahu, N., Szathmári, B., Varga, T., Wu, H., Yang, X., & Merényi, Z. (2023). Lessons on fruiting body morphogenesis from genomes and transcriptomes of agaricomycetes. *Studies in Mycology, 104*, 1–85. https://doi.org/10.3114/sim.2022.104.01

Naranjo-Ortiz, M. A., & Gabaldón, T. (2020). Fungal evolution: Cellular, genomic and metabolic complexity. *Biological Reviews, 95*(5), 1198–1232. https://doi.org/10.1111/brv.12605

Naz, T., Ullah, S., Nazir, Y., Li, S., Iqbal, B., Liu, Q., Mohamed, H., & Song, Y. (2023). Industrially important fungal carotenoids: Advancements in biotechnological production and extraction. *Journal of Fungi (Basel, Switzerland), 9*(5). https://doi.org/10.3390/jof9050578

Newsholme, P. (2021). Cellular and metabolic mechanisms of nutrient actions in immune function. *Nutrition and Diabetes, 11*(1). https://doi.org/10.1038/s41387-021-00162-3

Novak, B., Rainer, V., Sulyok, M., Haltrich, D., Schatzmayr, G., & Mayer, E. (2019). Twenty-eight fungal secondary metabolites detected in pig feed samples: Their occurrence, relevance and cytotoxic effects in vitro. *Toxins, 11*(9). https://doi.org/10.3390/TOXINS11090537

Nutrition: Micronutrient Intake, Imbalances, and Interventions. (2024, January 1). PubMed. https://pubmed.ncbi.nlm.nih.gov/37983324

Ochieng, P. E., Scippo, M. L., Kemboi, D. C., Croubels, S., Okoth, S., Kang'ethe, E. K., Doupovec, B., Gathumbi, J. K., Lindahl, J. F., & Antonissen, G. (2021). Mycotoxins in poultry feed and feed ingredients from sub-Saharan Africa and their impact on the production of broiler and layer chickens: A review. *Toxins, 13*(9). https://doi.org/10.3390/toxins13090633

Ojha, B., Singh, P. K., & Shrivastava, N. (2018). *Enzymes in the animal feed industry* (pp. 93–109). Elsevier eBooks. https://doi.org/10.1016/b978-0-12-813280-7.00007-4

Olmo, R., Wetzels, S. U., Armanhi, J. S. L., Arruda, P., Berg, G., Cernava, T., Cotter, P. D., Araujo, S. C., de Souza, R. S. C., Ferrocino, I., Frisvad, J. C., Georgalaki, M., Hansen, H. H., Kazou, M., Kiran, G. S., Kostic, T., Krauss-Etschmann, S., Kriaa, A., Lange, L., et al. (2022). Microbiome research as an effective driver of success stories in agrifood systems - a selection of case studies. *Frontiers in Microbiology, 13*, 834622. https://doi.org/10.3389/fmicb.2022.834622

Patience, J. F., Rossoni-Serão, M. C., & Gutiérrez, N. A. (2015). A review of feed efficiency in pig: Biology and application. *Journal of Animal Science and Biotechnology/Journal of Animal Science and Biotechnology, 6*(1). https://doi.org/10.1186/s40104-015-0031-2

Pawar, K. S., Singh, P. N., & Singh, S. K. (2023). Fungal alkaline proteases and their potential applications in different industries. *Frontiers in Microbiology, 14*, 1138401. https://doi.org/10.3389/fmicb.2023.1138401

Peles, F., Sipos, P., Győri, Z., Pfliegler, W. P., Giacometti, F., Serraino, A., Pagliuca, G., Gazzotti, T., & Pócsi, I. (2019). Adverse effects, transformation and channeling of aflatoxins into food raw materials in livestock. *Frontiers in Microbiology, 10*. https://doi.org/10.3389/FMICB.2019.02861

Peng, K., Jin, L., Niu, Y. D., Huang, Q., McAllister, T. A., Yang, H. E., Denise, H., Xu, Z., Acharya, S., Wang, S., & Wang, Y. (2018). Condensed tannins affect bacterial and fungal microbiomes and mycotoxin production during ensiling and upon aerobic exposure. *Applied and Environmental Microbiology, 84*(5). https://doi.org/10.1128/AEM.02274-17

Ravindran, V. (2013). Feed enzymes: The science, practice, and metabolic realities. *The Journal of Applied Poultry Research, 22*(3), 628–636. https://doi.org/10.3382/japr.2013-00739

Ravindran, V., & Abdollahi, M. R. (2021). Nutrition and digestive physiology of the broiler chick: State of the art and outlook. *Animals, 11*(10), 2795. https://doi.org/10.3390/ani11102795

Richter, F., Bindschedler, S., Calonne-Salmon, M., Declerck, S., Junier, P., & Stanley, C. E. (2022). Fungi-on-a-Chip: Microfluidic platforms for single-cell studies on fungi. *FEMS Microbiology Reviews, 46*(6). https://doi.org/10.1093/femsre/fuac039

Rizwanuddin, S., Kumar, V., Naik, B., Singh, P., Mishra, S., Rustagi, S., & Kumar, V. (2023). Microbial phytase: Their sources, production, and role in the enhancement of nutritional aspects of food and feed additives. *Journal of Agriculture and Food Research, 12*, 100559. https://doi.org/10.1016/j.jafr.2023.100559

Roik, A., Reverter, M., & Pogoreutz, C. (2022). A roadmap to understanding diversity and function of coral reef-associated fungi. *FEMS Microbiology Reviews, 46*(6). https://doi.org/10.1093/femsre/fuac028

Romero, J. J., Park, J., Joo, Y., Zhao, Y., Killerby, M., Reyes, D. C., Tiezzi, F., Gutierrez-Rodriguez, E., & Castillo, M. S. (2021). A combination of Lactobacillus buchneri and Pediococcus pentosaceus extended the aerobic stability of conventional and brown midrib mutants-corn hybrids ensiled at low dry matter concentrations by causing a major shift in their bacterial and fungal community. *Journal of Animal Science, 99*(8). https://doi.org/10.1093/jas/skab141

Rutkowski, D., Weston, M., & Vannette, R. L. (2023). Bees just wanna have fungi: A review of bee associations with nonpathogenic fungi. *FEMS Microbiology Ecology, 99*(8). https://doi.org/10.1093/femsec/fiad077

Salazar-Alekseyeva, K., Herndl, G. J., & Baltar, F. (2024). Influence of salinity on the extracellular enzymatic activities of marine pelagic fungi. *Journal of Fungi (Basel, Switzerland), 10*(2). https://doi.org/10.3390/jof10020152

Salem, H. M., Saad, A. M., Soliman, S. M., Selim, S., Mosa, W. F. A., Ahmed, A. E., Al Jaouni, S. K., Almuhayawi, M. S., Abd El-Hack, M. E., El-Tarabily, K. A., & El-Saadony, M. T. (2023). Ameliorative avian gut environment and bird productivity through the application of safe antibiotics alternatives: A comprehensive review. *Poultry Science, 102*(9). https://doi.org/10.1016/j.psj.2023.102840

Samtiya, M., Aluko, R. E., & Dhewa, T. (2020). Plant food anti-nutritional factors and their reduction strategies: An overview. *Food Production Processing and Nutrition, 2*(1). https://doi.org/10.1186/s43014-020-0020-5

Shan, S., Wang, W., Song, C., Wang, M., Sun, B., Li, Y., Fu, Y., Gu, X., Ruan, W., & Rasmann, S. (2019). The symbiotic bacteria Alcaligenes faecalis of the entomopathogenic nematodes Oscheius spp. exhibit potential biocontrol of plant- and entomopathogenic fungi. *Microbial Biotechnology, 12*(3), 459–471. https://doi.org/10.1111/1751-7915.13365

Sharma, S., Compant, S., Franken, P., Ruppel, S., & Ballhausen, M. B. (2021). It takes two to tango: A bacterial biofilm provides protection against a fungus-feeding bacterial predator. *Microorganisms, 9*(8). https://doi.org/10.3390/MICROORGANISMS9081566

Shurson, G. C., Urriola, P. E., & Schroeder, D. C. (2023). Biosecurity and mitigation strategies to control pig viruses in feed ingredients and complete feeds. *Animals: An Open Access Journal from MDPI, 13*(14). https://doi.org/10.3390/ani13142375

Stoev, S. D. (2024). Natural feed additives and bioactive supplements versus chemical additives as a safe and practical approach to combat foodborne mycotoxicoses. *Frontiers in Nutrition, 11*, 1335779. https://doi.org/10.3389/fnut.2024.1335779

Sun, W., & Shahrajabian, M. H. (2023). The application of arbuscular mycorrhizal fungi as microbial biostimulant, sustainable approaches in modern agriculture. *Plants (Basel, Switzerland), 12*(17). https://doi.org/10.3390/plants12173101

Swift, C. L., Louie, K. B., Bowen, B. P., Olson, H. M., Purvine, S. O., Salamov, A., Mondo, S. J., Solomon, K. V., Wright, A. T., Northen, T. R., Grigoriev, I. V., Keller, N. P., & O'Malley, M. A. (2021). Anaerobic gut fungi are an untapped reservoir of natural products. *Proceedings of the National Academy of Sciences of the United States of America, 118*(18). https://doi.org/10.1073/pnas.2019855118

Tanaka, T., Terauchi, Y., Yoshimi, A., & Abe, K. (2022). Aspergillus Hydrophobins: Physicochemical properties, biochemical properties, and functions in solid polymer degradation. *Microorganisms, 10*(8). https://doi.org/10.3390/microorganisms10081498

Urgessa, O. E., Koyamo, R., Dinka, H., Tefese, K., & Gemeda, M. T. (2024). Review on desirable microbial phytases as a poultry feed additive: Their sources, production, enzymatic evaluation, market size, and regulation. *International Journal of Microbiology, 2024*(1), 9400374. https://doi.org/10.1155/2024/9400374

Uwineza, C., Bouzarjomehr, M., Parchami, M., Sar, T., Taherzadeh, M. J., & Mahboubi, A. (2023). Evaluation of in vitro digestibility of Aspergillus oryzae fungal biomass grown on organic residue derived-VFAs as a promising ruminant feed supplement. *Journal of Animal Science and Biotechnology, 14*(1), 120. https://doi.org/10.1186/s40104-023-00922-4

Uwineza, C., Parchami, M., Bouzarjomehr, M., Taherzadeh, M. J., & Mahboubi, A. (2024). Recent developments in the application of filamentous fungus aspergillus oryzae in ruminant feed. *Animals: An Open Access Journal from MDPI, 14*(16). https://doi.org/10.3390/ani14162427

Venkatesh, N., & Keller, N. P. (2019). Mycotoxins in conversation with bacteria and fungi. *Frontiers in Microbiology, 10*(Mar). https://doi.org/10.3389/FMICB.2019.00403

Vitale, G. A., Coppola, D., Palma Esposito, F., Buonocore, C., Ausuri, J., Tortorella, E., & de Pascale, D. (2020). Antioxidant molecules from marine fungi: Methodologies and perspectives. *Antioxidants (Basel, Switzerland), 9*(12), 1–35. https://doi.org/10.3390/antiox9121183

Wan Mohtar, W. H. M., Wan-Mohtar, W. A. A. Q. I., Zahuri, A. A., Ibrahim, M. F., Show, P. L., Ilham, Z., Jamaludin, A. A., Abdul Patah, M. F., Ahmad Usuldin, S. R., & Rowan, N. (2022). Role of ascomycete and basidiomycete fungi in meeting established and emerging sustainability opportunities: A review. *Bioengineered, 13*(7–12), 14903–14935. https://doi.org/10.1080/21655979.2023.2184785

Wang, H., Li, P., Liu, X., Zhang, C., Lu, Q., Xi, D., Yang, R., Wang, S., Bai, W., Yang, Z., Zhou, R., Cheng, X., & Leng, J. (2019). The composition of fungal communities in the rumen of Gayals (*Bos frontalis*), Yaks (*Bos grunniens*), and Yunnan and Tibetan Yellow Cattle (*Bos taurs*). *Polish Journal of Microbiology, 68*(4), 505–514. https://doi.org/10.33073/PJM-2019-050

Wang, W., Tan, J., Nima, L., Sang, Y., Cai, X., & Xue, H. (2022). Polysaccharides from fungi: A review on their extraction, purification, structural features, and biological activities. *Food Chemistry: X, 15*, 100414. https://doi.org/10.1016/j.fochx.2022.100414

Wang, J., Deng, L., Chen, M., Che, Y., Li, L., Zhu, L., Chen, G., & Feng, T. (2024). Phytogenic feed additives as natural antibiotic alternatives in animal health and production: A review of the literature of the last decade. *Animal Nutrition (Zhongguo Xu Mu Shou Yi Xue Hui), 17*, 244–264. https://doi.org/10.1016/j.aninu.2024.01.012

Weiland, P., Dempwolff, F., Steinchen, W., Freibert, S. A., Tian, H., Glatter, T., Martin, R., Thomma, B. P. H. J., Bange, G., & Altegoer, F. (2023). Structural and functional analysis of the cerato-platanin-like protein Cpl1 suggests diverging functions in smut fungi. *Molecular Plant Pathology, 24*(7), 768–787. https://doi.org/10.1111/mpp.13349

Zain, M. E. (2011). Impact of mycotoxins on humans and animals. *Journal of Saudi Chemical Society, 15*(2), 129–144. https://doi.org/10.1016/j.jscs.2010.06.006

Zhang, P. (2022). Influence of foods and nutrition on the gut microbiome and implications for intestinal health. *International Journal of Molecular Sciences, 23*(17), 9588. https://doi.org/10.3390/ijms23179588

Zhen, Y., Wang, M., Gu, Y., Yu, X., Shahzad, K., Xu, J., Gong, Y., Li, P., & Loor, J. J. (2021). Biosorption of copper in pig manure using aspergillus and yeast: Characterization and its microbial diversity study. *Frontiers in Microbiology, 12*. https://doi.org/10.3389/FMICB.2021.687533

Zheng, L., Ma, Y., Zhang, Y., Meng, Q., Yang, J., Gong, W., Liu, Q., Cai, L., Hao, L., Wang, B., & Yang, Y. (2020). Distribution of zinc in mycelial cells and antioxidant and anti-inflammatory activities of mycelia zinc polysaccharides from Thelephora ganbajun TG-01. *Oxidative Medicine and Cellular Longevity, 2020*. https://doi.org/10.1155/2020/2308017

Zhu, X. M., Li, L., Bao, J. D., Wang, J. Y., Daskalov, A., Liu, X. H., Del Poeta, M., & Lin, F. C. (2023). The biological functions of sphingolipids in plant pathogenic fungi. *PLoS Pathogens, 19*(11 November). https://doi.org/10.1371/journal.ppat.1011733

Benefits of Fungal Feed Additives in Animals and Future Prospects

John Tsado Mathew, Abel Inobeme, Musah Monday, Yakubu Azeh, Elijah Yanda Shaba, Charles Oluwaseun Adetunji, Mohammed Jibrin Ndejiko, Isah Mohammed Chado, Amos Mamman, and Jonathan Hussaini

Abstract Fungal feed additives, derived from fungi such as *Saccharomyces cerevisiae* and *Aspergillus oryzae*, have garnered significant interest for their role in enhancing animal nutrition and health. These additives are rich in bioactive compounds such as enzymes, prebiotics, and antioxidants, which improve digestive efficiency, strengthen immune responses, and support overall growth in livestock. By optimizing nutrient absorption and promoting gut health, fungal additives enhance feed conversion ratios and animal performance. Their antimicrobial properties help reduce pathogenic infections, contributing to better animal welfare. Additionally, fungal feed additives mitigate the harmful effects of mycotoxins—common feed contaminants—thereby improving feed safety. Research indicates that incorporating these additives into animal diets can reduce the need for antibiotics, aligning with global efforts to combat antibiotic resistance. This not only supports sustain-

J. T. Mathew (✉) · M. Monday · Y. Azeh · A. Mamman · J. Hussaini
Department of Chemistry, Ibrahim Badamasi Babangida University, Lapai, Lapai, Niger State, Nigeria
e-mail: jmathew@ibbu.edu.ng

A. Inobeme
Department of Chemistry, Edo State University Uzairue, Uzairue, Edo State, Nigeria

E. Y. Shaba
Department of Chemistry, Federal University of Technology, Minna, Lapai, Niger State, Nigeria

C. O. Adetunji
Applied Microbiology, Biotechnology and Nanotechnology Laboratory, Department of Microbiology, Edo State University Uzairue, Uzairue, Edo State, Nigeria

M. J. Ndejiko
Department of Microbiology, Ibrahim Badamasi Babangida University, Lapai, Lapai, Niger State, Nigeria

I. M. Chado
Department of Biology, Ibrahim Badamasi Babangida University, Lapai, Lapai, Niger State, Nigeria

A. Gupta (ed.), *Prospects of Fungal Biotechnologies for Livestock Volume 1*, Fungal Biology, https://doi.org/10.1007/978-3-031-90994-8_4

able animal husbandry but also addresses public health concerns. However, the effectiveness of fungal feed additives can vary based on fungal strains, dosages, and animal species, highlighting the need for further research. A deeper understanding of their mechanisms of action is crucial for developing targeted and reliable supplementation strategies. Despite these challenges, fungal feed additives offer a sustainable solution for improving livestock productivity while delivering economic and environmental benefits, paving the way for advancements in modern animal husbandry.

Keywords Additives · antibiotic · animals · bioactive · feed · resistance

1 Introduction

The burgeoning global demand for animal protein has intensified the need for innovations in animal nutrition that enhance productivity while maintaining health and welfare standards. Among various strategies, the inclusion of fungal feed additives has emerged as a promising approach. Derived from fungi such as Saccharomyces cerevisiae (brewer's yeast) and Aspergillus oryzae, these additives are rich in bioactive compounds offering multifaceted benefits for livestock. They enhance digestive efficiency, boost immune function, and improve overall growth performance. This introduction explores the background of animal nutrition, the role of fungal feed additives, and their potential impact on modern animal husbandry, highlighting their contribution to sustainable and effective animal production practices (Balehegn et al., 2020).

Animal nutrition is a critical factor in the agricultural industry, directly influencing the growth, productivity, and health of livestock. Traditional feed components, such as grains, protein meals, forages, and vitamins, collectively provide the necessary nutrients for optimal animal development. However, optimizing these diets to ensure maximum efficiency and health benefits is a complex and ongoing challenge. To address this, various feed additives have been incorporated into animal diets over the years, including antibiotics, probiotics, and enzymes. These additives aim to enhance growth performance, improve feed conversion ratios, and bolster disease resistance. Antibiotics, for instance, have been used to promote growth and prevent disease, but their extensive use has raised significant concerns regarding antibiotic resistance. As a result, there has been a growing interest in alternative additives that can provide similar benefits without the associated risks. Probiotics, which introduce beneficial microorganisms to the gut, and enzymes, which aid in the digestion of feed components, are examples of such alternatives. These innovations in feed additives play a crucial role in modern animal husbandry, seeking to improve the overall health and productivity of livestock while addressing sustainability and safety concerns (Ponnampalam et al., 2022; Inobeme et al., 2024a).

While antibiotics have been effective in promoting growth and preventing disease, their extensive use has raised significant concerns regarding antibiotic resistance, posing a serious threat to both animal and human health. This alarming issue has spurred an urgent search for alternative feed additives that can provide similar benefits without the associated risks. In this context, fungal feed additives have garnered considerable attention due to their natural origin and potential health benefits. Derived from beneficial fungi such as Saccharomyces cerevisiae (brewer's yeast) and Aspergillus oryzae, fungal feed additives are rich in bioactive compounds, including enzymes, prebiotics, and antioxidants. These compounds can enhance digestive efficiency, improve nutrient absorption, and support immune function in livestock. Moreover, fungal additives exhibit antimicrobial properties, reducing the incidence of pathogenic infections and improving overall animal welfare. The natural origin of fungal additives makes them an attractive alternative, aligning with the growing demand for sustainable and environmentally friendly agricultural practices. As research into their benefits continues to expand, fungal feed additives hold promise for revolutionizing animal nutrition, providing a viable solution to the challenges posed by antibiotic resistance while promoting healthier and more productive livestock (Muteeb et al., 2023).

Fungal feed additives encompass a broad range of products derived from different fungal species. Among these, Saccharomyces cerevisiae (brewer's yeast) and Aspergillus oryzae are the most commonly used. These fungi are carefully cultivated and processed to extract beneficial compounds, including enzymes, prebiotics, and antioxidants, which are then incorporated into animal feeds. Saccharomyces cerevisiae is particularly valued for its high protein content and a rich array of vitamins and minerals. It produces various bioactive compounds that enhance the digestive process and boost the immune system. Aspergillus oryzae, known for its potent enzymatic capabilities, excels in breaking down complex carbohydrates, thus improving nutrient absorption and feed efficiency. The bioactive compounds derived from these fungi offer a range of health benefits for livestock. Enzymes aid in the breakdown of feed components, enhancing digestive efficiency, while prebiotics promote a healthy gut microbiome, crucial for nutrient absorption and immune function. Additionally, antioxidants help mitigate oxidative stress, supporting overall animal health. Through these mechanisms, fungal feed additives contribute to the improved growth, productivity, and well-being of livestock (Amara & El-Baky, 2023).

Saccharomyces cerevisiae, or brewer's yeast, is particularly valued for its high protein content, as well as its rich array of vitamins and minerals. This yeast produces various bioactive compounds that can enhance the digestive process and boost the immune system in livestock. These compounds include enzymes that aid in the breakdown of feed components, making nutrients more accessible and digestible. Additionally, Saccharomyces cerevisiae supports a healthy gut microbiome by promoting the growth of beneficial bacteria, which is essential for optimal nutrient absorption and immune function. Aspergillus oryzae, on the other hand, is renowned for its potent enzymatic capabilities. This fungus excels in breaking down complex carbohydrates into simpler sugars, improving the overall nutrient absorption and feed efficiency in animals. The enzymes produced by Aspergillus oryzae, such as amylases and proteases, play a crucial role in enhancing the digestive process, allowing animals to extract maximum nutritional value from their feed. By improving digestion and nutrient uptake, Aspergillus oryzae contributes to better growth performance and overall health in livestock. Together, these fungi provide a powerful combination of nutritional and health benefits when used as feed additives (Sampath et al., 2023).

The beneficial effects of fungal feed additives are attributed to their bioactive components, particularly enzymes. These enzymes play a pivotal role in breaking down complex feed components into simpler, more absorbable forms, thereby enhancing digestive efficiency. This enzymatic action is especially important for young animals or those with underdeveloped digestive systems, as it helps them to better utilize the nutrients in their feed. For example, enzymes such as amylases break down starches into simple sugars, while proteases hydrolyze proteins into amino acids. This improved digestion ensures that animals receive the maximum nutritional benefit from their feed, promoting better growth and overall health. Additionally, by enhancing nutrient absorption, fungal feed additives help to optimize feed conversion ratios, leading to more efficient and sustainable livestock production. In addition, the enzymatic activity provided by fungal feed additives is a key factor in their ability to improve the health and productivity of livestock (Velázquez-De et al., 2021; Mathew et al., 2024a).

Prebiotics, another key component of fungal feed additives, serve as food for beneficial gut bacteria, promoting a healthy microbiome. A balanced gut microbiome is essential for optimal nutrient absorption and immune function, playing a crucial role in the overall health and productivity of livestock. By providing a substrate for beneficial bacteria, prebiotics help to maintain a stable and diverse microbial population in the gut. This balance is vital for several reasons. First, it enhances nutrient absorption by supporting the digestive process, ensuring that animals can extract maximum nutritional value from their feed. Second, a healthy microbiome competes with pathogenic bacteria, reducing the incidence of gastrointestinal diseases and infections. This not only improves animal health but also reduces the need for antibiotics, contributing to efforts to combat antibiotic resistance. Generally, prebiotics in fungal feed additives help to create a robust and resilient gut environment, which is crucial for the well-being and productivity of livestock (Shehata et al., 2022).

Additionally, fungal additives often contain antioxidants, which play a crucial role in mitigating oxidative stress in animals. Oxidative stress occurs when there is an imbalance between the production of free radicals and the body's ability to neutralize them. This imbalance can lead to cellular damage and tissue inflammation, contributing to various health issues and impairments in growth. Antioxidants work by neutralizing free radicals, thereby preventing them from causing harm to cells and tissues. By maintaining cellular integrity and reducing inflammation, antioxidants support overall health and well-being in animals. In the context of animal nutrition, the inclusion of antioxidants in fungal feed additives offers a proactive approach to promoting optimal health and enhancing the resilience of livestock against oxidative damage (Ponnampalam et al., 2022; Adetuyi et al. 2024a).

In general, fungal feed additives represent a promising innovation in animal nutrition. By leveraging the natural bioactive compounds found in fungi, these additives can enhance growth performance, boost immune function, and improve overall animal health. As the industry seeks sustainable alternatives to traditional feed additives, further research into the specific mechanisms and benefits of fungal feed additives will be crucial. This not only holds potential economic advantages but also aligns with global efforts to promote more sustainable and ethical animal husbandry practices (Phillips et al., 2023).

2 Types of Fungal Feed Additives

Saccharomyces cerevisiae (brewer's yeast): Saccharomyces cerevisiae, commonly known as brewer's yeast, is a fungus renowned for its rich array of bioactive components. These include enzymes, prebiotics, and antioxidants, which collectively contribute to its beneficial effects on animal health and performance. Enzymes produced by Saccharomyces cerevisiae play a crucial role in enhancing digestion by breaking down complex feed components into simpler, more absorbable forms. This improved nutrient availability supports optimal growth and development in animals. Additionally, Saccharomyces cerevisiae acts as a prebiotic, promoting the growth of beneficial bacteria in the gut. A balanced gut microbiome is essential for supporting immune function and overall health. Furthermore, Saccharomyces cerevisiae contains antioxidants that help mitigate oxidative stress in animals. By neutralizing free radicals, these antioxidants maintain cellular integrity and support overall well-being. Common products derived from Saccharomyces cerevisiae include yeast cultures, yeast cell walls, and yeast extracts, which are incorporated into animal feeds to harness their beneficial properties and promote optimal performance and health (Tadioto et al., 2023).

Aspergillus oryzae Aspergillus oryzae, renowned for its potent enzymatic capabilities, is a fungus commonly utilized to enhance nutrient absorption in animals. This fungus produces a variety of enzymes that play a crucial role in breaking down complex carbohydrates, proteins, and fibers present in animal feed. Enzyme preparations derived from Aspergillus oryzae are specifically tailored to target and hydrolyze these complex feed components, thereby increasing their digestibility and availability to the animal. By breaking down these substrates into simpler forms, such as sugars and amino acids, Aspergillus oryzae enzymes facilitate more efficient nutrient absorption in the digestive tract. In addition to direct enzyme preparations, fermented feed additives produced through the action of Aspergillus oryzae are commonly used in animal nutrition. These additives undergo fermentation processes that enhance the bioavailability of nutrients in feed ingredients, further optimizing feed efficiency and nutrient utilization by animals. Generally, the enzymatic capabilities of Aspergillus oryzae contribute significantly to improving the nutritional value of animal feeds and promoting better growth performance in livestock (Corbu et al., 2023).

Other Fungal Species In addition to Saccharomyces cerevisiae and Aspergillus oryzae, various other fungal species contribute to the array of feed additives available for animal nutrition. Fungi such as Trichoderma spp., Penicillium spp., and Rhizopus spp. offer a diverse range of bioactive components and functions beneficial for livestock. Trichoderma spp. are known for their ability to produce enzymes such as cellulases and hemicellulases, which aid in the breakdown of plant cell walls, improving fiber digestion in animals. Penicillium spp. produce enzymes such as xylanases and pectinases, enhancing the utilization of fibrous feed ingredients. Additionally, Penicillium spp. are also sources of secondary metabolites with potential antimicrobial properties. Rhizopus spp. contribute to the fermentation process, producing organic acids and other metabolites that improve feed preservation and digestibility. Their role in fermented feed additives helps to enhance nutrient availability and support gut health in animals. Each of these fungal species offers unique benefits and applications in animal nutrition, providing a broad spectrum of options for formulating feed additives tailored to specific nutritional needs and production goals in livestock farming (Chuang et al., 2020; Mathew et al., 2024b).

3 Bioactive Compounds in Fungal Feed Additives

Fungal feed additives harbor an array of bioactive compounds beneficial for livestock. These include enzymes such as cellulases, hemicellulases, and ligninases, crucial for breaking down plant cell walls and enhancing digestibility. Mycotoxin-binding agents in fungi mitigate the adverse effects of mycotoxins in feed, safeguarding animal health. Moreover, secondary metabolites such as antimicrobial peptides and organic acids inhibit pathogenic microorganisms, promoting gut health and immunity. Fungal additives also offer vitamins, minerals, and amino acids essential for animal growth and metabolism. Their diverse bioactive repertoire contributes to improved feed efficiency, nutrient utilization, and overall animal performance in sustainable agriculture (Corbu et al., 2023) (Table 1).

4 Benefits of Fungal Feed Additives

Improved Nutrient Digestion and Absorption Fungal feed additives, notably those incorporating enzymes such as cellulases, xylanases, and proteases, play a vital role in breaking down complex plant fibers and proteins present in animal feed.

Table 1 Bioactive compounds commonly found in fungal feed additives

Compound	Function	Source	Benefits	Examples	References
Cellulases	Break down cellulose	Fungi	Enhance fiber digestion and nutrient release	Trichoderma reesei, Aspergillus niger	Ejaz et al. (2021), Inobeme et al. (2024b)
Xylanases	Hydrolyze xylan	Fungi	Improve digestibility of hemicellulose	Penicillium sp., Bacillus subtilis	Dahiya et al. (2024)
Proteases	Break down proteins	Fungi	Enhance protein digestion and amino acid release	Aspergillus oryzae, Bacillus licheniformis	Song et al. (2023)
Beta-glucans	Modulate immune response	Fungi, yeast	Stimulate the immune system, and improve gut health	Saccharomyces cerevisiae, Ganoderma lucidum	Zhong et al. (2023)
Mannan oligosaccharides	Prebiotic activity	Fungi, yeast	Promote the growth of beneficial gut bacteria	Candida utilis, Aspergillus awamori	Tanihiro et al. (2020), Mathew et al. (2024c)

These enzymes catalyze the hydrolysis of intricate molecular structures, rendering nutrients more accessible for digestion and absorption by animals. By enhancing nutrient availability, fungal enzymes facilitate more efficient utilization of essential nutrients, leading to improved growth rates, feed efficiency, and overall performance in livestock. Animals can extract greater nutritional value from their feed, resulting in enhanced metabolic processes and physiological functions. Moreover, the increased digestibility of feed components reduces the potential for undigested material to accumulate in the gastrointestinal tract, minimizing the risk of digestive disorders and improving gut health. Incorporating fungal feed additives with enzymatic activity into animal diets represents a strategic approach to optimize nutrient utilization and promote superior performance outcomes in livestock production (Sureshkumar et al., 2023).

Enhanced Gut Health Fungi possess a unique capacity to colonize the gastrointestinal tract and engage with the gut microbiota. Certain fungal species produce metabolites that play a pivotal role in promoting gut health by modulating microbial populations, bolstering the immune system, and preserving intestinal integrity. These beneficial effects collectively contribute to a reduction in the incidence of digestive disorders, such as diarrhea and enteritis, ultimately fostering the development of healthier and more resilient animals. By interacting with the gut microbiota, fungi help maintain a balanced microbial ecosystem, which is crucial for optimal digestion and nutrient absorption. Additionally, fungal metabolites can stimulate immune responses, enhancing the host's ability to ward off pathogens and maintain gut homeostasis. Furthermore, by preserving intestinal integrity, fungi contribute to the prevention of gut barrier dysfunction and associated inflammatory conditions. Generally, the ability of fungi to positively influence gut health underscores their significance as potential dietary supplements for improving animal well-being and performance in livestock production systems (Zhang et al., 2021; Inobeme et al., 2024c).

Mycotoxin Mitigation Fungal feed additives play a crucial role in mitigating the harmful effects of mycotoxins, toxic compounds produced by molds. Certain fungi possess the capability to bind mycotoxins within the gastrointestinal tract, preventing their absorption and reducing their adverse impact on animal health. This binding action helps prevent mycotoxins from entering the bloodstream and exerting toxic effects on vital organs and systems. Furthermore, some fungal species competitively exclude toxigenic fungi from colonizing feed materials. By outcompeting these harmful molds for nutrients and space, these beneficial fungi help minimize mycotoxin contamination in feedstuffs. This proactive approach reduces the risk of mycotoxin exposure to livestock and subsequently mitigates the associated health issues and performance losses. By employing fungal feed additives with mycotoxin-binding and competitive exclusion properties, producers can safeguard animal health and welfare while ensuring the quality and safety of feed supplies (Kępińska-Pacelik & Biel, 2021; Olorunsola et al., 2024).

Reduced Environmental Impact Fungal feed additives provide environmentally friendly solutions for livestock production by addressing key sustainability challenges. By enhancing nutrient utilization and reducing feed waste, these additives minimize the environmental footprint of animal agriculture. Improved feed efficiency translates to reduced feed input needed to achieve desired growth outcomes, resulting in lower greenhouse gas emissions, decreased nutrient excretion, and reduced land use associated with livestock farming. Additionally, by optimizing digestion and absorption, fungal feed additives contribute to fewer nutrients being lost in manure, thus reducing nutrient runoff and its environmental impacts on water bodies. Furthermore, the reduced need for additional feed ingredients helps conserve natural resources, such as land and water, while also lowering the energy and environmental costs associated with feed production and transportation. Overall, the incorporation of fungal feed additives into animal diets represents a sustainable approach to livestock farming, promoting environmental stewardship while supporting efficient and responsible production practices (Sousa et al., 2024).

Alternative Protein Source Certain fungal species have the remarkable ability to thrive on agricultural by-products or waste materials, transforming them into valuable sources of high-quality protein for animal feed. These fungal biomass-based feed additives present a sustainable alternative to conventional protein sources like soybean meal and fishmeal. By utilizing underutilized resources that would otherwise go to waste, fungal feed additives contribute to the efficient utilization of feed resources and help reduce dependency on traditional protein sources that may raise sustainability concerns, such as land and water use, deforestation, and overfishing. Moreover, the cultivation of fungi on agricultural residues can help reduce environmental pollution associated with waste disposal. Overall, incorporating fungal biomass into animal diets not only improves feed sustainability but also enhances the economic viability of livestock production while supporting environmentally responsible practices (Amara and El-Baky, 2023).

Adaptability to Various Diets and Production Systems: Fungal feed additives demonstrate remarkable versatility, seamlessly integrating into various animal diets and production systems. They are equally effective in conventional intensive farming setups and alternative systems like organic or free-range production. Whether fed to poultry, swine, ruminants, or aquaculture species, fungal additives consistently deliver benefits across different stages of production. Their adaptability stems from their diverse mechanisms of action, which enhance nutrient utilization, promote gut health, mitigate mycotoxin contamination, and boost immune function. This versatility makes fungal feed additives valuable tools for improving animal nutrition and health under diverse farming conditions. Furthermore, their compatibility with different dietary compositions and management practices ensures flexibility for producers seeking sustainable and effective solutions. By harnessing the benefits of fungal additives, farmers can optimize animal performance, minimize health risks, and meet the evolving demands of consumers and regulatory standards across a wide spectrum of livestock production systems (Chuang et al., 2020).

Enhanced Immune Function Certain fungal metabolites exhibit immunomodulatory effects, enhancing the animal's immune system by promoting the production of cytokines, antibodies, and other immune factors. These fungal feed additives bolster the animal's defense mechanisms against pathogens and diseases, resulting in improved overall health and welfare. Additionally, by boosting immune function, fungal feed additives reduce the reliance on antibiotic interventions in animal production. This supports the shift towards antibiotic-free animal farming, a trend driven by concerns over antimicrobial resistance and consumer demand for more sustainable and natural farming practices. By enhancing the animal's innate ability to resist diseases, fungal feed additives contribute to a healthier and more resilient livestock population while aligning with efforts to reduce the use of antibiotics in agriculture, thus promoting both animal and public health (Musah et al., 2024).

Conservation of Energy and Resources Harnessing the metabolic capabilities of fungi for feed additive production offers a sustainable alternative to chemical or synthetic methods. Fungi excel at converting substrates into biomass and bioactive compounds through fermentation, requiring minimal energy and resources. This efficiency translates into reduced environmental impact and resource consumption throughout the production process. By choosing fungal-based additives, producers contribute to the conservation of energy and resources in feed production. Additionally, these natural alternatives maintain product efficacy and quality, ensuring that animals receive the necessary nutritional benefits without compromising performance or health. Overall, leveraging fungal fermentation for feed additive production aligns with sustainable agriculture practices and supports efforts to minimize the ecological footprint of livestock farming (Umar et al., 2024).

Reduction of Odor and Ammonia Emissions Fungal feed additives containing enzymes play a crucial role in mitigating odor nuisance and environmental pollution linked with livestock farming. These enzymes facilitate the breakdown of organic matter in animal waste, reducing the production of odorous compounds such as hydrogen sulfide and ammonia. By accelerating the decomposition of manure, fungal additives help diminish odor emissions from animal facilities. Moreover, by reducing ammonia emissions, these additives contribute to better air quality and environmental sustainability. Ammonia can have detrimental effects on air quality, human health, and ecosystems, making its reduction imperative in livestock farming. Improved waste management through the use of fungal feed additives not only minimizes odor nuisance but also mitigates environmental pollution, supporting sustainable agricultural practices. Overall, incorporating these additives into animal diets can have significant positive impacts on both the local environment and broader sustainability efforts within the livestock industry (Janni, 2020).

In general, fungal feed additives offer a range of benefits in animal nutrition and production systems, including improved nutrient digestion and absorption, enhanced gut health, mycotoxin mitigation, reduced environmental impact, alternative protein sources, adaptability to various diets and production systems, enhanced immune function, conservation of energy and resources, and reduction of odor and ammonia emissions. Their multifaceted advantages make them valuable tools for optimizing animal health, welfare, and sustainability in modern livestock production (Azeh et al., 2023).

5 Impact on Mycotoxin Mitigation in Feed

Fungal feed additives can mitigate mycotoxin contamination in feed by competitively excluding toxigenic fungi, binding mycotoxins, and producing enzymes that degrade mycotoxins. However, efficacy varies based on factors such as fungal strain and dosage. Research is essential to optimize formulations for effective mycotoxin mitigation in animal feed (Table 2).

6 Variability in Efficacy of Fungal Feed Additives in Animal Nutrition

Fungal feed additives have gained attention in animal nutrition due to their potential to improve digestive efficiency, enhance immune function, and promote overall animal health. However, one of the challenges encountered in the use of fungal feed

Table 2 The impact of fungal feed additives on mycotoxin mitigation in animal feed

Fungal feed additive	Mycotoxin targeted	Mechanism of action	References
Saccharomyces cerevisiae	Aflatoxin B1	Adsorption onto yeast cell wall	Yiannikouris et al. (2021)
Aspergillus oryzae	Deoxynivalenol (DON)	Enzymatic degradation	Adetunji et al. (2023a)
Trichoderma spp.	Zearalenone	Metabolization	Mironenka et al. (2021)
Bacillus subtilis	Ochratoxin A	Competitive exclusion	Miri et al. (2024)
Lactobacillus spp.	Fumonisins	pH modulation	Adetunji et al. (2023b)

additives is the variability in their efficacy across different conditions and production systems. This variability can arise from several factors, including strain selection, production methods, dosage levels, and interactions with other dietary components. Understanding and addressing this variability is crucial for optimizing the efficacy and maximizing the benefits of fungal feed additives in animal nutrition (Abdelli et al., 2021; Adetuyi et al., 2024b).

Strain Selection Indeed, the efficacy of fungal feed additives can vary significantly based on the specific strains utilized in their production. Different strains may exhibit variations in their ability to produce essential bioactive compounds, enzymes, and metabolites, ultimately influencing their effectiveness in animal nutrition. Conducting research to identify and select strains with the highest levels of beneficial components is paramount for enhancing the consistency and efficacy of fungal feed additives. By pinpointing strains that demonstrate superior production of desired compounds, such as digestive enzymes or growth-promoting metabolites, producers can ensure more reliable and potent formulations. This emphasis on strain selection not only improves the performance of fungal feed additives but also contributes to the advancement of precision nutrition in animal agriculture. Through ongoing research efforts, the industry can continue to refine and optimize fungal additive formulations to better meet the evolving needs of livestock production (Abdul-Hakim et al., 2023).

Production Methods The cultivation and processing methods employed for fungal feed additives significantly impact their composition and efficacy. Variables such as fermentation conditions, substrate composition, and processing techniques directly influence the production of bioactive compounds and enzymes crucial for enhancing animal nutrition. Optimizing these production methods is essential for maximizing both the yield and quality of beneficial components. For instance, controlling fermentation parameters such as temperature, pH, and aeration can enhance enzyme activity and metabolite production. Likewise, selecting appropriate substrate materials rich in nutrients and bioavailable compounds can enhance fungal growth and bioactive compound synthesis. Furthermore, refining processing techniques can help concentrate and preserve the bioactive components of fungal additives, ensuring their efficacy throughout storage and administration. By prioritizing the refinement of cultivation and processing methods, producers can consistently deliver high-quality fungal feed additives, ultimately optimizing animal health and performance while minimizing resource wastage. Regular monitoring and refinement of these methods based on emerging research findings are essential for continued improvement in efficacy and sustainability (Corbu et al., 2023).

Dosage Levels The dosage levels of fungal feed additives play a crucial role in determining their efficacy and potential impact on animal health. While higher doses may offer greater benefits in certain scenarios, excessive supplementation can lead to waste and even adverse effects on animals. Optimizing dosage levels involves considering factors such as the species of livestock, their production stages, and the composition of their diets. Different animals may require varying amounts of fungal additives to achieve optimal results. Additionally, the nutritional composition of the diet can influence the absorption and utilization of the additives. By carefully determining the appropriate dosage levels for specific circumstances, producers can maximize the benefits of fungal feed additives while minimizing the risk of negative outcomes. This approach ensures efficient utilization of resources and promotes the overall health and performance of the animals. Regular monitoring and adjustments may be necessary to maintain optimal dosage levels under changing conditions (Inobeme et al., 2023a).

Interactions with Other Dietary Components Absolutely, the effectiveness of fungal feed additives can indeed be influenced by interactions with other dietary components present in animal feeds. Some feed ingredients may either inhibit or enhance the activity of enzymes produced by fungal additives. This can directly affect their ability to improve nutrient digestion and absorption in animals. For instance, certain compounds in feed ingredients might interfere with enzyme function or compete for substrate binding sites, thereby reducing the efficacy of fungal additives. Conversely, synergistic interactions between fungal enzymes and certain feed components can enhance nutrient breakdown and utilization. Understanding these complex interactions is vital for formulating balanced diets that maximize the benefits of fungal feed additives. By carefully selecting and combining feed ingredients, producers can mitigate any negative effects and optimize the synergistic relationships between fungal enzymes and other dietary components, ultimately improving animal health and performance (Sureshkumar et al., 2023).

Environmental Conditions Indeed, environmental conditions play a significant role in determining the efficacy of fungal feed additives. Factors such as temperature, humidity, and pH can directly impact fungal growth and enzyme activity. Different fungal species and strains may have varying optimal conditions for growth and enzyme production. For example, some fungi thrive in higher temperatures and acidic environments, while others prefer lower temperatures and neutral pH levels. When designing production processes and formulating diets containing fungal feed additives, it is crucial to consider these environmental factors. Failure to do so could result in suboptimal growth of the fungi or reduced enzyme activity, ultimately diminishing the effectiveness of the additives. By optimizing environmental conditions to suit the specific requirements of the fungal species being used, producers can ensure maximum efficacy of the feed additives and achieve desired outcomes in terms of animal health and performance (Mathew et al., 2023a).

Animal Health and Physiology Understanding how the health and physiology of animals impact the efficacy of fungal feed additives is crucial for optimizing their use in various production scenarios. Stress or digestive disorders can alter the response to supplementation, leading to variable outcomes compared to healthy animals. Factors such as altered gut microbiota, immune function, and nutrient absorption may influence the effectiveness of fungal additives. Additionally, stress-induced changes in metabolism and physiology could affect the utilization of fungal components by the animal. By considering these factors, producers can tailor supplementation strategies to better meet the specific needs of their livestock, ultimately maximizing the benefits of fungal feed additives in enhancing animal health and performance (Inobeme et al., 2023b).

7 Challenges and Future Research Directions

Effective management of hazardous waste faces several challenges, including the complexity of waste composition, the environmental persistence of contaminants, and the need for sustainable treatment technologies (Table 3). Future research should focus on developing advanced chemical treatments, such as innovative advanced oxidation processes, to degrade complex pollutants effectively. Additionally, integrating circular economy principles can promote waste minimization and resource recovery, transforming waste into valuable products. Emphasizing green chemistry approaches will further enhance the sustainability of hazardous waste management practices.

Table 3 Outlining challenges and future research directions for fungal feed additives in animal

Challenges	Potential solutions	Future research directions	References
Limited efficacy	Genetic modification to enhance efficacy	Investigate novel fungal strains with higher efficacy	Salazar-Cerezo et al. (2023)
Regulatory hurdles	Collaboration with regulatory agencies	Develop standardized testing protocols for fungal additives	Laulund et al. (2017), Adetuyi et al. (2024b)
Safety concerns	Rigorous toxicity testing	Study long-term effects of fungal additives on animal health	
Cost-effectiveness	Optimization of production processes	Explore cost-effective fermentation techniques	Mathew et al. (2022)

8 Future Prospects in Sustainable Animal Husbandry

Future prospects in sustainable animal husbandry are shaped by a growing awareness of environmental, ethical, and health concerns associated with conventional farming practices. As we look ahead, several key trends and innovations are likely to define the trajectory of sustainable animal agriculture.

Alternative Protein Sources The demand for animal protein is escalating due to global population growth and evolving dietary habits. However, conventional livestock farming strains resources and harms the environment. Consequently, interest in alternative protein sources such as plant-based proteins, cultured meat, and insect protein is growing. These alternatives present an opportunity to mitigate the environmental impact of animal agriculture while satisfying the nutritional demands of a burgeoning population. Plant-based proteins require fewer resources and produce fewer greenhouse gas emissions compared to conventional livestock farming. Cultured meat offers a more sustainable method of meat production, requiring less land and water and generating fewer greenhouse gas emissions. Insect protein is a highly efficient source of protein, requiring minimal resources and emitting fewer greenhouse gases. Embracing these alternative protein sources can promote sustainability and address the challenges facing traditional animal agriculture (Sajid et al., 2023).

Precision Livestock Farming Advancements in technology, including sensors, artificial intelligence, and data analytics, are transforming livestock management practices. Precision livestock farming enables real-time monitoring of individual animals' health, behavior, and performance, facilitating targeted interventions and optimized resource allocation. By leveraging these technologies, farmers can identify health issues early, provide timely care, and adjust management practices accordingly. This approach not only improves animal welfare but also enhances productivity and reduces environmental impacts. With precision farming, resources such as feed, water, and medication can be allocated more efficiently, minimizing waste and environmental pollution. Additionally, by optimizing feed conversion and minimizing disease prevalence, precision livestock farming contributes to more sustainable and economically viable farming practices. Overall, the integration of technology in livestock management represents a significant step forward in achieving both production efficiency and environmental sustainability in animal agriculture (Papakonstantinou et al., 2024).

Nutritional Optimization Optimizing animal diets is essential for sustainable animal husbandry, aiming to improve efficiency and health while minimizing resource use. Tailored feed formulations are developed to maximize nutrient utilization, reduce waste, and decrease reliance on scarce resources like land and water. Innovative feed additives, such as fungal feed additives, prebiotics, and probiotics, play a crucial role in enhancing animal nutrition and health. Fungal feed additives, derived from fungi such as Saccharomyces cerevisiae and Aspergillus oryzae, contain bioactive compounds that promote digestion, support immune function, and improve growth performance. Prebiotics serve as food for beneficial gut bacteria, promoting a healthy microbiome and enhancing nutrient absorption. Probiotics introduce beneficial microorganisms to the gut, further improving digestion and reducing the incidence of gastrointestinal disorders. By incorporating these innovative feed additives into animal diets, farmers can optimize nutrient utilization, promote animal health and welfare, and contribute to more sustainable and efficient livestock production systems (Akintan et al., 2024).

Circular Economy Practices Adopting circular economy principles is pivotal in addressing waste and resource depletion in animal agriculture. Repurposing waste streams, like manure and food by-products, as inputs for other agricultural processes reduces reliance on external resources and minimizes environmental pollution. By implementing regenerative agriculture practices such as rotational grazing and cover cropping, farmers can enhance soil health, sequester carbon, and promote biodiversity. Rotational grazing involves moving livestock between different pasture areas, allowing for natural vegetation regeneration and soil recovery. Cover cropping involves planting crops during off-seasons to protect soil from erosion, fix nitrogen, and improve soil structure. These practices not only enhance ecosystem services but also contribute to climate change mitigation efforts by sequestering carbon in the soil. Overall, integrating circular economy principles and regenerative agriculture practices into animal farming systems fosters sustainability, resilience, and environmental stewardship in agriculture (Adami & Schiavon, 2021).

Animal Welfare Standards Consumer awareness and concern for animal welfare are driving a shift towards ethically raised animal products. In response, future prospects in sustainable animal husbandry will focus on implementing higher welfare standards and transparent production practices throughout the supply chain. This involves ensuring animals have adequate space, access to outdoor environments, and humane handling practices. Farmers and producers are increasingly seeking certification from animal welfare organizations to demonstrate their commitment to ethical practices. Labeling schemes that provide clear information about the welfare standards met during production are also gaining popularity among consumers. By prioritizing animal welfare and transparency, sustainable animal husbandry not only meets consumer expectations but also promotes healthier and happier animals. In general, aligning production practices with higher welfare standards not only satisfies consumer preferences but also contributes to the long-term viability and ethical integrity of the animal agriculture industry (Alonso et al., 2020; Mathew et al., 2023b).

Climate-Resilient Breeding Climate change presents formidable challenges to livestock production, exacerbating issues such as heat stress, disease susceptibility, and water scarcity. In response, breeding programs are pivotal for developing climate-resilient livestock breeds capable of thriving in evolving environmental conditions. These programs prioritize traits such as heat tolerance, disease resistance, and feed efficiency to ensure the long-term sustainability and resilience of animal populations. By selectively breeding animals with these desirable traits, farmers can mitigate the negative impacts of climate change on livestock production. Additionally, genetic technologies, including genomic selection and marker-assisted breeding, expedite the breeding process by identifying and selecting animals with desired genetic markers associated with climate resilience. This enables farmers to efficiently develop livestock breeds better suited to withstand the challenges posed by climate change. Ultimately, investing in climate-resilient breeding programs is essential for ensuring the continued productivity and viability of livestock production systems in the face of climate variability and change (Cheng et al., 2022).

Collaboration and Knowledge Sharing Achieving sustainable animal husbandry necessitates collaborative efforts across various sectors, including government, industry, academia, and civil society. Collaboration platforms and knowledge-sharing initiatives play a vital role in facilitating this collaboration by enabling stakeholders to exchange best practices, research findings, and innovative solutions. Government agencies can provide regulatory frameworks and incentives to promote sustainable practices, while industry stakeholders contribute expertise and resources for implementing these practices on the ground. Academic institutions conduct research to develop evidence-based solutions, while civil society organizations advocate for ethical and sustainable practices. Through collaboration, stakeholders can collectively identify challenges, share experiences, and develop practical solutions that promote sustainability, resilience, and equitable outcomes in animal agriculture. By leveraging the diverse expertise and perspectives of different sectors, collaborative efforts can drive positive change and create a more sustainable future for animal husbandry (Zamiri & Esmaeili, 2024).

In addition, future prospects in sustainable animal husbandry are driven by a convergence of technological, environmental, and social factors. By embracing alternative protein sources, adopting precision farming techniques, optimizing nutrition, promoting circular economy practices, prioritizing animal welfare, investing in climate-resilient breeding, and fostering collaboration and knowledge sharing, the livestock industry can transition towards more sustainable and resilient production systems that meet the needs of current and future generations (Galanakis, 2024).

9 Conclusion

Fungal feed additives represent a promising avenue for enhancing animal nutrition, health, and productivity in livestock farming. Through their rich array of bioactive components such as enzymes, prebiotics, and antioxidants, fungal additives offer multifaceted benefits for animals. These additives aid in improving digestive efficiency, supporting immune function, and enhancing overall growth performance. Moreover, fungal feed additives contribute to sustainable animal husbandry practices by promoting efficient nutrient utilization, reducing reliance on antibiotics, and minimizing environmental impacts. By optimizing animal diets and incorporating innovative feed additives, farmers can achieve better production outcomes while prioritizing animal welfare and environmental stewardship. However, further research is needed to fully understand the mechanisms of action and optimal dosage levels of fungal feed additives in different livestock species and production systems. Additionally, efforts to standardize production methods and ensure product quality will be crucial for maximizing the effectiveness of these additives. Overall, the continued exploration and adoption of fungal feed additives hold great promise for advancing the sustainability, resilience, and productivity of animal agriculture, ultimately benefiting both animals and the environment.

Acknowledgments We want to appreciate the management of Ibrahim Badamasi Babangida University, Lapai, Niger State, Nigeria for providing an enabling environment for this work.

References

Abdelli, N., Solà-Oriol, D., & Pérez, J. F. (2021). Phytogenic feed additives in poultry: Achievements, prospective, and challenges. *Animals, 11*(12), 3471. https://doi.org/10.3390/ani11123471

Abdul-Hakim, B. N., Xuan, N. J., & Oslan, S. N. H. (2023). A comprehensive review of bioactive compounds from lactic acid bacteria: Potential functions as functional food in dietetics and the food industry. *Food, 12*(15), 2850. https://doi.org/10.3390/foods12152850

Adami, L., & Schiavon, M. (2021). From circular economy to circular ecology: A review on the solution of environmental problems through circular waste management approaches. *Sustainability, 13*(2), 925. https://doi.org/10.3390/su13020925

Adetunji, C. O., Olaniyan, O. T., Olugbenga, S. M., Shariati, M. A., Osarenkhoe, O. O., Odoh, U. E., Oyedara, O. O., Rebezov, M., Ajayi, O. O., Babaeva, M., Inobeme, A., Mathew, J. T., Adetunji, J. B., Ogechukwu, H. U., Bodunrinde, R. E., & Adeniyi, M. J. (2023a). Pharmacology, photochemical, and healing effectiveness of *Chromolaena odorata* (L.) R.M. King and H. Robinson. In T. Pullaiah (Ed.), *Phytochemical composition and pharmacy of medicinal plants* (Vol. 2, pp. 461–482). Apple Academic Press, Co-published with CRC Press (Taylor & Francis).

Adetunji, C. O., Olaniyan, O. T., Shariati, M. A., Inobeme, A., Mathew, J. T., Olugbenga, S. M., Adetunji, J. B., Hefft, D. I., Rebezov, M., Ajayi, O. O., Gribkova, V., Bodunrinde, R. E., & Adeniyi, M. J. (2023b). Phytochemical constituents and medicinal effectiveness of *Buchholzia coriacea* Engl. In T. Pullaiah (Ed.), *Phytochemical composition and pharmacy of medicinal plants* (Vol. 2, pp. 483–498). Apple Academic Press, Co-published with CRC Press (Taylor & Francis).

Adetuyi, B. O., Mathew, J. T., Inobeme, A., Falana, Y. O., Adetunji, C. O., Shahnawaz, M., Oyewole, O. A., Eniola, K. I. T., Yerima, M. B., & Popoola, O. A. (2024a). Tyres, bitumen Wear and plastic bottles, other single used plastic as major sources of microplastic. In *Microplastic pollution* (Vol. 1, pp. 141–160). Springer. https://doi.org/10.1007/978-981-99-8357-5_9

Adetuyi, B. O., Mathew, J. T., Inobeme, A., Falana, Y. O., Adetunji, C. O., Shahnawaz, M., Oyewole, O. A., Eniola, K. I. T., & Yerima, M. B. (2024b). Sources, uses and transport of secondary microplastics from land to marine environment. In *Microplastic pollution* (Vol. 1, pp. 35–49). Springer. https://doi.org/10.1007/978-981-99-8357-5_3

Akintan, O., Gebremedhin, K. G., & Uyeh, D. D. (2024). Animal feed formulation—Connecting technologies to build a resilient and sustainable system. *Animals, 14*(10), 1497. https://doi.org/10.3390/ani14101497

Alonso, M. E., González-Montaña, J. R., & Lomillos, J. M. (2020). Consumers' concerns and perceptions of farm animal welfare. *Animals, 10*(3), 385. https://doi.org/10.3390/ani10030385

Amara, A. A., & El-Baky, N. A. (2023). Fungi as a source of edible proteins and animal feed. *Journal of Fungi, 9*(1), 73. https://doi.org/10.3390/jof9010073

Azeh, Y., Musah, M., Mathew, J. T., Badeggi, U. M., Estuyankpa, M. B., Muhammad, Z. D., Oluwaseyi, D. S., Umar, M. T., Abdulkarim, M. A., Muhammad, A. I., Ibrahim, A. A., Ismail, H., Muhammad, K. T., Umar, G., & Ahmed, T. (2023). In situ synthesis of polyaniline nanohybrid and formulation of polyaniline/carboxymethyl cellulose/ethylene glycol nanocomposite: Study of its conducting and antibacterial properties. *Journal of Materials and Polymer Science, 3*(4), 1–9. https://doi.org/10.47485/2832-9384.1041

Balehegn, M., Duncan, A., Tolera, A., Ayantunde, A. A., Issa, S., Karimou, M., Zampaligré, N., André, K., Gnanda, I., Varijakshapanicker, P., Kebreab, E., Dubeux, J., Boote, K., Minta, M., Feyissa, F., & Adesogan, A. T. (2020). Improving adoption of technologies and interventions for increasing supply of quality livestock feed in low- and middle-income countries. *Global Food Security, 26*, 100372. https://doi.org/10.1016/j.gfs.2020.100372

Cheng, M., McCarl, B., & Fei, C. (2022). Climate change and livestock production: A literature review. *Atmosphere, 13*(1), 140. https://doi.org/10.3390/atmos13010140

Chuang, W. Y., Hsieh, Y. C., & Lee, T. T. (2020). The effects of fungal feed additives in animals: A review. *Animals, 10*(5), 805. https://doi.org/10.3390/ani10050805

Corbu, V. M., Gheorghe-Barbu, I., Dumbravă, A. Ş., Vrâncianu, C. O., & Şesan, T. E. (2023). Current insights in fungal importance—A comprehensive review. *Microorganisms, 11*(6), 1384. https://doi.org/10.3390/microorganisms11061384

Dahiya, S., Rapoport, A., & Singh, B. (2024). Biotechnological potential of lignocellulosic biomass as substrates for fungal xylanases and its bioconversion into useful products: A review. *Fermentation, 10*(2), 82. https://doi.org/10.3390/fermentation10020082

de Sousa, R., Bragança, L., da Silva, M. V., & Oliveira, R. S. (2024). Challenges and solutions for sustainable food systems: The potential of home hydroponics. *Sustainability, 16*(2), 817. https://doi.org/10.3390/su16020817

Ejaz, U., Sohail, M., & Ghanemi, A. (2021). Cellulases: From bioactivity to a variety of industrial applications. *Biomimetics, 6*(3), 44. https://doi.org/10.3390/biomimetics6030044

Galanakis, C. M. (2024). The future of food. *Food, 13*(4), 506. https://doi.org/10.3390/foods13040506

Inobeme, A., Ajai, I. A., Inobeme, J., Adetunji, C. O., Obar, A., Mathew, J. T., Jacob, J. O., & Nwakife, N. C. (2023a). Superabsorbent polymers for the development of nanofiltration. In *Properties and applications of superabsorbent polymers* (pp. 157–170). Springer. https://doi.org/10.1007/978-981-99-1102-8_7

Inobeme, A., Adetunji, C. O., Mathew, J. T., Ajai, A.I., n Inobeme, J. and Oyedolapo, B. (2023b). Bioremediation of chlorinated compounds. Springer Nature Switzerland AG 2023 M. P. Shah (ed.), Modern approaches in waste bioremediation, 101–115. https://doi.org/10.1007/978-3-031-24086-7_5.

Inobeme, A., Mathew, J. T., AJai, I. A., Adetunji, C. O., Mann, A., Jacob, O. J., Shaba, E. Y., Okonkwo, O. S., Edema, G., Efosa, J., Osarenren, E. B., Evbuomwan, L., & Tedunjaiye, O. H. (2024a). MXenes for catalysis and electrocatalysis. In *MXenes: Next generation 2D materials: Fundamentals and applications* (p. 231). Wiley.

Inobeme, A., Mathew, J. T., Adetunji, C. O., Agbugui, M. O., Osarenren, E. J., Oti, C., Ngonso, B. F., Inobeme, J., Oyedolapo, M. B., Bernard, E., & Jibrin, N. A. (2024b). Biocomposites with graphene derivatives. In *Advances in biocomposites and their applications* (pp. 149–166). Elsevier. https://doi.org/10.1016/B978-0-443-19074-2.00005-8

Inobeme, A., Ajai, A. I., Adetunji, C. O., Maliki, M., Mathew, J. T., Mann, A., Osarenre, J. E., Inobeme, J., Adekoya, M., & Eziukwu, C. (2024c). Oxidation and advanced oxidation processes in pharmaceutical wastewater treatment. In *Development in wastewater treatment research and processes: Innovative trends in removal of refractory pollutants from pharmaceutical wastewater* (pp. 157–169). Elsevier. https://doi.org/10.1016/B978-0-323-99278-7.00008-0

Janni, K. (2020). Reflections on Odor Management for Animal Feeding Operations. *Atmosphere, 11*(5), 453. https://doi.org/10.3390/atmos11050453

Kępińska-Pacelik, J., & Biel, W. (2021). Mycotoxins—Prevention, detection, impact on animal health. *PRO, 9*(11), 2035. https://doi.org/10.3390/pr9112035

Laulund, S., Wind, A., Derkx, P. M. F., & Zuliani, V. (2017). Regulatory and safety requirements for food cultures. *Microorganisms, 5*(2), 28. https://doi.org/10.3390/microorganisms5020028

Mathew, J. T., Adetunji, C. O., Inobeme, A., Musah, M., Shaba, E. Y., Azeh, Y., Abulude, F. O., & Mamman, A. (2022). *Organic compounds in atmospheric aerosols: Atmospheric aerosols properties, sources and detection* (pp. 43–59). Nova Science Publishers.

Mathew, J. T., Musah, M., Azeh, Y., & Muhammed, M. (2023a). Adsorptive removal of selected toxic metals from pharmaceutical wastewater using Fe_3O_4/ZnO nanocomposite. *Dutse Journal of Pure and Applied Sciences, 9*(4a), 236–248. https://doi.org/10.4314/dujopas.v9i4a.22

Mathew, J. T., Musah, M., Azeh, Y., & Muhammed, M. (2023b). Kinetic study of heavy metals removal from pharmaceutical wastewater using geopolymer/Fe_3O_4 nanocomposite. *Bima Journal of Science and Technology, 7*(4), 152–163. https://doi.org/10.56892/bima.v7i4.539

Mathew, J. T., Inobeme, A., Abdulkadir, A., Etsuyankpa, M. B., Shaba, E. Y., Musah, M., Azeh, Y., Adetunji, C. O., Ibrahim, A. M., Tanko, E., & Mamman, A. (2024a). Properties and characteristics of Nanocoatings for medicinal applications. In *Sustainable approach to protective Nanocoatings* (pp. 203–227). IGI Global; publishing Tomorrow's Research Today. https://doi.org/10.4018/979-8-3693-3136-1.ch007

Mathew, J. T., Adetunji, C. O., Inobeme, A., Musah, M., Azeh, Y., Otori, A. A., & Mamman, A. (2024b). Nutritional qualities of agricultural crops after application of Nanobioherbicides. In *Handbook of agricultural biotechnology* (Vol. 3, pp. 189–204). Wiley. https://doi.org/10.1002/9781119836513.ch9

Mathew, J. T., Inobeme, A., Adetuyi, B. O., Falana, Y. O., Adetunji, C. O., & Shahnawaz, M. (2024c). Application of microplastics in toiletry products. In M. Shahnawaz, C. O. Adetunji, M. A. Dar, & D. Zhu (Eds.), *Microplastic pollution* (pp. 73–84). Springer. https://doi.org/10.1007/978-981-99-8357-5_5

Miri, B. Y., Benabdallah, A., Chentir, I., Djenane, D., Luvisi, A., & De Bellis, L. (2024). Comprehensive insights into ochratoxin a: Occurrence, analysis, and control strategies. *Food, 13*(8), 1184. https://doi.org/10.3390/foods13081184

Mironenka, J., Różalska, S., Soboń, A., & Bernat, P. (2021). Trichoderma harzianum metabolites disturb Fusarium culmorum metabolism: Metabolomic and proteomic studies. *Microbiological Research, 249*, 126770. https://doi.org/10.1016/j.micres.2021.126770

Musah, M., Matthew, J. T., Azeh, Y., Badeggi, U. M., Muhammad, A. I., Abu, L. M., Okonkwo, P. T., & Muhammad, K. T. (2024). Preparation and characterization of adsorbent from waste shea (Vitellaria paradoxa) nut Shell. *FUDMA Journal of Sciences, 8*(2), 338–344. https://doi.org/10.33003/fjs-2024-0802-2370

Muteeb, G., Rehman, M. T., Shahwan, M., & Aatif, M. (2023). Origin of antibiotics and antibiotic resistance, and their impacts on drug development: A narrative review. *Pharmaceuticals (Basel, Switzerland), 16*(11), 1615. https://doi.org/10.3390/ph16111615

Olorunsola, A., Adetunji, C. O., Olaniyan, O., Adetunji, J. B., Akinbo, K., Adetuyi, B. O., Inobeme, A., Mathew, J. T., & Ajenifujah-Solebo, S. O. (2024). Application of nanotechnology for the production of biopesticides, bioinsecticides, bioherbicides, mosquito Repellants and biofungicides. In *Handbook of agricultural biotechnology* (pp. 293–309). Wiley. https://doi.org/10.1002/9781119836483.ch11

Papakonstantinou, G. I., Voulgarakis, N., Terzidou, G., Fotos, L., Giamouri, E., & Papatsiros, V. G. (2024). Precision livestock farming technology: Applications and challenges of animal welfare and climate change. *Agriculture, 14*(4), 620. https://doi.org/10.3390/agriculture14040620

Phillips, C. J. C., Hosseintabar-Ghasemabad, B., Gorlov, I. F., Slozhenkina, M. I., Mosolov, A. A., & Seidavi, A. (2023). Immunomodulatory effects of natural feed additives for meat chickens. *Life, 13*(6), 1287. https://doi.org/10.3390/life13061287

Ponnampalam, E. N., Kiani, A., Santhiravel, S., Holman, B. W. B., Lauridsen, C., & Dunshea, F. R. (2022). The importance of dietary antioxidants on oxidative stress, meat and milk production, and their preservative aspects in farm animals: Antioxidant action, animal health, and product quality—Invited review. *Animals, 12*(23), 3279. https://doi.org/10.3390/ani12233279

Sajid, Q. U. A., Asghar, M. U., Tariq, H., Wilk, M., & Płatek, A. (2023). Insect meal as an alternative to protein concentrates in poultry nutrition with future perspectives (an updated review). *Agriculture, 13*(6), 1239. https://doi.org/10.3390/agriculture13061239

Salazar-Cerezo, S., de Vries, R. P., & Garrigues, S. (2023). Strategies for the development of industrial fungal producing strains. *Journal of Fungi, 9*(8), 834. https://doi.org/10.3390/jof9080834

Sampath, V., Sureshkumar, S., & Kim, I. H. (2023). The efficacy of yeast supplementation on monogastric animal performance—A short review. *Life, 13*(10), 2037. https://doi.org/10.3390/life13102037

Shehata, A. A., Yalçın, S., Latorre, J. D., Basiouni, S., Attia, Y. A., Abd El-Wahab, A., Visscher, C., El-Seedi, H. R., Huber, C., & Hafez, H. M. (2022). Probiotics, prebiotics, and phytogenic substances for optimizing gut health in poultry. *Microorganisms, 10*(2), 395. https://doi.org/10.3390/microorganisms10020395

Song, P., Zhang, X., Wang, S., Xu, W., Wang, F., Fu, R., & Wei, F. (2023). Microbial proteases and their applications. *Frontiers in Microbiology, 14*, 1236368. https://doi.org/10.3389/fmicb.2023.1236368

Sureshkumar, S., Song, J., Sampath, V., & Kim, I. (2023). Exogenous enzymes as zootechnical additives in monogastric animal feed: A review. *Agriculture, 13*(12), 2195. https://doi.org/10.3390/agriculture13122195

Tadioto, V., Giehl, A., Cadamuro, R. D., Guterres, I. Z., dos Santos, A. A., Bressan, S. K., Werlang, L., Stambuk, B. U., Fongaro, G., & Silva, I. T. (2023). Bioactive compounds from and against yeasts in the one health context: A comprehensive review. *Fermentation, 9*(4), 363. https://doi.org/10.3390/fermentation9040363

Tanihiro, R., Sakano, K., Oba, S., Nakamura, C., Ohki, K., Hirota, T., Sugiyama, H., Ebihara, S., & Nakamura, Y. (2020). Effects of yeast mannan which promotes beneficial Bacteroides on the intestinal environment and skin condition: A randomized, double-blind, placebo-controlled study. *Nutrients, 12*(12), 3673. https://doi.org/10.3390/nu12123673

Umar, A., Mubeen, M., Ali, I., Iftikhar, Y., Sohail, M. A., Sajid, A., Kumar, A., Solanki, M. K., Kumar Divvela, P., & Zhou, L. (2024). Harnessing fungal bio-electricity: A promising path to a cleaner environment. *Frontiers in Microbiology, 14*, 1291904. https://doi.org/10.3389/fmicb.2023.1291904

Velázquez-De, L. B. S., Hernández-Domínguez, E. M., Villa-García, M., Díaz-Godínez, G., Mandujano-Gonzalez, V., Mendoza-Mendoza, B., & Álvarez-Cervantes, J. (2021). Exogenous enzymes as zootechnical additives in animal feed: A review. *Catalysts, 11*(7), 851. https://doi.org/10.3390/catal11070851

Yiannikouris, A., Apajalahti, J., Siikanen, O., Dillon, G. P., & Moran, C. A. (2021). Saccharomyces cerevisiae cell wall-based adsorbent reduces aflatoxin B1 absorption in rats. *Toxins, 13*(3), 209. https://doi.org/10.3390/toxins13030209

Zamiri, M., & Esmaeili, A. (2024). Methods and technologies for supporting knowledge sharing within learning communities: A systematic literature review. *Administrative Sciences, 14*(1), 17. https://doi.org/10.3390/admsci14010017

Zhang, L., Zhan, H., Xu, W., Yan, S., & Ng, S. C. (2021). The role of gut mycobiome in health and diseases. *Therapeutic Advances in Gastroenterology, 14*, 17–28. https://doi.org/10.1177/17562848211047130

Zhong, X., Wang, G., Li, F., Fang, S., Zhou, S., Ishiwata, A., Tonevitsky, A. G., Shkurnikov, M., Cai, H., & Ding, F. (2023). Immunomodulatory effect and biological significance of β-glucans. *Pharmaceutics, 15*(6), 1615. https://doi.org/10.3390/pharmaceutics15061615

Mycotoxins in Food and Feed: Relevance to Human and Animal Health

Susheela Biranjia-Hurdoyal

Abstract Numerous foodstuffs such as cereals, maize, nuts, and dried fruits have been found to be commonly contaminated with fungi and mycotoxins. Animal products such as eggs, milk, and meat have also been reported to be positive for some mycotoxins. The level of contamination differs across the world as the production of mycotoxins is impacted by environmental factors such as temperature, pH, and humidity. Agricultural products can be contaminated with mycotoxin-producing fungi before or after harvest. Plants affected by droughts and severe weather conditions are more susceptible to pathogenic fungi. The most prevalent mycotoxins are produced by the fungi of *Aspergillus*, *Penicillium,* and *Fusarium*. The major isolated mycotoxins are aflatoxins, ochratoxin A, patulin, trichothecenes, fumonisins, deoxynivalenol, and zearalenone. The effects of mycotoxins on human and animal health can range from acute to severe symptoms. Mycotoxin-contaminated foods have economic consequences for both consumers and producers, as food security and food safety are also impacted. This chapter focuses on the occurrence of mycotoxins in food and feeds that eventually enter the food chain. The effects of mycotoxins on humans or livestock and the limitations of international standards are also discussed.

Keywords Animal health · Human health · Feeds · Food · Mycotoxins

1 Introduction

Mycotoxins are secondary metabolites produced by several genera of fungi and are found as contaminants in a variety of crops consumed by animals and humans. Previous studies have highlighted the connection between mycotoxins,

S. Biranjia-Hurdoyal (✉)
Department of Health Sciences, Faculty of Medicine and Health Sciences, University of Mauritius, Reduit, Mauritius
e-mail: s.biranjia@uom.ac.mu

A. Gupta (ed.), *Prospects of Fungal Biotechnologies for Livestock Volume 1*,
Fungal Biology, https://doi.org/10.1007/978-3-031-90994-8_5

contaminated food, and health problems. Up to 400 mycotoxins have been identified, of which 30 have been repeatedly reported as potential threats to the health of humans and animals. The main mycotoxins are aflatoxins (AF), ochratoxin (OT), deoxynivalenol (DON), T-2, Th-2, fumonisins (FUM), zearalenone (ZEN), and patulin (Nji et al., 2022). Recently, some emerging mycotoxins, such as fusaproliferin, moniliformin, beauvericin, and NX-2 toxin have also been identified (Agriopoulou et al., 2020; Mahato et al., 2021). Some mycotoxins are synthesized by several fungi, and some fungal species produce multiple mycotoxins. Thus, a food product can be positive for more than one mycotoxin (Agriopoulou et al., 2020).

Some mycotoxins have been extensively studied for their toxic and carcinogenic properties, their ability to disrupt physiological processes, and their ability to dysregulate the immune system (da Rocha et al., 2014). Mycotoxins represent health hazards for both developing and developed countries. The contamination of food and feeds by mycotoxins causes major economic losses due to damaged crops and poor animal productivity, affecting both the agricultural sector and the animal industry which consequently poses a risk for food safety and food security (Bryden, 2012). Reduced productivity and increased susceptibility to diseases affect animal welfare and the economy associated with animal husbandry (FAO, 2020).

The production of mycotoxins is influenced by the hot and humid climate. Tropical zones of the world are mostly affected, and contaminations are exacerbated by changes in weather patterns, environmental factors, and improper agricultural practices, such as pre-harvest and post-harvest conditions and crop storage (Marchese et al., 2018). The production of mycotoxins is further affected by the bioavailability of micronutrients, damage by insects, pest attacks, and the use of pesticides and fertilizers which complicates the scenario (Milani, 2013). Control of mycotoxins can be achieved by preventing their synthesis at all stages of food production and elimination from the food chain. Conventional cooking and food processing could not completely eliminate mycotoxins from food products (Bullerman & Bianchini, 2007).

2 Types of Mycotoxin in Food and Animal Feeds

Food and animal feed has often been reported to be contaminated by different types of mycotoxins. Feed supply is a critical aspect of the animal production industry. If the security of feed supply is compromised, animal production is affected (Bryden, 2012). Animal feeds include a variety of products that cater to the nutritional needs of the animal and also the substances required for their optimum health, welfare, and production. The products mostly used are corn, soyabean meal, coffee/rice husk, food waste, and agricultural by-products (Torok et al., 2021; Jalilzadeh-Amin et al., 2023; Wilkinson & Young, 2020). Contaminated feed affects the quality and safety of feeds, which can eventually be a risk for animals and humans (FAO, 2020). Feed spoilage by fungi causes deterioration of feed quality, organoleptic changes, and loss of nutrients (Cegielska-Radziejewska et al., 2013).

Feed safety has been reported to be affected by:

(i) homegrown feeds such as cereals, pulses, forage crops, and forages
(ii) usage of commercial feed
(iii) feed preparation and storage on the farm
(iv) the feed of livestock from which food for humans will be obtained, e.g., meat, milk, eggs

Good agricultural practices must be maintained during the production of feed or feed ingredients. Pastures, cereals, and forage crops could be exposed to biological, chemical, and physical hazards that could eventually cause damage to animals. The biological hazards include bacteria, viruses, fungi, and other pathogens. Pesticides, fertilizer, and antibiotic residues would constitute chemical contamination, while physical hazards would include foreign materials, broken machinery parts, and sharp materials such as needles (FAO, 2020).

3 Mycotoxins Commonly Reported

3.1 Aflatoxins (AF)

AFs are produced by *Aspergillus* species such as *A. flavus, A. parasiticus, A. omius,* and *A. pseudotamarri* (Alshannaq & Yu, 2017; Pereira et al., 2019). Several AFs have been reported and the most abundant ones are AFB_1, AFB_2, AFG_1, and AFG_2. The metabolized derivatives of AF are AFM_1 and AFM_2 which are important contaminants in food products from animals fed feeds contaminated with AF. Such food products include milk, eggs, and dairy products (da Rocha et al., 2014; Jiang et al., 2021; Pereira et al., 2019). AFs have been reported to contaminate animal feed, leading to economic losses (Jiang et al., 2021). WHO has classified AFB1 and AFM1 as group 1 and group 2B of human carcinogens (possibly carcinogenic to humans), respectively (IARC, 1993). The action level set by the US Food and Drug Administration has been detailed in Tables 1 and 2.

The level of total aflatoxins (B1, B2, G1, and G2) in animal feed or animal feed ingredients has also been defined by the US Food and Drug Administration (Table 2). The action level depends on the animal to be fed and the feed ingredient being used (WHO, 2023). The action level is much lower for immature animals compared to mature ones.

Table 1 Maximum level of aflatoxins permitted in food

Toxin	Action level	Recipients	Reference
AFM1	0.50 μg/kg	Liquid milk	FDA (2019)
Total aflatoxins	20 μg/kg	In foods intended for human consumption	

Table 2 Action level of aflatoxin in animal feed and animal feed ingredients

Intended use	Animal feed and animal feed ingredient	Action level
Finishing (i.e., feedlot) beef cattle	Corn and peanut products	300 μg/kg
Products beef cattle, swine, or poultry (irrespective of age or breeding status)	Cottonseed meal	300 μg/kg
Finished swine of 100 pounds or more	Corn and peanut products	200 μg/kg
Breeding beef cattle, breeding swine, or mature poultry	Corn and peanut products	100 μg/kg
Immature animals	Corn, peanut products, and other animal and food ingredients, but excluding cottonseed meal	20 μg/kg
Dairy animals and other animal species (including wildlife), or other uses not specified in this table; or, when the intended use is not known, 20 ppb corn, peanut products, cottonseed meal, and other animal food and food ingredients	Corn, peanut products, cottonseed meal, and other animal food and food ingredients	20 μg/kg

The European Commission has reported the action level for AFM1 as 0.05 μg/kg in liquid milk, AFB1 of 20 μg/kg in all feedstuffs, 10 μg/kg in complete feeds, and 5 μg/kg in complete feeds for dairy animals (European Commission Regulation, 2006).

3.2 *Fumonisins (FUM)*

The main producers of FUM are *F. verticillioides* and *F. proliferatum* (Xu et al., 2022). *A. niger* is also known as a producer of FUM. The B-series FUM which includes the FUM B_1, B_2, B_3, and B_4 are the most important ones (da Rocha et al., 2014). FB_1 is mostly encountered and has been classified as a group 2B carcinogen (possible human carcinogen) (Xu et al., 2022; da Rocha et al., 2014). The carcinogenic property of FUM has been associated with its ability to disrupt sphingolipid biosynthesis (Riley & Merrill Jr, 2019). The maximum acceptable levels of FUM have been detailed in Table 3 (Nji et al., 2022; Chen et al., 2021; FDA, 2001).

3.3 *Ochratoxins (OT)*

OTs are essentially synthesized by species of Aspergillus and Penicillium such as A. ochraceus, A. carbonarius, P. verrucosum, and P. nordicum. The most important OTs are ochratoxin A (OTA) and ochratoxin B (OTB) (Pereira et al., 2019).

Table 3 Maximum acceptable levels of fumonisins

Mycotoxin	US µg/kg	EU µg/kg
FUM	All foods—2000 to 4000 Feed for horses—5000 Feed for swine—10,000 Feed for beef cattle and poultry—50,000	Unprocessed maize for humans—4000 Maize for direct human consumption—1000 Processed-maize-based food for babies, infants, and young children—200 Feed for swine and rabbits—5000 Feed for poultry, lambs, and calves—20,000 Feed for adult ruminants—50,000

OTB is less toxic and co-occurs with OTA in foods and feeds. Cereal is the most common food contaminated with OTA. Other foodstuffs include alcoholic beverages, dried vine fruits, coffee, cocoa, chocolate, meat, and milk (Eskola et al., 2019). OTA was previously reported mainly as a field mycotoxin. However, recently, it has been observed to cause important storage contamination. The mycotoxin has been classified into Group 2B. Although the exact mechanism for toxicity is not well understood, OT has been linked to nephrotoxicity, immunotoxicity, neurotoxicity, and carcinogenicity (Perši et al., 2014). Food processing could not completely deactivate OT and mycotoxin could be transferred from raw materials to end-products such as in beer and wine (Bullerman & Bianchini, 2007). Furthermore, animals fed with OTA-contaminated feed had accumulated OTA in the meat intended for human consumption (Perši et al., 2014). The European Union has reported the maximum acceptable levels of OTA in raw cereal grains at 5 µg/kg, products derived at 3 µg/kg, and dried fruits at 10 µg/kg (Nji et al., 2022).

3.4 Trichothecenes (TRC)

TRCs include up to 200 structurally related toxins and are produced by some filamentous fungal species such as *Fusarium*, *Myrothecium*, *Cephalosporium*, *Trichochecium*, and *Stachybotrys*. The compounds are divided into four groups A, B, C, and D. Type A and B toxins are the most important (Table 4). T-2 and HT-2 from type A are the most toxic of the groups (Ferrigo et al., 2016).

Type A toxins are heat stable and are not destroyed during food processing (Meneely et al., 2023). T-2 and HT-2 toxins are commonly found in cereal-based foods, soybeans, rice, and wheat. DON and its derivatives have been identified in wheat straw used for animal feed and bedding. The grain ears that are used in animal feed contain the highest concentration of DON. The dehulling of grains reduces the amount of DON in food used by man, but increases the concentration in feed (Polak-Śliwińska & Paszczyk, 2021; EFSA, 2017).

Table 4 Types and examples of trichothecenes

TRC	Examples	Produced by	Limits by US FDA (μg/kg)	References
A	T-2 toxin (T-2) HT-2 toxin (HT-2) Neosolaniol (NEO) Diacetoxyscirpenol (DAS) monoacetoxyscirpenol (MAS) Verrucarol (VER) Scirpentriol (SCP)	*F. langsethiae* *F. sporotrichioides* *F. poae* *F. equiseti* *F. acumninatum*	The T-2 and HT-2 toxins are not listed	Petersen (2018)
B	Deoxynivalenol (DON) and its acetyl derivatives: 3-acetyldeoxynivalenol (3-AcDON), 15-acetyldeoxynivalenol (15-AcDON), nivalenol (NIV), and fusarenon X (FUS X).	*F. culmorum* *F. graminearum* *F. cerealis*	for DON 1000 μg/kg in finished wheat products 10,000 μg/kg in grains and grain by-products 5000 μg/kg in grains and grain by-products for swine	Piec et al. (2016), Agriopoulou et al. (2020), Krska et al. (2014), Nji et al. (2022)

Table 5 Limits of ZEN

Produced by	Maximum permissible limits (μg/kg) by the EU	Reference
F. graminearum *F. culmorum* *F. crookwellense F. semitectum* *F. equiseti*	100–200—unprocessed cereals 75—processed cereals 20—processed cereal foods 50—cereal snacks	EFSA (2011)

3.5 Zearalenone (ZEN)

ZEN is produced by the *Fusarium* genus (Table 5). The mycotoxin has a worldwide distribution and is more prevalent in the storage area compared to the field. ZEN production is affected by the storage conditions of the food products, and a moisture of more than 30–40% increases its production. Therefore, ZEN is an important contaminant in temperate climates. Food items commonly associated with ZEN are grains- rice, millet, maize, wheat, oats, barley, and spices (Tola & Kebede, 2016; EFSA, 2011).

IARC classified ZEN as a Group 3 carcinogen. However, it is a matter of concern since mycotoxin is a non-steroidal estrogenic compound that affects the reproductive system and mimics the effects of the female estrogen hormone (WHO, 1993; Calori-Domingues et al., 2016).

3.6 Patulins

The mycotoxin patulin is produced by several fungi such as *Penicillium expansum Bysochlamis nívea, A. clavatus, Penicillium patulum Penicillium crustosum.* Patulin is a major contaminant in the fruit industry as mycotoxin is detected primarily from fruits such as apples, pears, peaches, grapes, low-acid fruit juices, and fruit concentrates. The limit recommended by the US FDA is 50 µg/kg while for the EU it is 10–50 µg/kg (Agriopoulou et al., 2020). For food intended for babies, babies, and young children, the limit is 10 µg/kg (Nji et al., 2022). If mycotoxin is not removed during fruit juice processing, patulin is transferred to the processed juice.

4 Co-contamination of Mycotoxins

The co-occurrence of mycotoxins in foods and feeds has been commonly reported (Weaver et al., 2021). The molecules could have either an additive, antagonistic, or synergistic effect. Although many combinations of mycotoxins have been reported based on food and region of the world, AF/FUM, DON/ZEN, AF/OTA, and FUM/ZEN were the most commonly encountered (Smith et al., 2016). In vivo studies concluded that AFM1 cytotoxicity increased when another mycotoxin, such as OTA and ZEN, coexisted in the same environment. The combination of AFM1 and OTA had the greatest synergistic effect compared to other mycotoxins (Gao et al., 2016). Given that the cytotoxicity of one mycotoxin could be enhanced by the presence of another mycotoxin, it could affect food safety, especially in vulnerable populations such as babies and the elderly. It also requires a review of the maximum residue limits for mycotoxins such as AFM1 because the existing limits do not consider the potential risk of exposure to multiple mycotoxins (Gao et al., 2016). The co-occurrence of mycotoxins in animal feed causes a reduction in feed intake. DON and ZEN had a synergistic effect on feed intake compared to DON and ZEN alone (Jia et al., 2020). Exposure to co-contamination by *Fusarium* mycotoxins was associated with reduced digestibility in animals (Xu et al., 2022).

5 Emerging Fusarium Mycotoxins

In recent years, some fusarium mycotoxins, which have been defined as "emerging fusarium mycotoxins," have been increasingly reported. These mycotoxins are not yet routinely detected or legislatively regulated (Vaclavikova et al., 2013). Examples of emerging fusarium mycotoxins are fusaproliferin, moniliformin, beauvericin, NX-2 toxin, and enniatins (Table 6). There is currently a lack of toxicity data, and until now, the maximum limits of emerging mycotoxins have not been determined

Table 6 Emerging *fusarium* mycotoxins

Emerging *fusarium* mycotoxins	Produced by	References
Fusaproliferin	*F. proliferatum, F. subglutinans, F. verticillioides*	Agriopoulou et al. (2020) Bryła et al. (2022)
Beauvericin	*F. subglutinans, F. bulbicola F. denticulatum, F. lactis F. phyllophillum, F. pseudocircinatum F. succisae*	Calo' et al. (2003) Křížová et al. (2021)
Moniliformin	*F. subglutinans, F. temperatum, F. verticilloides, F. proliferatum*	Fremy et al. (2019)
Enniatins	Some species of the Fusarium, Alternaria, Halosarpheia, and Verticillium genera	Santini et al. (2012) Křížová et al. (2021)

(EFSA, 2014). Emerging mycotoxins are detected mainly in cereals, feeds for livestock animals, and forage feed (Agriopoulou et al., 2020; Santini et al., 2012).

Enniatins include a large group of related mycotoxins with 29 known analogs. Studies have reported enniatin and its metabolites from eggs and chicken liver that were fed foodstuffs positive for mycotoxin (Ivanova et al., 2014). The transfer of beauvericin and enniatins from food of animal origin to humans is probably marginal (EFSA, 2014), as the rates of chicken liver, skin, and thigh muscle contamination were found to be very low, ranging from 0.44% to 1.57% (Callebaut et al., 2012). Very low amounts of beauvericin, enniatins A and enniatins B have also been detected in sheep milk (Piątkowska et al., 2018).

6 Level of Mycotoxins in Food and Animal Feed

Several studies have reported the prevalence and levels of mycotoxins in food and animal feed. More than a third of the world's grain production is consumed by livestock animals and corn is the most common product used for animal feed (Torok et al., 2021). To address seasonal feed scarcity, farmers use food waste, commercial feed, and conserved seasonal surplus grass (Jalilzadeh-Amin et al., 2023). Farmers in developing countries often have no option other than to use contaminated products such as recycled bread for animal feed. In developed countries, the use of food waste such as bread, pastry, and cereals has been reported (Torok et al., 2021). Porks are mostly fed with household food waste, by-products from agricultural production, and discarded vegetables from marketplaces (Nyachoti et al., 2004). Other food or by-products known to be used in animal feeds are potato waste, rice husk, citrus by-products, jackfruit waste, groundnut, molasses, wheat bran, soybean meal, and coffee husk (Jalilzadeh-Amin et al., 2023; Wilkinson & Young, 2020).

Many countries do not have a national limit for mycotoxin regulation and follow international regulations. It is important to ensure that food and animal feed comply with mycotoxin legislation to avoid economic sanctions. Several countries have rejected imported products because of high levels of mycotoxins. The European Union rejected 579 shipments from Africa from 2005 to 2020 due to noncompliance with mycotoxin standards (Chilaka et al., 2022). The economic loss and negative effects on trade are difficult to assess as not all countries investigate the problem to the source. It should be noted that such sanctions further complicate the situation, since rejected-mycotoxin-contaminated foods are added to animal feed, which further accumulate mycotoxins in the food chain (Jalilzadeh-Amin et al., 2023).

Mycotoxin contamination has often been reported in common foods consumed by humans. Agricultural technology and climate change have increased the contamination rate (Pandey et al., 2023). A review from the Middle East and North Africa (MENA) concluded that mycotoxins such as AF, FUM, OTA, DON, ZEN, and the T-2 toxin occur commonly in animal feed. FUM and DON concentrations were detected at a very high concentration at 1240.01 µg/kg and 806.1 µg/kg respectively. Mycotoxins occurred in 28% of animal protein-based meals, 57% of silage, 46% of finished feeds, and 80% of dried bread (Jalilzadeh-Amin et al., 2023).

The Innovad 2022 mycotoxin feed survey reported that 14.8% of the feed samples exceeded EU limits. Don was the most prevalent (75%) followed by FUM (55%), ZEN (39%), AF (13%), T-2 (8%), and OTA (7%). The individual feed category was also compared and it was noted that 29.7% of ruminant feed exceeded EU limits compared to 10% of swine feed and 3.2% of poultry feed (Innovad, 2022). The World Mycotoxin Survey 2024 by DSM also reported a high percentage of mycotoxin in feed samples around the world (Figs. 1 and 2). The feed samples were from the Middle East and North Africa (MENA), Sub-Saharan Africa (SSA), Europe, Asia, North America, and South America. ZEN, DON, and FUM were detected with a higher prevalence. Furthermore, the average ppb for FUM and DON were the highest (DSM, 2024).

7 Impact of Mycotoxins on the Health of Animals and Humans

The presence of mycotoxins in food and animal feed presents a public health hazard in both developed and developing countries. Mycotoxin and its derivatives could eventually enter the food chain (Jalilzadeh-Amin et al., 2023). Human beings can be exposed to mycotoxins both directly and indirectly. Direct exposure occurs when contaminated crops or processed foods are consumed, while indirect exposure occurs during lactation, gestation, and from products obtained from animals fed mycotoxin-contaminated feeds (Marchese et al., 2018; Pereira et al., 2019).

Mycotoxin-contaminated foods cause acute and chronic toxicity in animals (Xu et al., 2022). Animals and humans exhibit varying degrees of susceptibility to mycotoxins. Non-ruminants, such as swine and poultry, were more susceptible to some mycotoxins compared to ruminants due to the difference in their digestive systems (Xu et al., 2022). Mycotoxins affect the intestine in a major way because the

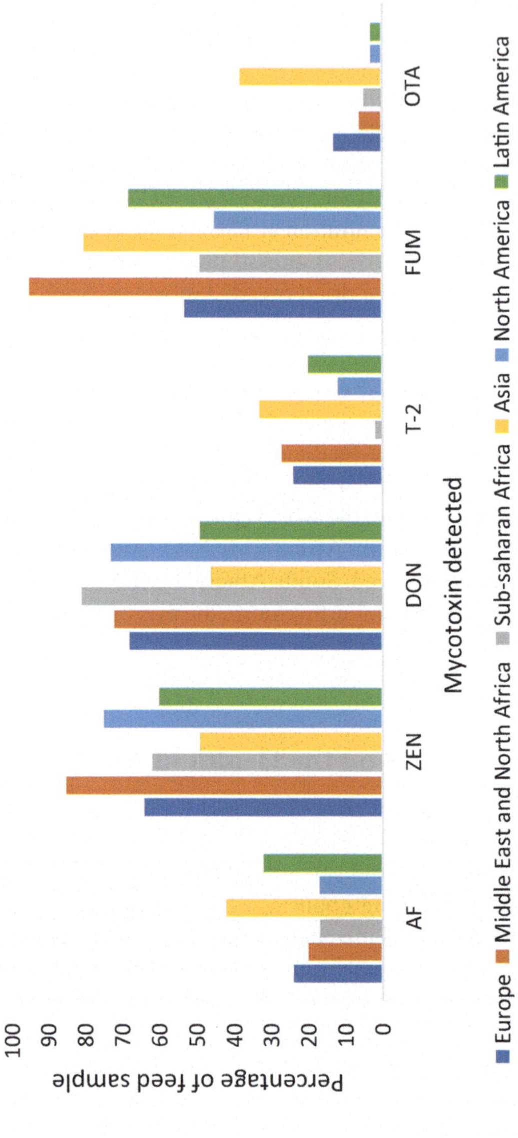

Fig. 1 Percentage of contaminated feed samples

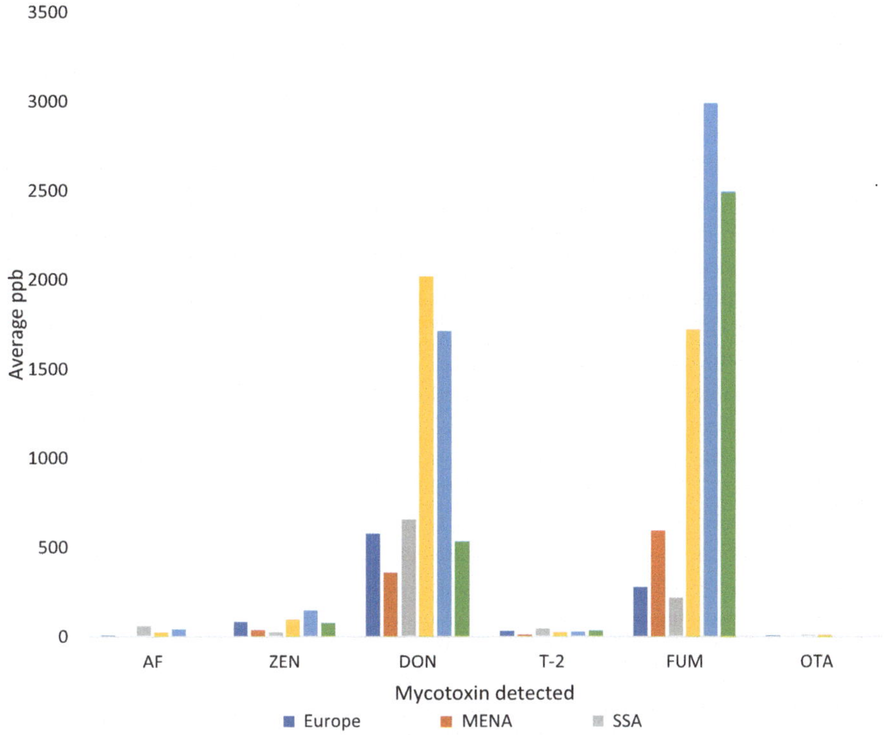

Fig. 2 Average ppb of the major mycotoxin in feed samples throughout the world

intestinal walls are exposed to absorbed mycotoxins. The observed impacts are reduced nutrient absorption, damaged intestinal barrier, and increased susceptibility to intestinal pathogens (Grenier & Applegate, 2013). The main consequences of mycotoxins in the livestock industry are disruptions in body systems such as the reproductive system and immune system, and metabolic disruptions with effects on the overall health and welfare of the animals. Therefore, leading to economic losses. Previous studies have also reported that a low level of mycotoxin exposure could cause poor feed intake, slow growth rate, reduced egg production, and reduced fertility (Bryden, 2012).

Animals such as poultry and pigs had a reduced ability to digest their feed after being exposed to the mycotoxins AF, DON, and FUM. Some studies reported that the negative effects were due to exposure to a higher concentration of mycotoxins while others reported that mycotoxins below maximum levels have similar effects as lactating cows had reduced digestibility (Gallo et al., 2020; Xu et al., 2022). The structure and functionality of the digestive system was found to be disrupted after exposure to mycotoxins. AFB1, OTA, FUM, and ZEN were found to affect the configuration of the intestine by reducing the height of the villus and increasing the depth of the crypt. The extent of damage varied with the concentration of mycotoxin and the animals under study, such as ducks, laying hens, and broiler chickens (Xu et al., 2022). Furthermore, in piglets, the co-occurrence of FUM and DON had a synergistic effect on the decreasing villus height and increasing crypt depth (Bracarense et al., 2012).

Aflatoxin affects monogastric animals more than ruminants, since the rumen microbiota metabolizes mycotoxins. Cereal grains constitute 85% of the swine diet. Given that grains are a very important source of mycotoxins, animals are exposed to contamination (Popescu et al., 2022). Exposure to AF causes impaired liver and kidney function, increased mortality, poor performance, increased susceptibility to diseases, and stress. Other most commonly reported symptoms have been slow growth rate, reproductive problems, refusal to eat, and an affected immune system (Lee et al., 2017). In broilers, AF affected the liver and immune system, and reduced productivity in broilers (Pereira et al., 2019).

Aflatoxin-contaminated feed consumed by dairy animals such as cows, sheep, and goats has been found to be a human health hazard, as the toxin can be transferred to milk (Jiang et al., 2021). AF has a wide range of effects in humans, which can be classified as acute aflatoxicosis and chronic sublethal aflatoxicosis. Acute intoxication occurs due to short exposure to a large amount of the toxin, whereas in chronic aflatoxicosis, exposure is continuous and the dose is low. The symptoms also differ, as acute aflatoxicosis is characterized by liver damage, jaundice, edema, hemorrhage, and death. Immunosuppression and cancer are more common in chronic cases (Marchese et al., 2018). In the last decade, cereal crops such as wheat, oats, rice, barley, and maize have been consistently found to be positive for AF and DON (Latham et al., 2023). In 2004, an outbreak of AF poisoning was reported and 125 people had died (CDC, 2004). Children who consumed AF-contaminated food were reported to have low weight and growth impairment (Gong et al., 2003). The role of AF in the genotoxic process, DNA damage, and carcinogenesis is well documented. Lipid peroxidation, mutation, and inhibition of DNA repair have also been associated with aflatoxin carcinogenesis (Li et al., 1993; Weng et al., 2017). Mutation in the p53 tumor suppressor gene was observed in patients from regions of high AF exposure and suffering from hepatocellular carcinoma (Li et al., 1993). Furthermore, AFB1 was found to increase reactive oxygen species production, activate mitochondrial ROS-dependent signal pathways, and oxidative stress, thus inducing apoptosis through the mitochondrial signal pathway (Liu & Wang, 2016). Aflatoxin has been associated with several cancers, such as liver cancer, lung cancer, and gastrointestinal cancer (Kucukcakan & Hayrulai-Musliu, 2015; Wang et al., 2016; Koshiol et al., 2017). AFB1 was classified as a Group 1 carcinogen due to its important role in liver cancer formation and AFM1 was classified by IARC as a human Group 2B carcinogen (IARC, 1993).

FUM causes a wide range of diseases in animals. The effects differ between animal species and involve several vital organs. Horses, swine, and rabbits are more sensitive to FUM. Toxicity in the liver and kidney has been reported in many species, carcinogenic effects in rodents, the vascular system and brain were affected in equines, and pulmonary edema syndrome was observed in porks (Riley et al., 2018). In horses, leukoencephalomalacia syndrome was accompanied by lethargy, blindness, reduced feeding intake, convulsions, and death. In pigs, pulmonary edema was associated with decreased feed consumption, dyspnea, weakness, cyanosis, and death (Marin et al., 2013; da Rocha et al., 2014). Previous studies have reported that fumonisin B_1 (FB_1), and its hydrolyzed derivatives HFB_1 or FB_1 aminopentol (AP_1) share structural similarities to the sphingoid bases (sphinganine and sphingosine) of sphingolipids. FB1 was found to inhibit ceramide synthase, an enzyme required to transfer fatty acids from fatty acyl-CoA to sphinganine and sphingosine. This inhibition alters multiple sphingolipid metabolites that impact many biochemical processes such as signaling targets, membrane structures, lipid structures, and perturbations were observed in the form of altered growth, altered gene expression, altered immune functions, and cancer, among others (Riley & Merrill Jr, 2019; Wang et al., 1991). FUM was also reported to affect the reproductive system by reducing gonadotropin levels, reducing litter size, inhibiting sperm production, and causing embryo death (Chen et al., 2021).

In vitro human models showed that FUM could cause organ toxicity by increasing the production of inflammatory cytokines, inducing apoptosis, reducing cell viability, increasing membrane leakage, and causing peroxidation of intestinal cell lines (Chen et al., 2021). FUM may also cause esophageal cancer, growth disturbances, and idiopathic congestive cardiopathy (Chen et al., 2021). Consumption of a high level of FUM during early pregnancy was associated with a high risk of neural tube defects (Missmer et al., 2006).

Both Type A and Type B trichothecenes have toxic effects. Lipophilic T-2 and TH-2 toxins are easily absorbed, metabolized, and excreted in urine and feces. The outcome-dependent factors include the length of exposure, the route, the dosage, the age, the gender, and the health of the animals (Meneely et al., 2023). Symptoms include food refusal, diarrhea, vomiting, low egg production, skin problems, abortion, and death. Toxins also affect protein synthesis and negatively regulate the immune system. Poultry are more sensitive to T-2 toxins and are associated with irritation of the intestine, necrosis of the proventricular mucosa, gizzard erosion, and abnormal growth of bone and cartilage (Pinton et al., 2012). Furthermore, T-2 toxicosis was associated with decreased egg production, low shell quality, and increased susceptibility to infections (Meneely et al., 2023). In pigs, various systems were affected, including the gastrointestinal tract, immune system, reproductive system, and nervous system, which were suppressed (Rafai et al., 1995).

DON's effect on animals includes a range of toxicological impacts through ribotoxic stress, disturbances in gut microbiota, and impaired nutrient absorption which lead to significant health issues. The intestinal tract is the primary organ where DON cause damage in the form of disruption of the intestinal barrier, reduced intestinal mucosal immunity, and dysbiosis. DON affects protein synthesis by interacting with ribosome subunits leading to ribotoxic stress, inflammation, and apoptosis (Mishra et al., 2019). DON at 10 mg/kg was found to affect the beneficial microbiota in broiler chicken. These microbes contribute to the digestion and absorption of nutrients (Wu et al., 2018). DON could induce anorexic responses in mouse and pig models. Feed intake was reduced by direct regulation of anorexigenic pathways in the central nervous system after crossing the blood-brain barrier in pigs or by regulating gut satiety hormones (Xu et al., 2022; Terciolo et al., 2018; Li et al., 2018). In animals, the consumption of contaminated animal feed leads to impaired immune function and acute gastrointestinal symptoms such as diarrhea, vomiting, and refusal to feed (Li et al., 2018).

The effects of DON on animals and humans are similar. Symptoms ranged from weight loss, vomiting, skin irritation, and refusal to eat. The acetyl derivatives of DON have also been increasingly highlighted for their potential to affect consumer health. The toxicity of the acetylated derivatives (3-ADON and 15-ADON) is considered similar to that of DON (JECFA, 2001). Multiple outbreaks have been reported in China after the consumption of cereal grains contaminated with DON (Han et al., 2014). Consumption of DON-contaminated cereals caused vomiting and acute gastroenteritis (Bhat et al., 1989).

Effects of patulin, ZEN, FUM, and OTA on animals and humans are described in Table 7.

The effects of emerging fusarium mycotoxins have been described in Table 8.

Table 7 Effects of patulin, ZEN, FUM, and OTA on animals and humans

Mycotoxin	Effect on animals	Effect on human beings	References
Patulin	Exert immunotoxic and neurotoxic effects Teratogenic properties also reported	Gastrointestinal disturbances, nausea, and vomiting Caused harm to the spleen, liver, immune system, and kidneys	Agriopoulou et al. (2020), Awuchi et al. (2022)
ZEN	Negative effect on the reproductive system Prepubertal swine are the most susceptible species Infertility, stillbirths, small litters, increase in secretion in the uterus, among others Associated with pituitary tumors, hyperkeratotic papilloma, and hepatocellular adenomas in mice	Causes premature puberty Decreased embryo survival Low fetal weight Decreased milk production reduced sperm count and viability Altering spermatogenesis	Calori-Domingues et al. (2016), Awuchi et al. (2022), Longobardi et al. (2023)
OTA	Nephrotoxic effects from continuous natural exposure to feeds The main target of OTA is the kidney, and it has been associated with endemic nephropathy in the kidney High doses of the toxin could cause damage to the liver, intestinal walls, and lymphoid tissues	Associated with endemic nephropathy Balkan endemic nephropathy, a fatal kidney disease	Stiborová et al. (2016), da Rocha et al. (2014) Marroquín-Cardona et al. (2014), Groopman et al. (2013)

Table 8 Emerging fusarium mycotoxins and their effects

Emerging fusarium mycotoxins	Effects	References
Fusaproliferin	Toxicity to chicken embryos and brine shrimp larvae Teratogenic properties; Macrocephaly, leg deformations	Agriopoulou et al. (2020), Bryła et al. (2022)
Beauvericin	Endocrine disrupting antagonistic effects at the androgen receptor Induce oxidative stress Interferes with smooth muscle contraction Interferes with steatosis	Calo' et al. (2003), Křížová et al. (2021)
Moniliformin	Inhibit enzymes of the Kreb cycle and affect ATP production In animals, it affected muscles and caused harm to the respiratory and cardiac system Associated with respiratory distress, myocardial loss, and lack of energy Increase oxidative stress	Fremy et al. (2019)
Enniatins	Cytotoxic on several types of mammalian and cancer cells at low doses Low toxicity in vivo	Santini et al. (2012), Křížová et al. (2021)

8 Migitation Strategies to Control Mycotoxins in Food and Feed

8.1 *Management Strategies*

The prevention of fungus and mycotoxin contamination has been reported to be impossible. However, comprehensive management strategies such as the mycotoxin management program (Fumagalli et al., 2021) and good agricultural practices (FAO, 2020) could reduce risks. The integrated system is based on prevention and control measures such as Good Agricultural Practices, Good Manufacturing Practices, Good Hygienic Practices, Quality Control, Hazard Analysis and Critical Control Points (HACCP) throughout the production, from the raw material to the consumer. The integrated mycotoxin management system is based on regulatory limits, monitoring programs, crop control, processing control, decontamination treatments, and education (Fumagalli et al., 2021).

Good agricultural practices should be applied during the production of feed and feed ingredients, which would otherwise have a negative impact on feed quality. Good agricultural practice aims to reduce the risks of biological, chemical, and physical contamination during production and storage. Some of the main elements of good agricultural practices include agricultural sites, seeds/propagation material, crop rotation, use of fertilizers, integrated pest management, plant protection products, harvesting, storage, transport, and personnel training (FAO and IFIF 2020). HACCP would also reduce the risk of mycotoxin in animal feed. Hazards should be identified at various stages, such as before harvesting feed, harvest, storage, feed mill, and livestock production. Necessary prevention actions such as soil irrigation, fertilization, integrated pest management control, and aeration of storage areas have been suggested for hazards such as low soil moisture/plant stress, insufficient soil nutrients, insect attacks, and humidity in storage areas, respectively (Fumagalli et al., 2021).

For effective prevention and control management, farmers should be educated about mycotoxin contaminants and risk factors for fungal contamination (Jalilzadeh-Amin et al., 2023).

8.2 Compliance with the Regulatory System

The JECFA, an international expert committee under the aegis of both the WHO and the Food and Agriculture Organization of the United Nations (FAO), is responsible for assessing the health risks of mycotoxins. The committee reviews published scientific information and other relevant data available to report updates or new findings on maximum tolerable intake levels, exposure margin, prevention or control measures, and detection methods. Individual local authorities and the Codex Alimentarius Commission refer to the report of JECFA to establish maximum levels of mycotoxins in food (WHO, 2023). Governmental organizations such as the European Food Safety Agency and non-government bodies including DSM BIOMIN and the World Mycotoxin Survey monitor the level of mycotoxin in food and animal feed (Latham et al., 2023).

8.3 Decontamination Strategies

Several methods have been used to detoxify foods contaminated with mycotoxins. Physical, chemical, and biological methods have been used for a long time. In recent years, novel technologies such as the use of irradiation, essential oils, cold plasma, and ozone treatment have been reported (Pandey et al., 2023). Recently, the use of feed additives has proven to be useful in mitigating the toxic effect of mycotoxins in feed. The findings appeared promising and important to improve food security (Kolawole et al., 2022).

9 Conclusions

The mycotoxin contamination of animal and human food is of significance as some of the mycotoxins pose risks to both animal and human health. A substantial percentage of animal feed and food meant for human consumption have been reported to have high contamination rates. The economic dynamics surrounding mycotoxin contamination is complex and influenced by agricultural practices and regulatory frameworks. The strict adherence to regulations may not always resolve mycotoxin contamination in food and animal feed. Many countries still depend on the subsistence economy, and contaminated crops might be diverted to local markets rather than regional or international ones. Farmers could also divert their contaminated crops destined for humans to animal feed to maintain food security. However, the production and use of safe feed might contribute to improving the situation by ensuring that the animals are provided with all the necessary nutrients, which will then improve their welfare, performance, and reproducibility, at the best cost. Management strategies such as HACCP, regular monitoring, and training of farmers would effectively contribute to addressing mycotoxin contamination. By fostering collaboration among stakeholders, it would be possible to improve feed safety, improve animal welfare, and protect public health from the adverse effects of mycotoxins.

References

Agriopoulou, S., Stamatelopoulou, E., & Varzakas, T. (2020). Advances in occurrence, importance, and mycotoxin control strategies: Prevention and detoxification in foods. *Food, 9*(2), 137. https://doi.org/10.3390/foods9020137

Alshannaq, A., & Yu, J.-H. (2017). Occurrence, toxicity, and analysis of major mycotoxins in food. *International Journal of Environmental Research and Public Health, 14*, 632.

Awuchi, C. G., Ondari, E. N., Nwozo, S., Odongo, G. A., Eseoghene, I. J., Twinomuhwezi, H., Ogbonna, C. U., Upadhyay, A. K., Adeleye, A. O., & Okpala, C. O. R. (2022). Mycotoxins' toxicological mechanisms involving humans, livestock and their associated health concerns: A review. *Toxins, 14*, 167. https://doi.org/10.3390/toxins14030167

Bhat, R. V., Beedu, S. R., Ramakrishna, Y., & Munshi, K. L. (1989). Outbreak of trichothecene mycotoxicosis associated with consumption of mould-damaged wheat production in Kashmir Valley, India. *Lancet, 1*, 35–37.

Bracarense, A. P. F. L., Lucioli, J., Grenier, B., Pacheco, G. D., Moll, W. D., Schatzmayr, G., et al. (2012). Chronic ingestion of deoxynivalenol and fumonisin, alone or in interaction, induces morphological and immunological changes in the intestine of piglets. *The British Journal of Nutrition, 107*, 1776–1786.

Bryden, W. L. (2012). Mycotoxin contamination of the feed supply chain: Implications for animal productivity and feed security. *Animal Feed Science and Technology, 173*(1–2), 134–158.

Bryła, M., Pierzgalski, A., Zapaśnik, A., Uwineza, P. A., Ksieniewicz-Woźniak, E., Modrzewska, M., & Waśkiewicz, A. (2022). Recent research on fusarium mycotoxins in maize-a review. *Food, 11*(21), 3465. https://doi.org/10.3390/foods11213465

Bullerman, L. B., & Bianchini, A. (2007). Stability of mycotoxins during food processing. *International Journal of Food Microbiology, 119*, 140–146.

Callebaut, F., Tangni, E. K., Debongnie, P., Stals, E., Huybrechts, B., Waegeneers, N., Delezie, E., Van Pamel, E., & Daeseleire, E. (2012). Carry-over of mycotoxins to animal products: Case study poultry. Centrum voor Onderzoek in Diergeneeskunde en Agrochemie-Centre d'Étude et de Recherches Vétérinaires et Agrochemiques; Brussels, Belgium Scientific Report 211/212 CODA-CERVA.

Calo', L., Fornelli, F., Nenna, S., Tursi, A., Caiaffa, M. F., & Macchia, L. (2003). Beauvericin cytotoxicity to the invertebrate cell line SF-9. *Journal of Applied Genetics, 44*, 515–520.

Calori-Domingues, M. A., Bernardi, C. M. G., Nardin, M. S., de Souza, G. V., dos Santos, F. G. R., de Abreu, S. M., da Gloria, E. M., dos Santos Dias, C. T., & de Camargo, A. C. (2016). Co-occurrence and distribution of deoxynivalenol, nivalenol and zearalenone in wheat from Brazil. *Food Additives & Contaminants. Part B, Surveillance, 9*, 142–151. https://doi.org/10.1080/19393210.2016.1152598

Cegielska-Radziejewska, R., Stuper-Szablewska, K., & Szablewski, T. (2013). Microflora and mycotoxin contamination in poultry feed mixtures from western Poland. *Annals of Agricultural and Environmental Medicine, 20*(1), 30–35.

Centers for Disease Control and Prevention. (2004). Outbreak of aflatoxin poisoning—Eastern and Central Provinces, Kenya, January–July 2004. *Morbidity and Mortality Weekly Report, 53*, 790–793.

Chen, J., Wen, J., Tang, Y., Shi, J., Mu, G., Yan, R., Cai, J., & Long, M. (2021). Research progress on fumonisin B1 contamination and toxicity: A review. *Molecules, 26*(17), 5238. https://doi.org/10.3390/molecules26175238

Chilaka, C. A., Obidiegwu, J. E., Chilaka, A. C., Atanda, O. O., & Mally, A. (2022). Mycotoxin regulatory status in Africa: A decade of weak institutional efforts. *Toxins, 14*(7), 442. https://doi.org/10.3390/toxins14070442

da Rocha, M. E. B., da Freire, F. C. O., Maia, F. E. F., Guedes, M. I. F., & Rondina, D. (2014). Mycotoxins and their effects on human and animal health. *Food Control, 36*, 159–165. https://doi.org/10.1016/j.foodcont.2013.08.021

DSM. (2024). REP_MTXsurvey_Q2_2024_EN_0724_62415.pdf (dsm.com) DSM Firmenich. Jan to June 2024. https://www.dsm.com/content/dam/dsm/anh/en/documents/REP _MTXsurvey_ Q2_2024_ EN_0724_62415.pdf

Eskola, M., Kos, G., Elliott, C. T., Hajšlová, J., Mayar, S., & Krska, R. (2019). Worldwide contamination of food-crops with mycotoxins: Validity of the widely cited 'FAO estimate' of 25%. *Critical Reviews in Food Science and Nutrition, 59*, 1–17. https://doi.org/10.1080/10408398.2019.1658570

European Commission Regulation. (2006). No 1881/2006 of 19 December 2006. Setting maximum levels for certain contaminants in foodstuffs. *Official Journal of the European Union, 364*, 5–24.

European Food Safety Authority. (2011). Scientific Opinion on the risks for human and animal health related to the presence of modified forms of certain mycotoxins in food and feed. *EFSA Journal, 9*, 2197. https://doi.org/10.2903/j.efsa.2011.2197

European Food Safety Authority. (2017). Human and animal dietary exposure to T-2 and HT-2 toxin. *EFSA Journal, 15*, 1–57.

European Food Safety Authority. EFSA Panel on Contaminants in the Food Chain. (2014). Scientific Opinion on the risks to human and animal health related to the presence of beauvericin and enniatins in food and feed. *EFSA Journal, 12*, 3802. https://doi.org/10.2903/j.efsa.2014.3802

FAO & IFIF (2020). Good practices for the feed sector- Implementing the Codex Alimentarius Code of Practice on Good Animal Feeding. FAO Animal production and health manual no.24. Rome. https://doi.org/10.4060/cb1761en

Ferrigo, D., Raiola, A., & Causin, R. (2016). *Fusarium* toxins in cereals: Occurrence, legislation, factors promoting the appearance and their management. *Molecules, 21*, 21050627. https://doi.org/10.3390/molecules21050627

Food and Drug Administration. (2001). *Guidance for industry: Fumonisin levels in human foods and animal feeds.*

Food and Drug Administration. (2019). *Sec. 683.100 action levels for aflatoxins in animal food.* https://www.fda.gov/media/121202/download.

Food and Agriculture Organization of the United Nations. (2020). https://openknowledge.fao.org/server/api/core/bitstreams/38b56761-c230-4ffb-83b5-d0a34c939a16/content.

Fremy, J. M., Alassane-Kpembi, I., Oswald, I. P., Cottrill, B., & Van Egmond, H. P. (2019). A review on combined effects of moniliformin and co-occurring *Fusarium* toxins in farm animals. *World Mycotoxin Journal, 12*, 281–291. https://doi.org/10.3920/WMJ2018.2405

Fumagalli, F., Ottoboni, M., Pinotti, L., & Cheli, F. (2021). Integrated mycotoxin management system in the feed supply chain: Innovative approaches. *Toxins (Basel), 13*(8), 572. https://doi.org/10.3390/toxins13080572

Gallo, A., Minuti, A., Bani, P., Bertuzzi, T., Cappelli, F. P., Doupovec, B., et al. (2020). A mycotoxin-deactivating feed additive counteracts the adverse effects of regular levels of fusarium mycotoxins in dairy cows. *Journal of Dairy Science, 103*, 11314–11331.

Gao, Y. N., Wang, Q., Li, S. L., Zhang, Y. D., & Zheng, N. (2016). Aflatoxin M1 cytotoxicity against human intestinal Caco-2 cells is enhanced in the presence of other mycotoxins. *Food and Chemical Toxicology, 96*, 79.

Gong, Y. Y., Egal, S., Hounsa, A., Turner, P. C., Hall, A. J., Cardwell, K. F., & Wild, C. P. (2003). Determinants of aflatoxin exposure in young children from Benin and Togo, West Africa: The critical role of weaning. *International Journal of Epidemiology, 32*, 556–562. https://doi.org/10.1093/ije/dyg109

Grenier, B., & Applegate, T. J. (2013). Modulation of intestinal functions following mycotoxin ingestion: Meta-analysis of published experiments in animals. *Toxins, 5*, 396–430. https://doi.org/10.3390/toxins5020396

Groopman, J. D., Kensler, T. W., & Wu, F. (2013). Mycotoxins—Occurrence and toxic effects. In *Encyclopedia of human nutrition* (Vol. 2, pp. 337–341) Elsevier Inc.

Han, Z., Nie, D., Ediage, E. N., Yang, X., Wang, J., Chen, B., Li, S., On, S. L. W., De Saeger, S., & Wu, A. (2014). Cumulative health risk assessment of co-occurring mycotoxins of deoxynivalenol and its acetyl derivatives in wheat and maize: Case study, Shanghai, China. *Food and Chemical Toxicology, 74*, 334–342. https://doi.org/10.1016/j.fct.2014.10.018

IARC. (1993). International Agency for Research on Cancer. Monographs on the evaluation of carcinogenic risks to humans. In *Some naturally occurring substances: Food items and constituents, heterocyclic aromatic amines and mycotoxins* (Vol. 56, pp. 1–609). World Health Organization.

Innovad mycotoxin feed survey. (2022). Infographic_Feed Survey 2022 (1)_0.pdf (innovad-global.com). https://innovad-global.com/sites/default/files/file-entity-browser/infographic_Feed%20Survey%202022%20%281%29_0.pdf

Ivanova, L., Fæste, C. K., Van Pamel, E., Daeseleire, E., Callebaut, A., & Uhlig, S. (2014). Presence of enniatin B and its hepatic metabolites in plasma and liver samples from broilers and eggs from laying hens. *World Mycotoxin Journal, 7*, 167–175. https://doi.org/10.3920/WMJ2013.1609

Jalilzadeh-Amin, G., Dalir-Naghadeh, B., Ahmadnejad-Asl-Gavgani, M., Fallah, A. A., & Mousavi Khaneghah, A. (2023). Prevalence and concentration of mycotoxins in animal feed in the Middle East and North Africa (MENA): A systematic review and meta-analysis. *Toxins (Basel), 15*(3), 214. https://doi.org/10.3390/toxins15030214

JECFA. (2001). Safety evaluation of certain mycotoxins in food prepared by the Fifty-Sixth Meeting of the Joint FAO/WHO Expert Committee on Food Additives. WHO FoodAdditives Series. 47.

Jia, R., Liu, W., Zhao, L., Cao, L., & Shen, Z. (2020). Low doses of individual and combined deoxynivalenol and zearalenone in naturally moldy diets impair intestinal functions via inducing inflammation and disrupting epithelial barrier in the intestine of piglets. *Toxicology Letters, 333*, 159–169.

Jiang, Y., Ogunade, I. M., Vyas, D., & Adesogan, A. T. (2021). Aflatoxin in dairy cows: Toxicity, occurrence in feedstuffs and milk and dietary mitigation strategies. *Toxins (Basel), 3*(4), 283. https://doi.org/10.3390/toxins13040283

Kolawole, O., Siri-Anusornsak, W., Petchkongkaw, A., Meneely, J., & Elliott, C. (2022). The efficacy of additives for the mitigation of aflatoxins in animal feed: A systematic review and network meta-analysis. *Toxins (Basel), 14*(10), 707. https://doi.org/10.3390/toxins14100707

Koshiol, J., Gao, Y. T., Dean, M., Egner, P., Nepal, C., Jones, K., Wang, B., Rashid, A., Luo, W., Van Dyke, A. L., et al. (2017). Association of aflatoxin and gallbladder cancer. *Gastroenterology, 153*, 488–494.

Křížová, L., Dadáková, K., Dvořáčková, M., & Kašparovský, T. (2021). Feedborne mycotoxins beauvericin and enniatins and livestock animals. *Toxins (Basel), 13*(1), 32. https://doi.org/10.3390/toxins13010032

Krska, R., Malachova, A., Berthiller, F., & van Egmond, H. P. (2014). Determination of T-2 and HT-2 toxins in food and feed: An update. *World Mycotoxin Journal, 7*, 131–142. https://doi.org/10.3920/WMJ2013.1605

Kucukcakan, B., & Hayrulai-Musliu, Z. (2015). Challenging role of dietary aflatoxin B1 exposure and hepatitis B infection on risk of hepatocellular carcinoma. *Open Access Macedonian Journal of Medical Sciences, 3*, 363.

Latham, R. L., Boyle, J. T., Barbano, A., Loveman, W. G., & Brown, N. A. (2023). Diverse mycotoxin threats to safe food and feed cereals. *Essays in Biochemistry, 67*(5), 797–809. https://doi.org/10.1042/EBC20220221

Lee, H. S., Lindahl, J., Nguyen-Viet, H., Khong, N. V., Nghia, V. B., Xuan, H. N., & Grace, D. (2017). An investigation into aflatoxin M1in slaughtered fattening pigs and awareness of aflatoxins in Vietnam. *BMC Veterinary Research, 13*, 1–7. https://doi.org/10.1186/s12917-017-1297-8

Li, D., Cao, Y., He, L., Wang, N. J., & Gu, J. R. (1993). Aberrations of p53 gene in human hepatocellular carcinoma from China. *Carcinogenesis, 14*, 169–173.

Li, X., Guo, Y., Zhao, L., Fan, Y., Ji, C., Zhang, J., et al. (2018). Protective effects of Devosia sp. ANSB714 on growth performance, immunity function, antioxidant capacity and tissue residues in growing-finishing pigs fed with deoxynivalenol contaminated diets. *Food and Chemical Toxicology, 121*, 246–251.

Liu, Y., & Wang, W. (2016). Aflatoxin B1 impairs mitochondrial functions, activates ROS generation, induces apoptosis and involves Nrf2 signal pathway in primary broiler hepatocytes. *Animal Science Journal, 87*, 1490–1500.

Longobardi, C., Damiano, S., Ferrara, G., Montagnaro, S., Meucci, V., Intorre, L., Bacci, S., Esposito, L., Piscopo, N., Rubino, A., Raffaele, A., Florio, S., & Ciarcia, R. (2023). Zearalenone (ZEN) and its metabolite levels in tissues of wild boar (*Sus scrofa*) from Southern Italy: A pilot study. *Toxins (Basel), 15*(1), 56. https://doi.org/10.3390/toxins15010056

Mahato, D. K., Kamle, M., Sharma, B., Pandhi, S., Devi, S., Dhawan, K., Selvakumar, R., Mishra, D., Kumar, A., Arora, S., Singh, N. A., & Kumar, P. (2021). Patulin in food: A mycotoxin concern for human health and its management strategies. *Toxicon, 198*, 12–23, ISSN 0041-0101. https://doi.org/10.1016/j.toxicon.2021.04.027

Marchese, S., Polo, A., Ariano, A., Velotto, S., Costantini, S., & Severino, L. (2018). Aflatoxin B1 and M1: Biological properties and their involvement in cancer development. *Toxins (Basel), 10*(6), 214. https://doi.org/10.3390/toxins10060214

Marin, S., Ramos, A. J., Cano-Sancho, G., & Sanchis, V. (2013). Mycotoxins: Occurrence, toxicology, and exposure assessment. *Food and Chemical Toxicology, 60*, 218–237. https://doi.org/10.1016/j.fct.2013.07.047

Marroquín-Cardona, A. G., Johnson, N. M., Phillips, T. D., & Hayes, A. W. (2014). Mycotoxins in a changing global environment—A review. *Food and Chemical Toxicology, 69*, 220–230. https://doi.org/10.1016/j.fct.2014.04.025

Meneely, J., Greer, B., Kolawole, O., & Elliott, C. (2023). T-2 and HT-2 toxins: Toxicity, occurrence and analysis: A review. *Toxins (Basel), 15*(8), 481. https://doi.org/10.3390/toxins15080481

Milani, J. M. (2013). Ecological conditions affecting mycotoxin production in cereals: A review. *Veterinární Medicína, 58*(8).

Mishra, S., Srivastava, S., Dewangan, J., Divakar, A., & Rath, S. K. (2019). Global occurrence of deoxynivalenol in food commodities and exposure risk assessment in humans in the last decade: A survey. *Critical Reviews in Food Science and Nutrition, 59*, 1–29. https://doi.org/10.1080/10408398.2019.1571479

Missmer, S. A., Suarez, L., Felkner, M., Wang, E., Merrill, A., Rothman, K., & Hendricks, K. A. (2006). Exposure to fumonisins and the occurrence of neural tube defects along the Texas–Mexico border. *Environmental Health Perspectives, 114*, 237–241. https://doi.org/10.1289/ehp.8221

Nji, Q. N., Babalola, O. O., Ekwomadu, T. I., Nleya, N., & Mwanza, M. (2022). Six main contributing factors to high levels of mycotoxin contamination in African foods. *Toxins (Basel), 14*, 318. https://doi.org/10.3390/toxins14050318

Nyachoti, C., Zijlstra, R., De Lange, C., & Patience, J. (2004). Voluntary feed intake in growing-finishing pigs: A review of the main determining factors and potential approaches for accurate predictions. *Canadian Journal of Animal Science, 84*, 549–566. https://doi.org/10.4141/A04-001

Pandey, A. K., Samota, M. K., Kumar, A., Silva, A. S., & Dubey, N. K. (2023). Mycotoxins in food commodities: Present status and future concerns. *Frontiers in Sustainable Food Systems, 7*. https://doi.org/10.3389/fsufs.2023.1162595

Pereira, S. C., Cunha, S. C., & Fernandes, J. O. (2019). Prevalent mycotoxins in animal feed: Occurrence and analytical methods. *Toxins (Basel), 11*(5), 290. https://doi.org/10.3390/toxins11050290

Perši, N., Pleadin, J., Kovačević, D., Scortichini, G., & Milone, S. (2014). Ochratoxin A in raw materials and cooked meat products made from OTA-treated pigs. *Meat Science, 96*, 203–210. https://doi.org/10.1016/j.meatsci.2013.07.005

Petersen, A. B. (2018). UNICEF and World Food programme. Expert advice on appropriate criteria and limits for contaminants in Ready to Use Therapeutic Foods.

Piątkowska, M., Sulyok, M., Pietruszka, K., & Panasiuk, Ł. (2018). Pilot study for the presence of fungal metabolites in sheep milk from first spring milking. *Journal of Veterinary Research, 62*, 167–172. https://doi.org/10.2478/jvetres-2018-0026

Piec, J., Pallez, M., Beyer, M., Vogelgsang, S., Hoffmann, L., & Pasquali, M. (2016). The Luxembourg database of trichothecene type B *F. graminearum* and *F. culmorum* producers. *Bioinformation, 12*, 1–3. https://doi.org/10.6026/97320630012001

Pinton, P., Guzylack-Piriou, L., Kolf-Clauw, M., & Oswald, I. P. (2012). The effect on the intestine of some fungal toxins: The trichothecenes. *Current Immunology Reviews, 8*, 193–208. https://doi.org/10.2174/157339512800671967

Polak-Śliwińska, M., & Paszczyk, B. (2021). Trichothecenes in food and feed, relevance to human and animal health and methods of detection: A systematic review. *Molecules, 26*(2), 454. https://doi.org/10.3390/molecules26020454

Popescu, R. G., Rădulescu, A. L., Georgescu, S. E., & Dinischiotu, A. (2022). Aflatoxins in feed: Types, metabolism, health consequences in swine and mitigation strategies. *Toxins, 14*(12), 853. https://doi.org/10.3390/toxins14120853

Rafai, P., Bata, A., Ványi, A., Papp, Z., Brydl, E., Jakab, L., Tuboly, S., & Túry, E. (1995). Effect of various levels of T-2 toxin on the clinical status, performance and metabolism of growing pigs. *The Veterinary Record, 136*, 485–489. https://doi.org/10.1136/vr.136.19.485

Riley, R. T., & Merrill, A. H., Jr. (2019). Ceramide synthase inhibition by fumonisins: A perfect storm of perturbed sphingolipid metabolism, signaling, and disease. *Journal of Lipid Research, 60*(7), 1183–1189. https://doi.org/10.1194/jlr.S093815

Riley, R. T., Edwards, S. G., Aidoo, K., Alexander, J., Bolger, M., Boon, P., Cressey, P., Doerge, D. R., Edler, L., Miller, J. D., et al. (2018). Fumonisins (addendum). In *WHO food additive series 74: Safety evaluation of certain contaminants in food* (pp. 415–573). World Health Organization.

Santini, A., Meca, G., Uhlig, S., & Ritieni, A. (2012). Fusaproliferin, beauvericin and enniatins: Occurrence in food-a review. *World Mycotoxin Journal, 5*, 71–81. https://doi.org/10.3920/WMJ2011.1331

Smith, M. C., Madec, S., Coton, E., & Hymery, N. (2016). Natural co-occurrence of mycotoxins in foods and feeds and their *in vitro* combined toxicological effects. *Toxins, 8*(4), 94. https://doi.org/10.3390/toxins8040094

Stiborová, M., Arlt, V. M., & Schmeiser, H. H. (2016). Balkan endemic nephropathy: An update on its aetiology. *Archives of Toxicology, 90*(11), 2595–2615. https://doi.org/10.1007/s00204-016-1819-3

Terciolo, C., Maresca, M., Pinton, P., & Oswald, I. P. (2018). Review article: Role of satiety hormones in anorexia induction by trichothecene mycotoxins. *Food and Chemical Toxicology, 121*, 701–714.

Tola, M., & Kebede, B. (2016). Occurrence, importance and control of mycotoxins: A review. *Cogent Food & Agriculture, 2*, 1–12. https://doi.org/10.1080/23311932.2016.1191103

Torok, V. A., Luyckx, K., & Lapidge, S. (2021). Human food waste to animal feed: Opportunities and challenges. *Animal Production Science, 62*, 1129–1139. https://doi.org/10.1071/AN20631

Vaclavikova, M., Malachova, A., Veprikova, Z., Dzuman, Z., Zachariasova, M., & Hajslova, J. (2013). "Emerging" mycotoxins in cereals processing chains: Changes of enniatins during beer and bread making. *Food Chemistry, 136*, 750–757. https://doi.org/10.1016/j.foodchem.2012.08.031

Wang, E., Norred, W. P., Bacon, C. W., Riley, R. T., & Merrill, A. H., Jr. (1991). Inhibition of sphingolipid biosynthesis by fumonisins. Implications for diseases associated with Fusarium moniliforme. *The Journal of Biological Chemistry, 266*, 14486–14490.

Wang, Y., Zhang, Z., Wang, H., Zhang, Y., Ji, M., Xu, H., Wang, C., Sun, Z., Gao, W., & Wang, S. L. (2016). miR-138-1 regulates aflatoxin B1-induced malignant transformation of BEAS-2B cells by targeting PDK1. *Archives of Toxicology, 90*, 1239–1249.

Weaver, A. C., Weaver, D. M., Adams, N., & Yiannikouris, A. (2021). Co-occurrence of 35 mycotoxins: A seven-year survey of corn grain and corn Silage in the United States. *Toxins, 13*, 516.

Weng, M. W., Lee, H. W., Choi, B., Wang, H. T., Hu, Y., Mehta, M., Desai, D., Amin, S., Zheng, Y., & Tang, M. S. (2017). AFB1 hepatocarcinogenesis is via lipid peroxidation that inhibits DNA repair, sensitizes mutation susceptibility and induces aldehyde-DNA adducts at p53 mutational hotspot codon 249. *Oncotarget, 8*, 18213–18226.

Wilkinson, J. M., & Young, R. H. (2020). Strategies to reduce reliance on soya bean meal and palm kernel meal in livestock nutrition. *Journal of Applied Animal Nutrition, 8*(2), 75–82. https://doi.org/10.3920/JAAN2020.0007

World Health Organisation. (2023). *Mycotoxins*. https://www.who.int/news-room/fact-sheets/detail/mycotoxins.

World Health Organization. (1993). *International Agency for Research on Cancer. IARC monographs on the evaluation of carcinogenic risks to humans* (Vol. 56, pp. 245–395). IARC Press: Aflatoxins.

Wu, S., Liu, Y., Duan, Y., Wang, F., Guo, F., Yan, F., et al. (2018). Intestinal toxicity of deoxynivalenol is limited by supplementation with Lactobacillus plantarum JM113 and consequentially altered gut microbiota in broiler chickens. *Journal of Animal Science and Biotechnology, 9*, 74.

Xu, R., Kiarie, E. G., Yiannikouris, A., et al. (2022). Nutritional impact of mycotoxins in food animal production and strategies for mitigation. *Journal of Animal Science and Biotechnology, 13*, 69. https://doi.org/10.1186/s40104-022-00714-2

Sustainable Utilization of Fungal Resources and Applications in Animal Health

Shambhavee Srivastav, Hiral Krishnatray, Prabhanshu Kumar, and Nahid Masood Siddiqui

Abstract The kingdom of fungi holds vast potential for addressing critical challenges in animal health and environmental sustainability. This paper discusses the diverse roles fungi play in ecosystems, biotechnology, and veterinary medicine. It highlights the profound impact of fungal metabolites, enzymes, and proteins in promoting ecological balance, enhancing animal nutrition, and increasing disease resistance. Through sustainable cultivation and harvesting techniques, fungal resources can be conserved while unlocking applications in pharmaceuticals, feed additives, and antimicrobial agents. Advancements in fungal genomics and bioinformatics enable the discovery of novel bioactive compounds with therapeutic and industrial applications. However, significant challenges such as toxicity concerns, regulatory barriers, and scalability must be addressed. This paper emphasizes the need for innovative solutions and interdisciplinary approaches to harness fungi's full potential in improving animal health and achieving sustainable development goals.

S. Srivastav · H. Krishnatray · P. Kumar · N. M. Siddiqui (✉)
Amity Institute of Biotechnology, Amity University, Noida, Noida, Uttar Pradesh, India
e-mail: nsiddiqui@amity.edu

A. Gupta (ed.), *Prospects of Fungal Biotechnologies for Livestock Volume 1*,
Fungal Biology, https://doi.org/10.1007/978-3-031-90994-8_6

Graphical Abstract

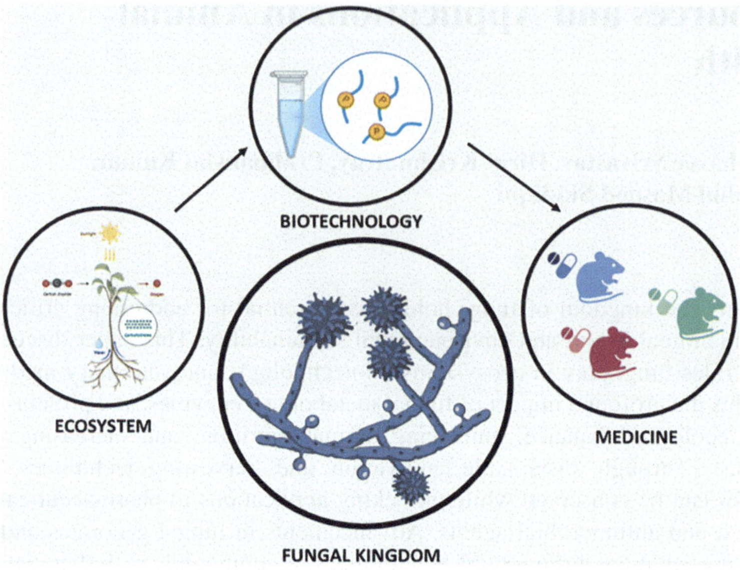

Keywords Fungi · Sustainability · Animal health · Biotechnology · Bioactive compounds

1 Introduction

As we advance into the twenty-first century, we can discern the burgeoning challenges faced by our world, encompassing issues of animal health, environmental degradation, and sustainable development. This multifaceted puzzle demands valuable paradigm-shifting solutions, which the "Fungal Kingdom" can provide. The kingdom enriches a spectrum of unparalleled wealth of biomaterials, enzymes, and bioactive compounds. With around 1.5 million species, the fungal kingdom offers transformative potential, driving the increasing use of fungal resources (Buckley, 2008).

Fungal biotechnology's impact on animal health has been profound (Roth et al., 2023). From the discovery of edible mushrooms like *Ganoderma lucidum* and *Cordyceps sinensis*to (Kumar et al., 2021), the development of life-changing pharmaceuticals such as cyclosporin, lovastatin, and penicillin, fungi have played a pivotal role in transforming healthcare (Prescott et al., 2023). Fungal-based feed additives and supplements enhance animal nutrition and resilience, ensuring improved livestock health while upholding safety and efficacy.

As we unravel the facets of this diverse community, we vouchsafe compounds and materials that uphold pervasive applications. Studies have elucidated the roles of laccases, cellulases, and proteases in biodegradation, biofuel production, and animal feeding (Janusz et al., 2020). Furthermore, polysaccharides, proteins, and polyketides exhibit promise in the biotechnology and pharmacology industries, demonstrating immunomodulatory properties. Antibacterial, antiviral, and anticancer drugs derived from fungal secondary metabolites also present a prospective substitute for traditional treatments (Keller, 2018).

In addition to their contributions to healthcare, fungi hold tremendous potential in improving animal welfare, human wellness, and ecological sustainability ("Fungus Focus", 2018). Emerging technologies like synthetic biology and nanotechnology—derived from fungal sources—have opened new avenues for biomedical innovation and the production of specialized biochemicals (Abd-Elsalam, 2023). Furthermore, the intricate interplay between fungal populations and the microbiome is crucial in maintaining ecological balance. Understanding the mechanisms that regulate this equilibrium is essential for preserving ecosystem resilience.

Recent advances in transcriptomics, proteomics, and fungal genomics have expanded our understanding of fungal biology. Bioinformatics tools have simplified the exploration of fungal genomes, uncovering novel gene clusters and biosynthetic pathways (Tran et al., 2019). Integrating these cutting-edge technologies promises to accelerate the discovery and development of new fungal-derived compounds with far-reaching applications.

As we delve deeper into the fungal kingdom, new solutions are emerging to address pressing environmental and health challenges. To bridge the knowledge gap regarding fungal-microbial interactions and their impact on ecological balance. Fully harnessing the potential of fungi can revolutionize healthcare, environmental sustainability, and biotechnology, offering solutions to some of the world's most urgent challenges.

This flowchart encompasses the roles of fungi in ecological balance, animal nutrition, disease resistance, their applications in pharmaceuticals and sustainability, challenges like toxicity and scalability, and innovative solutions like genomics, eco-friendly harvesting, and controlled fermentation.

2 Fungal Diversity and Ecology

Ecology and variety of fungi are essential to comprehending ecosystems and their applications in real life. Fungi are essential for plant development, the cycling of nutrients, and the stability of the ecosystem. A large portion of the fungal world is still unknown despite advancements in fungal categorization, providing intriguing possibilities for novel discoveries (Bahram & Netherway, 2021).

This chapter first looks at G.C. Ainsworth's classification approach. It next examines the role fungi play in plant symbiosis and nutrient cycling in various environments. To emphasize their ecological and biomedical significance, it ends by stressing their function in animal health through the use of probiotics, antibiotics, and bioactive substances (Fig. 1) (Latif et al., 2023).

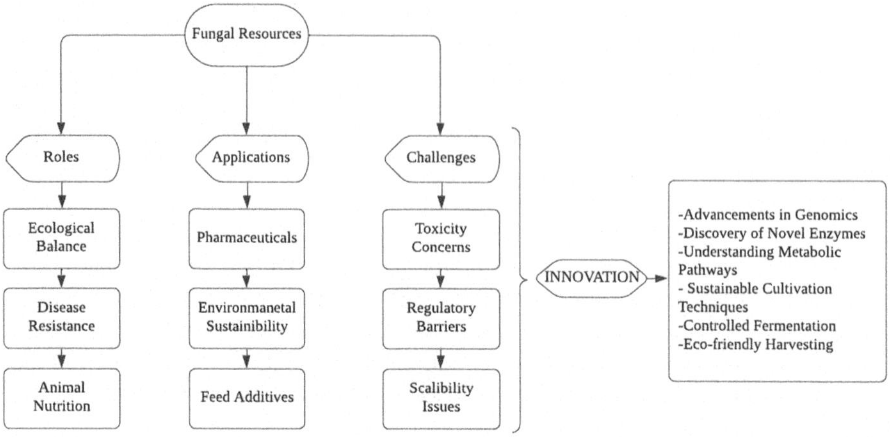

Fig. 1 Fungal resources and their roles, applications, challenges, and innovations

2.1 Classification of Fungi

Fungal taxonomy was revolutionized by G.C. Ainsworth's (1973) paradigm-shifting categorization system, which identified seven separate divisions: Ascomycota, Zygomycota, Glomeromycota, Blastocladiomycota, Neocallimastigomycota, and Basidiomycota. This framework, which offers a taxonomic roadmap, is still essential for modern fungal study. Despite over 120,000 identified fungal species, scientists estimate that millions more are undiscovered (Wu et al., 2019). Thus, understanding fungal classification is crucial for exploring the vast potential of the fungal kingdom.

2.2 Ecological Roles of Fungi

The complex interactions between fungi and their environments underscore their critical role in maintaining ecological balance. By understanding these relationships, researchers can better appreciate the importance of fungi in shaping ecosystem processes.

Fungi play a pivotal role in nutrient cycling, facilitating the breakdown of organic matter and releasing vital components into the environment. Through symbiotic connections, this mechanism preserves soil fertility and encourages plant development (Jagadesh et al., 2024).

Additionally, fungi mobilize nutrients and enhance their availability to plants. For example, *mycorrhizal fungi* form mutualistic associations that boost nutrient uptake in ecosystems like the Amazon rainforest, while *saprotrophic fungi* facilitate carbon release in Arctic tundra soils through heterotrophic processes (Teotia et al.,

2017). Fungal populations also influence soil nitrogen availability, impacting plant composition and ecological interactions (Jagadesh et al., 2024).

2.3 Fungal Habitats Relevant to Animal Health

Fungi are crucial for the environment, but they are also beneficial to animal health. Antibiotics, medicines, and bioactive substances with potential uses in animal health are abundant in fungi (Case et al., 2022). In aquaculture, *probiotic yeast* like *Sporidiobolus ruineniae* is effective in boosting animal health (Kanpiengjai et al., 2020).

Interestingly, substances generated from fungi have antibacterial qualities that can combat bacterial illnesses resistant to antibiotics. *Saccharomyces cerevisiae* and other mycotoxin-binding yeasts decrease toxicity in animal feed, promoting better growth rates and lowering mortality in pigs and poultry. Animals' resistance to illness is increased when immunological responses are stimulated by fungus *beta-glucans* (Mirseyed et al., 2024). Furthermore, antimicrobial peptides produced from fungi, like *penicillin*, have been employed in animal health to treat bacterial infections. It is important to continue researching the potential of fungus to improve animal health through probiotics, nutritional supplements, and antibacterial chemicals.

3 Sustainable Harvesting and Cultivation of Fungi

The ecology and variety of fungi must be thoroughly understood to develop sustainable wild harvesting methods. Nguyen et al.'s (2015) research underlines the significance of understanding the functions that fungi play in ecosystems and how human behavior influences fungal populations (Větrovský et al., 2019). Researching fungal biology and ecology in-depth is necessary to develop sustainable harvesting methods and conservation initiatives (s40694—020-00095-z). These studies guide best practices and establish the foundation for sustained conservation efforts.

3.1 Wild Harvesting Techniques

Understanding the biological interactions between fungi in their environment is essential for successful wild harvesting. Fungi are essential for preserving the equilibrium of ecosystems, affecting the cycling of nutrients, and sustaining plant communities, according to Allen et al. (2017) (Bahram & Netherway, 2021). It is crucial to acknowledge these ecological roles while developing harvesting techniques that safeguard fungal diversity.

For instance, Nguyen et al. (2015) emphasize the need to consider habitat fragmentation and fungal dispersion techniques when developing conservation plans (Wilson et al., 2015). According to the Horton et al. (2013), chanterelle mushroom (Cantharellus cibarius) depends on mycorrhizal relationships with different tree types. By integrating these ecological insights into regulatory measures like as quantity constraints and seasonal harvesting restrictions, we may ensure the preservation of fungal biodiversity while maintaining the health of the ecosystem. These regulations promote the maintenance of fungi's natural ecological roles while protecting against habitat degradation.

3.2 Cultivation Methods

While sustainable wild harvesting is critical, cultivation offers a complementary approach that can help alleviate pressure on natural fungal populations. Utilizing locally accessible substrates, such as wood wastes or agricultural waste, can make growing less resource-intensive and more sustainable. This method lessens the need for wild populations while allowing the ongoing use of fungus for culinary, medicinal, and ecological purposes. To prevent the misuse of fungal resources, May et al. (2019) emphasizes the necessity of enforcing regulated harvesting even in growing procedures (Davies et al., 2021). Reducing exploitation and guaranteeing a steady supply of fungal species for many businesses, ethical cultivation not only supports wild harvesting but also functions as a feasible conservation strategy.

3.3 Conservation and Sustainability Practices

Sustainable harvesting and cultivation methods must be integrated into a complete strategy for the long-term preservation of fungal ecosystems. Achieving long-term conservation success requires restoring habitat, raising public awareness, and including fungus in larger ecological management strategies. According to Allen et al. (2017), mycorrhizal fungi are essential to preserving ecosystem resilience, which is necessary to support biodiversity and guarantee the stability of ecosystems (Thirkell et al., 2017).

These initiatives are strengthened by cooperation between researchers, decision-makers, and conservation organizations. Incorporating fungal conservation into ecological management strategies guarantees that fungi are not disregarded in more general conservation debates, as May et al. (2019) notes (Lofgren & Stajich, 2021). Long-term sustainability is ensured by this integrated strategy, which not only safeguards fungal species but also contributes to preserving the equilibrium of the ecosystems they support.

4 Bioactive Compounds from Fungi

Bioactive chemicals found in fungi have been shown to provide a multitude of health advantages as well as commercial applications. The potential of fungal metabolites as key chemicals, additives, nutraceuticals, supplements, and functional components in food and pharmaceutical products has been thoroughly studied.

4.1 Secondary Metabolites

Secondary metabolites from fungi have a variety of pharmacological properties. These substances play a major role in the management and prevention of chronic non-communicable diseases (NCDs). Against different human tumor cell lines, for example, hypocriol A, hypocriol F, sesquiterpene strichocaranes E, and sesquiterpene strichocaranes F have exhibited cytotoxic action. This offers a bright future for creating cutting-edge anticancer treatments. Additionally, substances such as terrein have demonstrated impressive anti-inflammatory and antioxidant properties, which makes them appropriate for use in functional foods and medications (Goutam et al., 2017).

The interactions between the environment and the ecological niche of fungal species are closely related to the production of these secondary metabolites. By elucidating the mechanisms behind secondary metabolite formation, researchers can progress the development of novel drugs and industrial applications. Therefore, the focus of current research is on understanding the complex relationships between fungal secondary metabolites, their bioactive properties, and their applications.

4.2 Enzymes and Proteins

Fungi not only create a large number of secondary metabolites but also a large number of proteins and enzymes. These are essential to the degradation of complex polysaccharides and serve their purpose in several industrial applications. For instance, the capacity of cellulases and xylanases to degrade cellulose and hemicellulose, respectively, has led to their widespread use (De Souza & Kawaguti, 2021). Furthermore, fungal proteins have shown promise as viable substitutes for conventional protein sources. They facilitate better digestion, support intestinal health, and assist in controlling blood cholesterol levels.

Fungal proteins and enzymes have several biotechnological uses. The food, feed, and pharmaceutical sectors may find use for them. Government agencies like the US Food and Drug Administration (FDA) have declared edible fungus proteins safe for human consumption. These proteins can be included in diets regularly to offer health advantages and take the place of meals high in animal protein. Fungal proteins are intriguing as supplements for both plant and animal proteins due to their intrinsic bio-functional characteristics (Li et al., 2023).

4.3 Polysaccharides

The capacity of fungi to break down complex polysaccharides that include cellulose, hemicellulose, pectin, starch, and inulin is another important characteristic (Wang et al., 2020). They accomplish this by creating enzymes that break down polysaccharides (El-Gendi et al., 2021). Interestingly, anaerobic fungi are useful for biotechnological applications since they have evolved special ways to degrade crystalline cellulose. Fungal exopolysaccharides are used in food, textiles, cosmetics, and pharmaceuticals, among other sectors, because of their ability to retain moisture, improve viscosity, and stabilize materials (Osemwegie et al., 2020).

It will be essential to investigate the breakdown mechanisms of fungal polysaccharides to create innovative industrial processes and applications connected to health (Wang et al., 2022).

The goal of current research is to clarify the intricate relationships that exist between enzymes, fungal polysaccharides, and the environment to develop novel biotechnological solutions (El-Gendi et al., 2021).

5 Applications in Animal Health

Fungi have been shown in several studies to offer potential benefits for animal health, including their use as antibacterial agents, immunomodulators, and nutritional supplements (Corbu et al., 2023). Aspergillus, Penicillium, and Saccharomyces are among the promising species (Corbu et al., 2023). More research is needed to fully grasp the potential of fungi for animal health.

5.1 Antimicrobial Agents

The wide range of bioactive compounds that fungi create indicates that they may be useful as antimicrobial agents for treating a variety of illnesses. Antimicrobial substances generated from fungi offer a good substitute for traditional antibiotics. These compounds' antibacterial qualities have been the subject of several research. Penicillin, which is produced from the Penicillium species, is one famous example (Sułkowska-Ziaja et al., 2023). Furthermore, Aspergillus, Trichoderma, and Fusarium are examples of fungi that have antibacterial qualities against a variety of diseases (Sal et al., 2022).

These fungal antimicrobials operate through an array of techniques, including permeabilizing membranes, inhibiting protein synthesis, dissolving cell walls, and interfering with DNA replication (Hossain, 2024). As a result, they present encouraging chances for managing plant illnesses, curing bacterial infections, stopping food from spoiling, and creating novel antimicrobial treatments (Sar et al., 2023).

5.2 Immunomodulators

Fungi are appealing candidates for innovative immunotherapies because they include a variety of bioactive chemicals that can either activate or inhibit immunological responses (Venturella et al., 2021). They can control and improve immunological function because of their two immunomodulatory qualities.

Some fungi, like the Reishi mushroom (Ganoderma lucidum), can stimulate the generation of cytokines and activate immune cells, which increases resistance to infections (Ekiz et al., 2023). On the other hand, certain fungi, such as Cordyceps sinensis, have immunosuppressive properties that can be used to control T- and B-cell activity and regulate inflammatory responses. This combined effect provides new therapeutic options for autoimmune diseases (Das et al., 2021).

5.3 Growth Promoters

In multiple industries including agriculture, animal husbandry, and human health, fungi are important for increasing sustainability and production. For example, mycorrhizal fungi encourage plant development and nutrient absorption, whereas Trichoderma species lessen plant illnesses and improve soil health (Tyśkiewicz et al., 2022). Additionally, fungi probiotics support animal intestinal health, while medicinal mushrooms such as chaga and reishi boost immunity and general health (Ganoderma lucidum). We can enhance health outcomes and promote sustainable development by utilizing the advantageous features of fungi.

5.4 Anthelmintics

Fungi offer a promising solution to anthelmintic resistance, with various species and their metabolites exhibiting potent anthelmintic properties. Research has demonstrated that certain fungal compounds, such as chitinases and proteases, have been shown via research to damage worm cell membranes and block vital enzymes (Javed et al., 2013).

Certain species, such as Aspergillus and Trichoderma, have demonstrated potential in the fight against parasitic nematodes in cattle and maybe humans.

By efficiently targeting parasites, these fungi aid in the control of infections (Guzmán-Guzmán et al., 2023). Their numerous strategies of immune response control, cell membrane rupture, and enzyme inhibition account for their effectiveness. This makes using fungus as a natural parasite infection treatment option appealing (Marcos et al., 2016). Fungal anthelmintics have a great deal of potential as a sustainable and natural substitute for manmade chemicals (Jayawardene et al., 2021). However, further investigation is necessary to thoroughly assess their safety and efficacy.

6 Fungi-Derived Products in Veterinary Medicine

Modern methods to enhance animal health have been made possible by fungal biotechnology, which has revolutionized veterinary care. By using the unique bioactive compounds present in fungi, veterinary medicine may expand the scope of treatments it offers, therefore addressing a multitude of animal health issues and improving the general health and performance of these animals (Hashem et al., 2023). These discoveries demonstrate how fungi might be creatively used to provide powerful treatments for animal health.

6.1 Pharmaceuticals

Drugs developed from fungi have greatly benefited veterinary practice by increasing the options for treating a variety of animal health conditions. For instance, itraconazole, which is produced from Trichophyton, works effectively for animals with dermatophytosis; dosages for cats vary from 3 to 5 mg/kg/day (Moriello, 2004). In addition to antifungal medications, research is being done to create compounds derived from fungi as therapeutic agents to treat infections and diseases in veterinary patients (Alghuthaymi et al., 2021). There is hope for future developments in this growing field of medications made from fungi.

6.2 Nutraceuticals

Fungi, which enhance the immune system and prevent disease in animal care, are also important players in the nutraceutical business, building on the success of medicines (Chugh et al., 2022). Fungal products, such as those obtained from marine organisms, have the potential to enhance the health and welfare of animals (Ślusarczyk et al., 2021). Moreover, adding fungus to feed formulations raises the nutritional content and therapeutic potential for livestock and companion animals, emphasizing the double benefits of using fungi in animal husbandry (Amara & El-Baky, 2023).

7 Challenges in Fungal Resource Utilization

Harnessing fungal resources presents several difficulties, mostly about toxicity and safety, legal compliance, and industrial scalability (Umar et al., 2024). It is imperative to tackle these concerns to guarantee the secure and efficient use of goods obtained from fungi across diverse sectors.

7.1 Safety and Toxicity Concern

Notably, fungi's potential is offset by significant safety and toxicity concerns. Mycotoxin-producing fungus poses serious health dangers to both people and animals, hence extensive testing and risk assessment are required before the product is sold (Awuchi et al., 2021). Strict safety procedures and efficient detoxification techniques are essential for reducing the dangers connected to food and feed sources (Udomkun et al., 2017). In addition, the existence of harmful substances in some fungus-based products needs close observation to protect consumer health (Garnier et al., 2017). These worries highlight the necessity of effective quality control procedures (Garnier et al., 2017).

7.2 Regulatory Issues

The development and launch of goods derived from fungi into the market are complicated by regulatory obstacles in addition to safety considerations (Davies et al., 2021). Businesses looking to commercialize these products face challenges due to the many and complex regulatory frameworks that exist across different countries (Davies et al., 2021). It takes a lot of time and money to meet the strict safety and effectiveness criteria in regulatory reviews, which frequently delays licensure. This discrepancy emphasizes the necessity of unified regulatory guidelines to enable the safe introduction of advantageous fungal applications into the marketplace (Ng et al., 2022).

7.3 Scalability and Production Challenges

Furthermore, there are major obstacles to effectively using fungal resources, including scalability and production issues. Although culture on a laboratory scale produces encouraging results, transferring these techniques to an industrial scale presents difficulties because of things like reliable product quality, ideal growth conditions, and the availability of substrate (Segeritz & Vallier, 2017). It will need creative thinking and technological developments to get beyond these obstacles, improve farming practices, and increase production efficiency. Fungal-derived goods can satisfy expanding consumer needs while remaining sustainable by tackling these issues (El-Gendi et al., 2021).

8 Future Perspectives

Thanks to innovative technology, applications in human health, and sustainable practices, fungal research is set to make major strides shortly. This advancement has been made possible by breakthroughs in transcriptomics, proteomics, and fungal genomes, which have opened up new perspectives on the diverse range of uses and biology of fungi (Wijayawardene et al., 2023).

8.1 Emerging Technologies in Fungal Research

Building on this momentum, discoveries—particularly in the areas of transcriptomics, proteomics, and genome sequencing—have significantly increased our knowledge of fungal biology (Ball et al., 2020). Large-scale genomic and proteomic data processing has been made easier by the incorporation of bioinformatics tools, opening the door to the identification of new genes and metabolic pathways (Tariq et al., 2020). These cutting-edge techniques improve the mapping of the fungal genome and produce findings with broad biotechnological and medical applications (Ross & Santiago-Tirado, 2024). These technologies and their advancements are having a significant influence on human health that goes beyond fundamental research (Junaid et al., 2022).

8.2 Potential New Applications in Human Health

As the promise of fungi in novel medicinal applications becomes more widely acknowledged, scientists are delving further into their function in drug discovery and development (Prescott et al., 2023). Fungi that produce bioactive molecules can be a valuable source of new therapeutics, especially when looking for antibiotics that do not work as well (Sanchez & Demain, 2017). Furthermore, new treatment avenues are made possible by the understanding of the molecular mechanisms driving fungal bioactivity provided by transcriptomics and proteomics (Zamith-Miranda et al., 2021).

Another crucial area of growth is the intersection of sustainable practices with fungal research (Vassilev & De Oliveira Mendes, 2024). This area has emerged to provide solutions relating to environmental issues, ignite sustainable practices, and put forward cultivated work management with fungal resources (Vassilev & De

Oliveira Mendes, 2024). Fungi are essential parts of ecosystems that maintain good soil and nutrient cycling, which is the foundation of sustainable agriculture (Saha et al., 2023). Furthermore, understanding how different fungus species interact with their surroundings may lead to the development of bioremediation strategies that reduce pollution (Deshmukh et al., 2016). Through the use of fungal technology in combination with bioinformatics technologies, scientists may create environmentally sustainable solutions that bolster ecological resilience and showcase the advantageous applications of fungus (Tiwari & Park, 2024).

9 Conclusion

Fungi possess the capacity to bring about revolutionary changes in the realms of biotechnology, animal health, and environmental sustainability, among other critical global issues. The fungal kingdom, which has over 1.5 million species, provides an unmatched abundance of biomaterials, enzymes, and bioactive substances. Fungal biotechnology has recently produced medicines that can change people's lives, feed additives derived from fungi, and supplements that improve the resistance and nutrition of animals. Innovative technologies like as genome sequencing, proteomics, and transcriptomics in conjunction with sustainable techniques of growth and harvesting have created new opportunities for biomedical innovation and the manufacturing of specialized biochemicals.

References

Abd-Elsalam, K. A. (2023). Special issue: Fungal nanotechnology 2. *Journal of Fungi, 9*(5), 553. https://doi.org/10.3390/jof9050553

Ainsworth, G. C., Sparrow, F. K., & Sussman, A. S. (1973). Introduction and keys to higher taxa. In The fungi: an advanced treatise (Vol. 4B, pp. 247–279). Academic Press, New York

Alghuthaymi, M. A., Hassan, A. A., Kalia, A., Ahl, R. M. H. S. E., Hamaky, A. a. M. E., Oleksak, P., Kuca, K., & Abd-Elsalam, K. A. (2021). Antifungal nano-therapy in veterinary medicine: Current status and future prospects. *Journal of Fungi, 7*(7), 494. https://doi.org/10.3390/jof7070494

Allen, J. L., Leflaive, J., Bringuier, C., Ten-Hage, L., Chauvet, E., Cornut, J., & Danger, M. (2017). Allelopathic inhibition of primary producer growth and photosynthesis by aquatic fungi. Fungal Ecology, 29, 133–138. https://doi.org/10.1016/j.funeco.2017.07.001

Amara, A. A., & El-Baky, N. A. (2023). Fungi as a source of edible proteins and animal feed. *Journal of Fungi, 9*(1), 73. https://doi.org/10.3390/jof9010073

Awuchi, C. G., Ondari, E. N., Ogbonna, C. U., Upadhyay, A. K., Baran, K., Okpala, C. O. R., Korzeniowska, M., & Guiné, R. P. F. (2021). Mycotoxins affecting animals, foods, humans, and plants: Types, occurrence, toxicities, action mechanisms, prevention, and detoxification strategies—A revisit. *Food, 10*(6), 1279. https://doi.org/10.3390/foods10061279

Bahram, M., & Netherway, T. (2021). Fungi as mediators linking organisms and ecosystems. *FEMS Microbiology Reviews, 46*(2). https://doi.org/10.1093/femsre/fuab058

Ball, B., Langille, M., & Geddes-McAlister, J. (2020). Fun(gi)omics: Advanced and diverse technologies to explore emerging fungal pathogens and define mechanisms of antifungal resistance. *MBio, 11*(5). https://doi.org/10.1128/mbio.01020-20

Buckley, M. (2008). *The Fungal Kingdom: diverse and essential roles in earth's ecosystem.* https://doi.org/10.1128/aamcol.2nov.2007.

Case, N. T., Berman, J., Blehert, D. S., Cramer, R. A., Cuomo, C., Currie, C. R., Ene, I. V., Fisher, M. C., Fritz-Laylin, L. K., Gerstein, A. C., Glass, N. L., Gow, N. a. R., Gurr, S. J., Hittinger, C. T., Hohl, T. M., Iliev, I. D., James, T. Y., Jin, H., Klein, B. S., et al. (2022). The future of fungi: Threats and opportunities. *G3 Genes Genomes Genetics, 12*(11). https://doi.org/10.1093/g3journal/jkac224

Chugh, R. M., Mittal, P., Mp, N., Arora, T., Bhattacharya, T., Chopra, H., Cavalu, S., & Gautam, R. K. (2022). Fungal mushrooms: A natural compound with therapeutic applications. *Frontiers in Pharmacology, 13.* https://doi.org/10.3389/fphar.2022.925387

Corbu, V. M., Gheorghe-Barbu, I., Dumbravă, A. Ş., Vrâncianu, C. O., & Şesan, T. E. (2023). Current insights in fungal importance—A comprehensive review. *Microorganisms, 11*(6), 1384. https://doi.org/10.3390/microorganisms11061384

Das, G., Shin, H., Leyva-Gómez, G., Del Prado-Audelo, M. L., Cortes, H., Singh, Y. D., Panda, M. K., Mishra, A. P., Nigam, M., Saklani, S., Chaturi, P. K., Martorell, M., Cruz-Martins, N., Sharma, V., Garg, N., Sharma, R., & Patra, J. K. (2021). Cordyceps spp.: A review on its immune-stimulatory and other biological potentials. *Frontiers in Pharmacology, 11.* https://doi.org/10.3389/fphar.2020.602364

Davies, C. R., Wohlgemuth, F., Young, T., Violet, J., Dickinson, M., Sanders, J., Vallieres, C., & Avery, S. V. (2021). Evolving challenges and strategies for fungal control in the food supply chain. *Fungal Biology Reviews, 36*, 15–26. https://doi.org/10.1016/j.fbr.2021.01.003

De Souza, T. S. P., & Kawaguti, H. Y. (2021). Cellulases, hemicellulases, and pectinases: Applications in the food and beverage industry. *Food and Bioprocess Technology, 14*(8), 1446–1477. https://doi.org/10.1007/s11947-021-02678-z

Deshmukh, R., Khardenavis, A. A., & Purohit, H. J. (2016). Diverse metabolic capacities of fungi for bioremediation. *Indian Journal of Microbiology, 56*(3), 247–264. https://doi.org/10.1007/s12088-016-0584-6

Ekiz, E., Oz, E., El-Aty, A. A., Proestos, C., Brennan, C., Zeng, M., Tomasevic, I., Elobeid, T., Çadırcı, K., Bayrak, M., & Oz, F. (2023). Exploring the potential medicinal benefits of Ganoderma lucidum: From metabolic disorders to coronavirus infections. *Food, 12*(7), 1512. https://doi.org/10.3390/foods12071512

El-Gendi, H., Saleh, A. K., Badierah, R., Redwan, E. M., El-Maradny, Y. A., & El-Fakharany, E. M. (2021). A comprehensive insight into fungal enzymes: Structure, classification, and their role in Mankind's challenges. *Journal of Fungi, 8*(1), 23. https://doi.org/10.3390/jof8010023

Fungus focus. (2018). *Nature Ecology & Evolution, 2*(11), 1675. https://doi.org/10.1038/s41559-018-0721-1

Garnier, L., Valence, F., & Mounier, J. (2017). Diversity and control of spoilage fungi in dairy products: An update. *Microorganisms, 5*(3), 42. https://doi.org/10.3390/microorganisms5030042

Goutam, J., Sharma, G., Tiwari, V. K., Mishra, A., Kharwar, R. N., Ramaraj, V., & Koch, B. (2017). Isolation and characterization of "Terrein" an antimicrobial and antitumor compound from endophytic fungus Aspergillus terreus (JAS-2) associated from Achyranthus aspera Varanasi, India. *Frontiers in Microbiology, 8.* https://doi.org/10.3389/fmicb.2017.01334

Guzmán-Guzmán, P., Kumar, A., De Los Santos-Villalobos, S., Parra-Cota, F. I., Del Carmen Orozco-Mosqueda, M., Fadiji, A. E., Hyder, S., Babalola, O. O., & Santoyo, G. (2023). Trichoderma species: Our best fungal allies in the biocontrol of plant diseases—A review. *Plants, 12*(3), 432. https://doi.org/10.3390/plants12030432

Hashem, A. H., Attia, M. S., Kandil, E. K., Fawzi, M. M., Abdelrahman, A. S., Khader, M. S., Khodaira, M. A., Emam, A. E., Goma, M. A., & Abdelaziz, A. M. (2023). Bioactive compounds and biomedical applications of endophytic fungi: A recent review. *Microbial Cell Factories, 22*(1). https://doi.org/10.1186/s12934-023-02118-x

Horton, T. R., Hayward, J., Tourtellot, S. G., & Taylor, D. L. (2013). Uncommon ectomycorrhizal networks: richness and distribution of Alnus-associating ectomycorrhizal fungal communities. *New Phytologist, 198*(4), 978–980. https://doi.org/10.1111/nph.12313

Hossain, T. J. (2024). Methods for screening and evaluation of antimicrobial activity: A review of protocols, advantages, and limitations. *European Journal of Microbiology and Immunology, 14*(2), 97–115. https://doi.org/10.1556/1886.2024.00035

Jagadesh, M., Dash, M., Kumari, A., Singh, S. K., Verma, K. K., Kumar, P., Bhatt, R., & Sharma, S. K. (2024). Revealing the hidden world of soil microbes: Metagenomic insights into plant, bacteria, and fungi interactions for sustainable agriculture and ecosystem restoration. *Microbiological Research, 285*, 127764. https://doi.org/10.1016/j.micres.2024.127764

Janusz, G., Pawlik, A., Świderska-Burek, U., Polak, J., Sulej, J., Jarosz-Wilkołazka, A., & Paszczyński, A. (2020). Laccase properties, physiological functions, and evolution. *International Journal of Molecular Sciences, 21*(3), 966. https://doi.org/10.3390/ijms21030966

Javed, S., Hamid, R., Khan, M., Ahmad, M., Ahmad, M., Abdin, M., & Musarrat, J. (2013). Chitinases: An update. *Journal of Pharmacy and Bioallied Sciences, 5*(1), 21. https://doi.org/10.4103/0975-7406.106559

Jayawardene, K. L. T. D., Palombo, E. A., & Boag, P. R. (2021). Natural products are a promising source for anthelmintic drug discovery. *Biomolecules, 11*(10), 1457. https://doi.org/10.3390/biom11101457

Junaid, S. B., Imam, A. A., Balogun, A. O., De Silva, L. C., Surakat, Y. A., Kumar, G., Abdulkarim, M., Shuaibu, A. N., Garba, A., Sahalu, Y., Mohammed, A., Mohammed, T. Y., Abdulkadir, B. A., Abba, A. A., Kakumi, N. a. I., & Mahamad, S. (2022). Recent advancements in emerging technologies for healthcare management systems: A survey. *Healthcare (Basel, Switzerland), 10*(10), 1940. https://doi.org/10.3390/healthcare10101940

Kanpiengjai, A., Khanongnuch, C., Lumyong, S., Kummasook, A., & Kittibunchakul, S. (2020). Characterization of Sporidiobolus ruineniae A45.2 cultivated in tannin substrate for use as a potential multifunctional probiotic yeast in aquaculture. *Journal of Fungi, 6*(4), 378. https://doi.org/10.3390/jof6040378

Keller, N. P. (2018). Fungal secondary metabolism: Regulation, function and drug discovery. *Nature Reviews Microbiology, 17*(3), 167–180. https://doi.org/10.1038/s41579-018-0121-1

Kumar, K., Mehra, R., Guiné, R. P. F., Lima, M. J., Kumar, N., Kaushik, R., Ahmed, N., Yadav, A. N., & Kumar, H. (2021). Edible mushrooms: A comprehensive review on bioactive compounds with health benefits and processing aspects. *Food, 10*(12), 2996. https://doi.org/10.3390/foods10122996

Latif, A., Shehzad, A., Niazi, S., Zahid, A., Ashraf, W., Iqbal, M. W., Rehman, A., Riaz, T., Aadil, R. M., Khan, I. M., Özogul, F., Rocha, J. M., Esatbeyoglu, T., & Korma, S. A. (2023). Probiotics: Mechanism of action, health benefits and their application in food industries. *Frontiers in Microbiology, 14*. https://doi.org/10.3389/fmicb.2023.1216674

Li, K., Qiao, K., Xiong, J., Guo, H., & Zhang, Y. (2023). Nutritional values and bio-functional properties of fungal proteins: Applications in foods as a sustainable source. *Food, 12*(24), 4388. https://doi.org/10.3390/foods12244388

Lofgren, L. A., & Stajich, J. E. (2021). Fungal biodiversity and conservation mycology in light of new technology, big data, and changing attitudes. *Current Biology, 31*(19), R1312–R1325. https://doi.org/10.1016/j.cub.2021.06.083

Marcos, C. M., De Oliveira, H. C., De Cássia M Antunes De Melo, W., De Fátima Da Silva, J., Assato, P. A., Scorzoni, L., Rossi, S. A., De Paula E Silva, A. C. A., Mendes-Giannini, M. J. S., & Fusco-Almeida, A. M. (2016). Anti-immune strategies of pathogenic fungi. *Frontiers in Cellular and Infection Microbiology, 6*. https://doi.org/10.3389/fcimb.2016.00142

May, T. W., Cooper, J. A., Dahlberg, A., Furci, G., Minter, D. W., Mueller, G. M., Pouliot, A., & Yang, Z. (2019). Recognition of the discipline of conservation mycology. *Conservation Biology, 33*, 733–736. https://doi.org/10.1111/cobi.13228

Mirseyed, P. S., Kheirabadi, S., Anbarteh, R., & Ghaffari, M. H. (2024). Assessment of mycotoxin sequestration efficacy in Saccharomyces cerevisiae by-products cultured in wheat bran and whey protein medium. *Scientific Reports, 14*(1). https://doi.org/10.1038/s41598-024-53633-9

Moriello, K. A. (2004). Treatment of dermatophytosis in dogs and cats: Review of published studies. *Veterinary Dermatology, 15*(2), 99–107. https://doi.org/10.1111/j.1365-3164.2004.00361.x
Ng, J. Y., Kim, M., & Suri, A. (2022). Exploration of facilitators and barriers to the regulatory frameworks of dietary and herbal supplements: A scoping review. *Journal of Pharmaceutical Policy and Practice, 15*(1). https://doi.org/10.1186/s40545-022-00447-7
Nguyen, N. H., Song, Z., Bates, S. T., Branco, S., Tedersoo, L., Menke, J., Schilling, J. S., & Kennedy, P. G. (2015). FUNGuild: An open annotation tool for parsing fungal community datasets by ecological guild. *Fungal Ecology, 20*, 241–248. https://doi.org/10.1016/j.funeco.2015.06.006
Osemwegie, O. O., Adetunji, C. O., Ayeni, E. A., Adejobi, O. I., Arise, R. O., Nwonuma, C. O., & Oghenekaro, A. O. (2020). Exopolysaccharides from bacteria and fungi: Current status and perspectives in Africa. *Heliyon, 6*(6), e04205. https://doi.org/10.1016/j.heliyon.2020.e04205
Prescott, T. a. K., Hill, R., Mas-Claret, E., Gaya, E., & Burns, E. (2023). Fungal drug discovery for chronic disease: History, new discoveries and new approaches. *Biomolecules, 13*(6), 986. https://doi.org/10.3390/biom13060986
Ross, R. L., & Santiago-Tirado, F. H. (2024). Advanced genetic techniques in fungal pathogen research. *mSphere, 9*(4). https://doi.org/10.1128/msphere.00643-23
Roth, M. G., Westrick, N. M., & Baldwin, T. T. (2023). Fungal biotechnology: From yesterday to tomorrow. *Frontiers in Fungal Biology, 4*. https://doi.org/10.3389/ffunb.2023.1135263
Saha, S., Paul, D., Poudel, T. R., Basunia, N. M., Hasan, T., Hasan, M., Li, B., Reza, R., Haque, A. R., Hanif, M. A., Sarker, M., Roberts, N. J., Khoso, M. A., Wu, H., & Shen, H. (2023). Biofertilizer science and practice for agriculture and forestry: A review. *Journal of Applied Biology & Biotechnology*. https://doi.org/10.7324/jabb.2023.148741
Sal, E., Stemler, J., Salmanton-García, J., Falces-Romero, I., Kredics, L., Meyer, E., Würstl, B., Lass-Flörl, C., Racil, Z., Klimko, N., Cesaro, S., Kindo, A. J., Wisplinghoff, H., Koehler, P., Cornely, O. A., & Seidel, D. (2022). Invasive Trichoderma spp. infections: Clinical presentation and outcome of cases from the literature and the FungiScope® registry. *Journal of Antimicrobial Chemotherapy, 77*(10), 2850–2858. https://doi.org/10.1093/jac/dkac235
Sanchez, S., & Demain, A. L. (2017). Bioactive products from fungi. In *Springer eBooks* (pp. 59–87). https://doi.org/10.1007/978-3-319-51639-4_3
Sar, T., Kiraz, P., Braho, V., Harirchi, S., & Akbas, M. Y. (2023). Novel perspectives on food-based natural antimicrobials: A review of recent findings published since 2020. *Microorganisms, 11*(9), 2234. https://doi.org/10.3390/microorganisms11092234
Segeritz, C., & Vallier, L. (2017). Cell culture. In *Elsevier eBooks* (pp. 151–172). https://doi.org/10.1016/b978-0-12-803077-6.00009-6
Ślusarczyk, J., Adamska, E., & Czerwik-Marcinkowska, J. (2021). Fungi and algae as sources of medicinal and other biologically active compounds: A review. *Nutrients, 13*(9), 3178. https://doi.org/10.3390/nu13093178
Sułkowska-Ziaja, K., Trepa, M., Olechowska-Jarząb, A., Nowak, P., Ziaja, M., Kała, K., & Muszyńska, B. (2023). Natural compounds of fungal origin with antimicrobial activity—Potential cosmetics applications. *Pharmaceuticals (Basel, Switzerland), 16*(9), 1200. https://doi.org/10.3390/ph16091200
Tariq, M. U., Haseeb, M., Aledhari, M., Razzak, R., Parizi, R. M., & Saeed, F. (2020). Methods for proteogenomics data analysis, challenges, and scalability bottlenecks: A survey. *IEEE Access, 9*, 5497–5516. https://doi.org/10.1109/access.2020.3047588
Teotia, P., Kumar, M., Prasad, R., Kumar, V., Tuteja, N., & Varma, A. (2017). Mobilization of micronutrients by mycorrhizal fungi. In *Springer eBooks* (pp. 9–26). https://doi.org/10.1007/978-3-319-53064-2_2
Thirkell, T. J., Charters, M. D., Elliott, A. J., Sait, S. M., & Field, K. J. (2017). Are mycorrhizal fungi our sustainable saviours? Considerations for achieving food security. *Journal of Ecology, 105*(4), 921–929. https://doi.org/10.1111/1365-2745.12788
Tiwari, P., & Park, K. (2024). Advanced fungal biotechnologies in accomplishing sustainable development goals (SDGs): What do we know and what comes next? *Journal of Fungi, 10*(7), 506. https://doi.org/10.3390/jof10070506

Tran, P. N., Yen, M., Chiang, C., Lin, H., & Chen, P. (2019). Detecting and prioritizing biosynthetic gene clusters for bioactive compounds in bacteria and fungi. *Applied Microbiology and Biotechnology, 103*(8), 3277–3287. https://doi.org/10.1007/s00253-019-09708-z

Tyśkiewicz, R., Nowak, A., Ozimek, E., & Jaroszuk-Ściseł, J. (2022). Trichoderma: The current status of its application in agriculture for the biocontrol of fungal phytopathogens and stimulation of plant growth. *International Journal of Molecular Sciences, 23*(4), 2329. https://doi.org/10.3390/ijms23042329

Udomkun, P., Wiredu, A. N., Nagle, M., Müller, J., Vanlauwe, B., & Bandyopadhyay, R. (2017). Innovative technologies to manage aflatoxins in foods and feeds and the profitability of application—A review. *Food Control, 76*, 127–138. https://doi.org/10.1016/j.foodcont.2017.01.008

Umar, A., Mubeen, M., Ali, I., Iftikhar, Y., Sohail, M. A., Sajid, A., Kumar, A., Solanki, M. K., Divvela, P. K., & Zhou, L. (2024). Harnessing fungal bio-electricity: A promising path to a cleaner environment. *Frontiers in Microbiology, 14*. https://doi.org/10.3389/fmicb.2023.1291904

Vassilev, N., & De Oliveira Mendes, G. (2024). Soil fungi in sustainable agriculture. *Microorganisms, 12*(1), 163. https://doi.org/10.3390/microorganisms12010163

Venturella, G., Ferraro, V., Cirlincione, F., & Gargano, M. L. (2021). Medicinal mushrooms: Bioactive compounds, use, and clinical trials. *International Journal of Molecular Sciences, 22*(2), 634. https://doi.org/10.3390/ijms22020634

Větrovský, T., Kohout, P., Kopecký, M., Machac, A., Man, M., Bahnmann, B. D., Brabcová, V., Choi, J., Meszárošová, L., Human, Z. R., Lepinay, C., Lladó, S., López-Mondéjar, R., Martinović, T., Mašínová, T., Morais, D., Navrátilová, D., Odriozola, I., Štursová, M., et al. (2019). A meta-analysis of global fungal distribution reveals climate-driven patterns. *Nature Communications, 10*(1). https://doi.org/10.1038/s41467-019-13164-8

Wang, B., Hu, S., Yu, X., Jin, L., Zhu, Y., & Jin, F. (2020). Studies of cellulose and starch utilization and the regulatory mechanisms of related enzymes in fungi. *Polymers, 12*(3), 530. https://doi.org/10.3390/polym12030530

Wang, W., Tan, J., Nima, L., Sang, Y., Cai, X., & Xue, H. (2022). Polysaccharides from fungi: A review on their extraction, purification, structural features, and biological activities. *Food Chemistry: X, 15*, 100414. https://doi.org/10.1016/j.fochx.2022.100414

Wijayawardene, N., Boonyuen, N., Ranaweera, C., De Zoysa, H., Padmathilake, R., Nifla, F., Dai, D., Liu, Y., Suwannarach, N., Kumla, J., Bamunuarachchige, T., & Chen, H. (2023). OMICS and other advanced technologies in mycological applications. *Journal of Fungi, 9*(6), 688. https://doi.org/10.3390/jof9060688

Wilson, M. C., Chen, X., Corlett, R. T., Didham, R. K., Ding, P., Holt, R. D., Holyoak, M., Hu, G., Hughes, A. C., Jiang, L., Laurance, W. F., Liu, J., Pimm, S. L., Robinson, S. K., Russo, S. E., Si, X., Wilcove, D. S., Wu, J., & Yu, M. (2015). Habitat fragmentation and biodiversity conservation: Key findings and future challenges. *Landscape Ecology, 31*(2), 219–227. https://doi.org/10.1007/s10980-015-0312-3

Wu, B., Hussain, M., Zhang, W., Stadler, M., Liu, X., & Xiang, M. (2019). Current insights into fungal species diversity and perspective on naming the environmental DNA sequences of fungi. *Mycology: an International Journal on Fungal Biology, 10*(3), 127–140. https://doi.org/10.1080/21501203.2019.1614106

Zamith-Miranda, D., Da Silva, R. P., Couvillion, S. P., Bredeweg, E. L., Burnet, M. C., Coelho, C., Camacho, E., Nimrichter, L., Puccia, R., Almeida, I. C., Casadevall, A., Rodrigues, M. L., Alves, L. R., Nosanchuk, J. D., & Nakayasu, E. S. (2021). Omics approaches for understanding biogenesis, composition and functions of fungal extracellular vesicles. *Frontiers in Genetics, 12*. https://doi.org/10.3389/fgene.2021.648524

Aquaculture Production and Management Practices of Farmed Fish in Nigeria

Marian Onwude Agbugui, Abel Inobeme, Elizabeth Princess Okhamafe, Fran Ehizokhale Abhulimen, Grace Abe, John Chiwuzulum Odozie, and John Temitope Ogbiti

Abstract Aquaculture in Nigeria has over time emerged as an essential component in the agricultural sector. This sector contributes significantly to food security, serves as a means of employment, and to the overall economic growth of the nation. This paper provides in-depth information on the breeding techniques, culture measures feed and feeding techniques, health management practices as well as processing methods in the Nigerian aquaculture industry. Diverse methods are used to breed different species of fish—tilapia, clarias, and heterobranchus; diverse culture practices for different fish species including ponds, tanks, and cage cultures for tilapia, catfish, heterobranchus, Heterotis, and knife fish are used. Despite the aforementioned, the sector has faced a few challenges. The cost of feeding and raising table-size fish using imported feeds is not economically satisfying as 70% of the cost of procuring feed goes into feeding and raising the fish to maturity. Ponds with inadequate water supply face serious poisoning; water monitoring and management and diseases to maintain optimal fish health are common challenges. Common dis-

M. O. Agbugui (✉) · E. P. Okhamafe
Department of Biological Sciences, Edo State University Uzairue, Uzairue, Edo State, Nigeria
e-mail: agbugui.marian@edouniversity.edu.ng

A. Inobeme
Department of Chemistry, Edo State University Uzairue, Uzairue, Edo State, Nigeria

F. E. Abhulimen
North Somerset Council, Weston-super-Mare, UK

G. Abe
Department of Aquatic Bio-resources, Bio-resources Development Centre, Kano, Nigeria

J. C. Odozie
Department of Agricultural and Development Economics, Ajayi Crowther University, Oyo Town, Oyo, Nigeria

J. T. Ogbiti
Department of Computer Science, Edo State University Uzairue, Uzairue, Edo State, Nigeria

A. Gupta (ed.), *Prospects of Fungal Biotechnologies for Livestock Volume 1*, Fungal Biology, https://doi.org/10.1007/978-3-031-90994-8_7

eases such as bacterial parasites and fungal infections pose significant threats to fish populations. This sector is also posed with environmental threats and limitations. The study also highlights the need for infusing a combination of various approaches and methods to improve the already existing management practices. Enhanced breeding methods, sustainable local and hybrid feeds, improved disease control mechanisms, environmental sustainability measures, government support and policies, and farmer education are pivotal in addressing these challenges and totally obtaining the full potential of aquaculture in Nigeria.

Keywords Culture · Feeding · Breeding · Diseases · Challenges · Fish species

1 Introduction

Aquaculture is the carefully managed rearing of aquatic species, including fish, crabs, mollusks, algae, and other valuable organisms like aquatic plants. Another name for aquaculture is aquafarming.

Fish, shellfish, and aquatic plant breeding, growing, and harvesting are also referred to as aquaculture. Though comparable methods can be used in non-commercial contexts to replace wild stocks, restore habitats, and rebuild populations of threatened and endangered species, aquaculture primarily provides food and commercial items. Nigerian aquaculture has become an important sector for food production, income generation, and employment. With its vast coastline, abundant water resources, and favorable climatic conditions, Nigeria has the potential to become a major aquaculture producer in Africa. In addition to being important providers of protein and other critical nutrients, fish and seafood also contribute significantly to economic growth and food security. Unfortunately, overfishing, habitat degradation, and falling fish stocks are just a few of the issues facing Nigeria's capture fishery industry, which has historically provided the majority of the country's fish supply. The production of fish through aquaculture has become more popular as a substitute source of fish due to this. The United Nations Food and Agriculture Organization (FAO, 2023a) claims that aquaculture production in Nigeria has been growing at an average annual rate of 10% over the past decade. In 2020, Nigeria produced an estimated 136,000 tons of fish from aquaculture, making it one of the largest producers of farmed fish in Africa (FAO, 2023a). The majority of aquaculture production in Nigeria is based on tilapia and catfish, which are the most commonly cultivated species in the country. Other species farmed in Nigeria include Nile tilapia *Orechromis niloticus, Clarias gariepinus* (African catfish), and *Heterotis niloticus* (Nile perch).

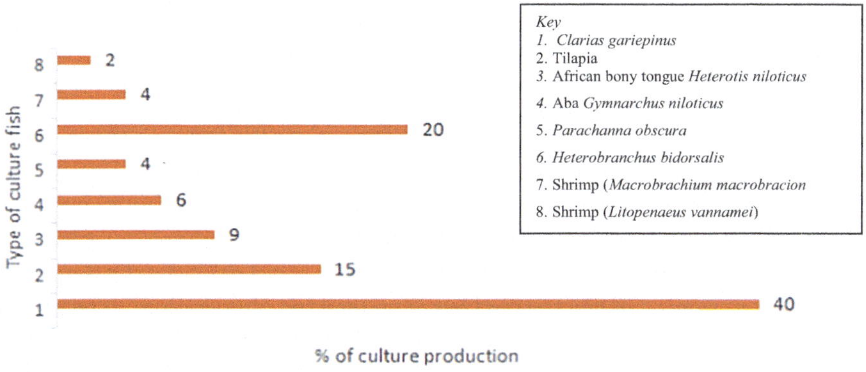

Fig. 1 Types of culture fish in Nigeria and the percentage rate of production

2 Economic Importance of Aquaculture

Aquaculture contributes significantly to Nigeria's economy, providing employment opportunities, generating revenue, and contributing to food security. The sector could lessen the nation's need for imported seafood and fish.

3 Key Fish Species Farmed in Nigeria

Below is a list of fishes that have the potential to be cultured in Nigeria and the percentage rate of culture (Fig. 1).

However, the waters of Nigeria are home to various fish species that grow successfully in the wild.

4 Fish Breeding and Hatchery Management

This sector is important in the technology of catfish breeding. It is the establishment of a system that carries the burden of reproduction. Reproduction is carried out in three different ways, namely: natural spawning, artificial reproduction, and the ecological and physiological method of obtaining offspring.

Breeding techniques for major farmed fish species include:

Selective breeding
Hybridization

In the hatchery operations, the steps involved include:

Egg collection
Fertilization
Larval rearing

5 Types of Aquaculture Systems in Nigeria

Aquaculture systems can be classified based on the kind of operations to be carried out. Culture systems have been classified into four basic categories:

A. Type of culture structure: The container that holds the aquaculture organisms is described by the culture structure. Ponds, tanks, raceways, cages, and pens are typical aquaculture facilities
B. Water exchange: This refers to the control over the flow of water into the system or the quantity of water exchanged. Water exchange levels are often classified as static (open), semi-closed, or recirculating (closed).
C. Intensity of culture: This refers to both the capacity of natural productivity to sustain the crop and the quantity of aquaculture organisms per unit area or water volume. There are three common classifications for cultural intensity: extensive, semi-intense, and intensive.
D. Fish farming techniques: The quantity of fish species raised is indicated by the assortment of fish species raised in one pond.

There is no hard and fast rule over how aquaculture organisms are kept, aquaculture production could be a combination of many structures and criteria. The above-listed methods could be combined for aquaculture production. A pond system could be extensive and static, used to grow major Heterotis in earthen ponds; semi-intensive and semi-closed, used to grow silver perch in Australian ponds; semi-closed and intensive, used to grow shrimp in Asian ponds; or open and intensive, used to grow Atlantic salmon in Canadian sea cages.

Aquaculture systems are categorized based on the following standards:

(a) Different kinds of raising facilities, such as tanks, raceways, ponds, cages, and enclosures
(b) Open or closed water exchange
(c) The level of cultural intensity (high, intense, semi-intensive, and extensive)
(d) Aquaculture techniques (polyculture and monoculture)

5.1 Culture Structure Type

5.1.1 Earthen Ponds

Typically vast in size, earthen ponds are man-made lakes, reservoirs, or dams that are used by fishermen to raise a variety of fish species over a long period. In Nigeria, they are typically located on the periphery of settlements. In regions where the soil has a low capacity to hold water, sandbags are utilized as deck support to strengthen the pond's bottom (Fig. 2). Earthen ponds are incredibly resilient and efficient in these types of environments.

Fig. 2 Type of culture structure

5.1.2 Raceways

Raceways are the forerunners of high-volume production. They usually consist of two to three raceway sets in series built in parallel sets of thin channels in sequential blocks as shown in Fig. 2b.

5.1.3 Concrete or Fiberglass

These are tanks that are either rectangular, circular, or even oval. Water enters the tank through a nozzle that causes a rotational circulation inside, and it exits the tank through a bottom drain or standpipe that is encircled by a screen. Tanks may be constructed above or in the ground. For this type of pond, Due to the center limit's self-cleaning ability, which in practice achieves the water vortex formed by flow rates and may be suitable for the selection of fish for culture, the tank hygiene appears to be better. There is a maximum stocking rate, water changing routine, aeration capacity, and feeding regime as shown in Fig. 2d.

5.1.4 Floating Cages, Pens, and Enclosure

Modern cages with floating structures and a net hanging below are known as floating cages. They could be circular, square, or rectangular. Floating cages can be made of several thousand cubic meters and intended for usage in open water, or they can be small and weak. When fish reach their market size, which is in the months or years leading up to it, they are grown in cages. *Heterotis niloticus* and shrimp have been successfully housed in floating cages (Fig. 2f).

5.2 Water Exchange

This describes the amount of water exchanged during production can be looked at through two main systems:

5.2.1 Static Systems

Numerous nationwide aquaculture production systems make use of the traditional pond culture methods. The ponds are said to be static because no water exchange is carried out during the period of culture, though periodic topping of water can be carried out due to evaporation. Static pond culture usually cares for a large biomass and variety of fish to be cultured. It also takes into consideration the major problems in maintaining water quality and quantity. A great amount of supplementary feeds are employed to maintain productivity.

5.2.2 Open Systems

In an open system, the water body's whole aquatic ecosystem serves as a fish farm. For example, when using cages and pens, the culture organisms are contained or protected inside the farm in a large body of water (such as a lake, river, or ocean) to allow natural flows and processes to maintain water quality. Natural tides and water currents in these cages maintain the water quality, artificial feeding may be provided to increase productivity, and artificial aeration is not necessary.

5.2.3 Closed or Re-circulating System

Closed or re-circulating systems are systems with minimal amount of water during their production circle. Water is not frequently exchanged during the season or time of production. Water can be added periodically to curb the rate of evaporation and maintain water quality. In recirculating systems, connection with the ambient environment and the original water source is minimal. Water quality in completely closed tank systems with intensive culture is much more difficult to maintain than in systems in which there is a regular 5% or more replacement per day. Even with some limited water exchange each day, water quality within a recirculating tank system will only be maintained by artificial manipulation.

5.3 Fish Intensification or Culture Intensity

The different densities of organisms per unit volume or per unit area are referred to as aquaculture intensity. It is the quantity of different fish species that are kept in an enclosure. It contrasts the cultural levels of a given species with those of comparable culturable species. The densities of species from various groups are not compared, nevertheless. It is called intensity because it views density which is the total number of individuals that are as culturally significant. In other words, the more organisms that are cultivated, the more inputs are needed to the framework. Thus, to ensure that the cultured organisms develop to a suitable size, the intensity of culture will take into account all system inputs. For instance, intensive culture is defined as the culture of *Pomadasys jubelini* at 1000 kg/m^3 of water in a recirculating system or the culture of shrimp in ponds at 50 individuals per m^2 (1–2 kg/m^3). Intensive aquaculture systems can be found in raceways, cages, ponds, streams, and seas.

The goal of intensification is to produce the maximum amount of food, labor, and water from a small area in the shortest amount of time. Fish only eat the naturally occurring food that is available, frequently entire protein diets.

6 Fish Farming Methods

There are three major types of fish farming methods, namely:

(i) Monoculture
(ii) Polyculture
(iii) Integrated fish farming

6.1 *Monoculture Fish Farming Method*

This fish farming method involves the culture or production of a particular fish species. It is when only one species of fish or shrimp/prawn is cultured in a water body, e.g., only tilapia, catfish, gymnarchus, and heterotis culture in a pond. This method is highly applicable for commercial purposes under the intensive management system. In this method, feeding fish is mainly on artificially formulated diets and not on the natural food from the water body. Hence, through the application of a well-balanced diet/food, fish with high market prices are cultured. In Nigeria, monoculture fish production is the most versatile and most practiced form of fish production. Most Nigerian fish markets are stocked through a monoculture system.

6.2 *Polyculture*

The polyculture fish farming method is when more than one species of fish is cultured in a water body. This method takes into consideration the optimal use of the natural food items present in the water body as well as artificial feed. Fish of different species that grow in different shapes and sizes depending on different food and feeding habits are stocked in a water body. The advantage of the polyculture system is that a higher yield is obtained from polyculture than from monoculture, e.g., polyculture of Clarias and tilapia, and Heterotis with Gymnarchus, etc. together. This system of culture is effortlessly practiced in many Nigerian farms.

6.3 *Integrated Fish Farming*

A system known as "integrated fish farming" produces or raises fish with other agricultural farming practices. These agricultural activities which include raising fish, crops, and cattle, are typically interconnected such that the wastes and byproducts from one farming activity (subsystem) serve as valuable inputs for the development and upkeep of another subsystem, ensuring full use of the farm's water and land resources and maximizing farm output at the lowest possible cost of labor and

Fig. 3 Integrated fish farming systems in Nigeria

materials. The foundation of integrated fish farming is the idea that "waste" is nothing more than a misplaced resource that may be turned into a profitable component for another product (FAO, 2023a). The fundamental tenets of integrated farming include conservation, which includes making full use of farm wastes, and harnessing the synergistic impacts of connected farm activities (Kadi, 2024). In Nigerian fish farming systems, integrated farming is an ancient tradition. Integrated farming has proven successful for a few farms. These include:

(i) Rice-Fish Farming: the culture of fish with rice farm (Fig. 3)
(ii) Duck-Fish Farming System: the culture of fish with raising dicks (Fig. 3)
(iii) Chicken-Fish Farming System: raising chicken above a fish farm (Fig. 3)
(iv) Pig-Fish Farming: farming of fish with piggery (Fig. 3)
(v) Aquaponics (Fig. 4)

6.4 Aquaponics

Hydroponics, which grows plants in water in a symbiotic environment, and aquaculture, which raises aquatic creatures like fish, crayfish, or prawns in tanks, are combined to create aquaponics. Excretions from the animals being reared in aquaculture might build up in the water and cause increased toxicity. An aquaponic system uses water from an aquaculture system to feed byproducts to a hydroponic system, where nitrifying bacteria convert them into nitrates and nitrites that plants can use as nutrients. The water is then recycled back into the aquaculture system.

Fig. 4 Aquaponics showing lettuce grown above fish cages

The aquaculture component of aquaponics is used to raise aquatic animals, while the hydroponics component is used to cultivate plants. Unused feed or caring for fish-like pets can produce aquatic effluents, which build up in water. Aquatic effluents, resulting from uneaten feed or rearing fish, end up in the water since most aquaculture systems use closed-system recirculation. High amounts of effluent-rich water are hazardous to aquatic life, but it also includes nutrients that are vital to plant growth. Let us talk about the key components of the aquaponics industry.

6.4.1 Water Usage

There is no discharge or exchange of water in the aquaponic system. What is obtainable is the efficient recirculation and reuse of water that depends on the presence and relationship between the animals and the plants to maintain a stable aquatic environment that experiences a minimum of fluctuation in ambient nutrient and oxygen levels. This allows for aquaponic production of both crops and fish in areas where water or fertile land is scarce (FAO, 2020).

To function properly, an aquaponic system needs a variety of live components.

Fish and other water organisms, plants, and bacteria are the three main living components. Additionally, some systems have extra live components like worms.

6.4.2 Plants

A deep water culture hydroponics system in which plants are grown without the use of a soil medium, straight into the wastewater-rich water. Because the roots do not have to spread outward to support the weight of the plant, plants can be planted closer together. Many plants are suitable for aquaponic systems, though which ones work for a specific system depends on the maturity and stocking density of the fish. These factors influence the concentration of nutrients from the fish effluent, and how much of those nutrients are made available to the plant roots via bacteria.

While tomatoes, cucumbers, and peppers have higher nutrient requirements and will only thrive in mature aquaponic systems with high stocking densities of fish, green leafy crops with low to medium nutrient requirements are well adapted to aquaponic systems. Common salad plants, such as cucumbers, shallots, tomatoes, lettuce, chiles, capsicum, red salad onions, and snow peas, are among the most successful aquaponics plants.

6.4.3 Fish

The most popular aquatic animal cultivated in aquaponics is freshwater fish, while freshwater crayfish and prawns are also occasionally utilized. Saltwater aquaponics is a subset of aquaponics that uses saltwater fish. Both warm- and cold-water fish species are well-suited to aquaculture systems. In actuality, perch, catfish, and tilapia are frequently utilized.

6.4.4 Bacteria

One of the most crucial processes in an aquaponics system is nitrification, or the aerobic conversion of ammonia into nitrates. This process lessens the toxicity of the water for fish and enables the plants to take the generated nitrate molecules for nutrition. As a byproduct of their metabolism, fish excrete ammonia, which is gradually released into the water through their gills. However, since fish can die from exposure to elevated ammonia concentrations (usually between 0.5 and 1 ppm), the water must be filtered.

Fish toxicity in the water is effectively reduced by nitrates, which are more readily absorbed by plants even if they can take ammonia from the water to a certain extent. Healthy populations of Nitrosomonas can cooperate to convert ammonia into other nitrogenous substances that convert ammonia to nitrites and Nitrobacter, which turns nitrites into nitrates (Ahmed & Turchini, 2021).

7 Nutrition and Feeding Practices

Fish will eat fish. However, fish will consume natural food from within the environment or palatable compounded feed (supplementary feeds). The natural food could be microscopic plants (phytoplankton), microscopic animals (zooplankton), plant origin (duckweed, water hyacinth, water lettuce, grasses), insects, crustaceans, copepods, and mollusks are examples of natural foods. For fish to grow well, fast, and healthy, a balanced diet has to be consumed. In the absence of sufficient quantity and quality of natural foods, artificial diets called feed must be provided in the right proportion to meet the nutritional requirement of any fish in question.

Supplementary feeds for chickens and pigs are sometimes suitable for feeding fish. Other forms include kitchen waste, offals, oil seed cakes, cassava leaves, and rice bran.

Artificial feeds are a nutritionally rich mixture of well-compounded ingredients in mash or pellet form that could be fed to fish. Mash feeds are good for fries and pellets (0.8 m to 1 mm) for fingerlings, juveniles (2 to 3 mm), and adults (4.5 mm) depending on pellet sizes. The cost of purchasing the raw materials and formulating these artificial feed is quite expensive thus alternate cheaper sources are often sorted from the wild.

The following benefits of artificial feeds, particularly the commercially compounded ration, outweigh their potential cost in fish culture:

1. Allow for high stocking densities, or stocking maximization, particularly in systems with polycultures.
2. Because food is always available, they encourage fish to grow more quickly.
3. A high yield of fish in relation to stocking density is assured.
4. A fish farmer may observe the health and behavior of the fish while they are being fed.

Generally, in fish culture, if other factors regarding to growth (water temperature, water quality) of fish are met, then the quality of feeds, and the feeding method should produce to a large extent nutritional sufficient fish. The nutritional requirement of fish varies within fish species. Some fish as said earlier are herbivores, carnivores, and some omnivores. The carnivorous fish will feed on mainly protein, making their protein requirement very high. Depending on the type of fish and its age limit, different dietary requirements for crude protein (CP) are needed for quick and healthy fish growth.

The general needs are as follows:

(a) semi-intensive culture using natural feeds (20–25% CP)
(b) intense culture for fingerlings (40–48% CP)
(c) intensive culture for juveniles and adult fish (30–35% CP)

7.1 *Feedstuffs and Feeding in Fish Culture*

In fish culture, feedstuffs are divided into two main categories: protein supplements and energy feedstuffs.

1. Energy feedstuffs: These are feedstuffs with less than 20% crude protein and are primarily derived from plants. They have a high carbohydrate content. Cassava, wheat offal, rice bran, corn, guinea corn, etc. are a few examples.
2. Protein supplements: Feedstuffs with at least 20% crude protein are classified as such. Either plant or animal resources are used to make them. Animal-derived

proteins are of a higher caliber than plant-derived proteins. Fish meals, fish meal, bone meal, blood meal, and crayfish are a few types of animal protein sources used in fish culture. Plant protein resources include things like soybean meal, cottonseed cake, and Bambara nut cake, a type of groundnut cake. The main obstacle to using legumes as animal feed, particularly when they are fresh, is the existence of harmful and anti-nutritional ingredients (Agbugui et al., 2020). If the amount of hazardous substances in some wild plants from Nigeria is decreased or eliminated, their proximate composition indicates that they could be used successfully in the production of animal feed (Agbugui et al., 2020).

7.2 *Choice of Feedstuffs*

The type of feed ingredient fed to fish should be a supplement to that in the natural waters. The feed can be fed as a single ingredient or combined ingredients. A combination of ingredients will make better-quality supplemental feed. The combination could be either plant and animal sources, two plant sources, or two animal sources. Many times, a combination of one plant and one animal source provides a good nutritional supplement that also reduces the cost of production for the fish farmer. Fish would grow well on a feed containing 20–30% crude protein, of which 7–10% of the protein is from animal sources (Agbugui et al., 2020). The Pearson's Square Method is an easy and reliable method of formulation feed for fish.

The recent improvement and need for balanced diets in fish culture is evident in the size of fish in today's Nigerian market. The market globally is in quest for healthy nutritious fish. The invention and development of locally-made affordable diets have supported the industry to expand to its current level. Adequate nutrition has long been recognized as a critical factor in promoting normal growth, thus generating good body composition/flesh quality, and sustaining the health of fish (Abe & Agbugui, 2023), hence proper feeding and nutrition aim to improve the following areas of fish culture:

(i) Immunity and disease resistance of cultured organisms to reduce disease-related economic losses
(ii) Provide essential nutrients to cultured fish species
(iii) Income and economic value. Good and healthy fish commands high market value, the ultimate purpose of fish husbandry
(iv) Optimal growth and development of fish within a limited time

The farming and culture of fish are key in the agricultural industry with economic prospects to reduce poverty, reduce nutritional deficiencies as well as improve the quality of feeding. Fish culture offers the Nigerian populace a major source of cheap, quality, rich, and nutritionally satisfying source of protein in diets. Nearly 40% of the consumption of fish as protein hovers around 11.2 and 14.3 kg/person / year (Ogunji & Wuertz, 2023). Yet the Nigerian per-capita protein intake is still lower than the Sub-Saharan average. The Nigerian fish farm and market is one of the fastest-growing farming businesses, increasing its contribution from the local to the global market from 0.07% in 1995 to 0.21% in 2022 thereby turning Nigeria into the second largest producer in Africa. Nigeria is home to various native and some imported fish species. The native species include catfish, tilapia, *Heterotis sp, Oreochromis niloticus, O. mossambicus, O. aureus, Sarothrodon galilaeus* and *S. melanotheron,* and *Coptodon rendilli* which are mainly cultured. The catfish species include the African catfish *Clarias gariepinus*, and the sharp-tooth catfish *Heterobranchus longifillis*. Heteroclarias is a hybrid of the two catfish species. The species contributing to over 90% of the fish production in Nigeria include *Clarias gariepinus*, Torpedo-shaped catfishes (*Heterobranchus longifilis*, hybrid Heteroclarias), carp fishes, tilapia (mainly *Oreochromis niloticus*), Aba *Gymnarchus niloticus*, and Nile perch (Ogunji & Wuertz, 2023). In general, the wide acceptability, taste, nutritional value, high price, fast growth, culture capacity, and robustness of the African catfish and tilapia are the main reasons for their predominant position in Nigerian aquaculture. The African catfish is one of the most commercially important freshwater fish species in Africa, In Nigeria, it is the most popular fish on the local market. The successful rate of production of the African catfish has placed Nigeria as the largest producer of catfish in Africa and in the world. Aquaculture production in Nigeria has risen from about 2000 tons in 1960 to more than 300,000 tons by 2015 to 275,645 tons in 2023 (Oboh, 2022; Ajayi et al., 2022). Other widely and successful culturable fish species in the Nigerian farming industry are the African bony tongue *Heterotis niloticus* and the obscure snakehead *Parachanna obscura*. The Heterotis is herbivorous while the snake head is omnivorous. Indeed, both species compete favorably with other carnivorous fish species, both species are highly valued, the growth rate is 3–4 kg in 6–9 months and the current supply does not meet the market demand thus, this diversification in the aquaculture sector is well encouraged.

8 Government and Private Sector Support for Aquaculture

To achieve increased domestic fish production from all sources on a sustainable and renewable basis to the level of self-sufficiency and fish export in the medium to long term, the Nigerian government developed a National Aquaculture Strategy, a nationwide sectorial document (FAO, 2024). A few cross-sectoral elements pertaining to fish farms and the handling and transportation of live fish are included in the strategy. Eight primary strategic objectives are listed to meet the overall goal:

(i) To greatly increase the output of fisheries and aquaculture by means of private sector investment, working with the public sector and all fisheries sub-sector operators.
(ii) Enhance the socioeconomic conditions of fishing communities by making fishing inputs, equipment, facilities, and credit more accessible.

(iii) Adopt a purposeful policy of reducing imports of fish and fisheries products by 25% annually over a five-year period.
(iv) Create and implement a national network for the diagnosis, control, and prevention of fish diseases.
(v) Encourage and strengthen organizations related to fishing for the best possible contribution to fisheries research and development.
(vi) Establish an efficient national safety and quality assurance system that can safeguard consumer health and increase foreign exchange earnings through the export of fish and fishery products.
(vii) Place an emphasis on value addition in fish processing.
(viii) Develop effective local and international marketing of fish and fisheries.

Source: (FAO, 2023b)

To boost employment and increase fish production, the Nigerian government has also revealed plans to build six additional "fish farming estates" in the vicinity of Lagos (Okai, 2020). Additionally, two estates have already been built in the Bwari and Kwali area councils, according to the Federal Capital Territory's Director of Fisheries (Okai, 2020). For the numerous federal employees who work in the nation's capital, these estates offer a beneficial investment opportunity as well as a practical way to help supply the enormous local demand for fish. State governments in several regions of Nigeria have previously built similar estates with the guarantee of federal government backing to aid in the production of fish, a significant element of the Nigerian diet, considering the population's rapid growth. The creation of jobs was another important factor. They were founded in regions with abundant, naturally occurring water supplies, and the local governments pledged to provide infrastructural and financial support to the farmers (Okai, 2020). A public-private partnership agreement in Lagos State for the construction of a fish processing factory to add value to the produce of the state's fish farmers was another encouraging development. Initially, the processing factory will concentrate on processing tilapia and catfish, but eventually, it will be able to handle high-value species such as pangasius, carp, barramundi, and other similar species. Once the processing plant is operational, the middlemen's monopoly on setting pricing will be eliminated. Off-taker agreements with the processing plant will provide farmers access to markets and fair pricing (Okai, 2020).

In Nigeria, the practice of aquaculture by rural people has evolved from traditional fish culture methods in tidal pools and flood plains to diverse production systems. This paper aims to assess trends in aquaculture production and marketed output over time using available data for Nigeria. Of the total global fish production in 2018, aquaculture production contributed 46% (FAO, 2020). African countries contributed 2.7% of this amount (Adeleke et al., 2020). Aquaculture production in Africa has increased over time by 20-fold between 1995 and 2018 (FAO, 2020). Nigeria together with nine other countries including Egypt and Ghana contributes 93% of the total regional contribution. Similarly, aquaculture production in Nigeria experienced growth of 81.71% between 1995 and 2011 as demonstrated in Fig. 5.

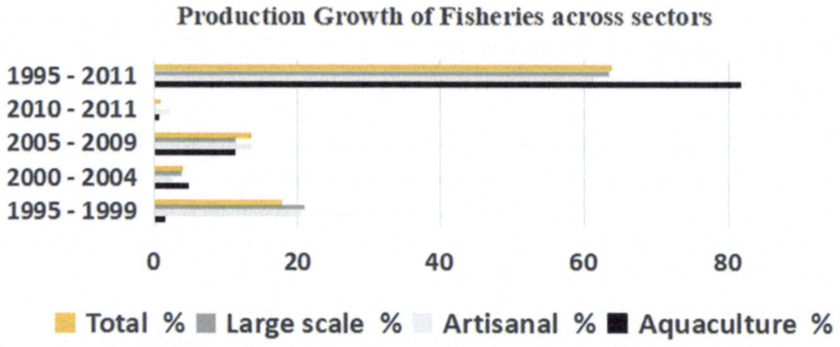

Fig. 5 Output production across time and sectors. (Estimates from NAERLS (2021))

9 Major Challenges Faced by Fish Farmers

9.1 Water Quality Issues

Water is the primary resource used in aquaculture; hence frequent water quality testing is essential. As a result, water quality and quantity are essential for the successful production of fish, and the water used for fish culture should be free of chemicals that are harmful to fish and within an acceptable pH range of 6.5–9.0. Fish live in water, and the quality of that water affects both the fish's suitability for use and its ability to support the health of aquatic farm animals (Akinnao, 2023). Aquaculture is primarily burdened by environmental pollution because low water quality in fish ponds leads to declining fish health and ultimately reduced output rates (Ahmed & Turchini, 2021). The use of *"Recirculating aquaculture systems (RAS)":* Maintaining a healthy aquatic environment requires paying attention to the state of several water parameters, including turbidity, pH, alkalinity, hardness, ammonia, nitrite, nitrate, and Biochemical Oxygen Demand (BOD). Given its impact on fish output, water quality is one of the most neglected components of pond management (Ali & Mishra, 2022). Culturists can achieve maximum fish output in their ponds to a larger extent if they are adequately educated and informed about water quality management procedures.

9.2 Equity and Access

Access to aquatic resources and equity are two of the main continuous issues facing an aquaculture business. It has been difficult to provide fair access, thus markets, funding, and resources for small-scale aquaculture producers have still not received the attention they require (FAO, 2023c). In contrast to catch fisheries, aquaculture necessitates intentional human interference in the production of the organisms, leading to yields that surpass those obtained alone from the natural environment. According to FAO (2023c), some examples of these interventions include stocking water with seeds (fingerlings), fertilizing the water, feeding the organisms, and preserving the water's quality.

9.3 Environmental Impacts

Mitigated communities may be disproportionately impacted by aquaculture's negative environmental effects, which include habitat destruction, water pollution, and disease transmission. Aquaculture typically uses materials that have the potential to pollute, such as nitrogen and phosphorus, which are utilized in diets and the chemical upkeep of aquaculture ponds (Jeanson et al., 2022). When these substances are discharged into the environment, they can cause algal blooms because they both overstimulate the growth of algae and lower oxygen concentrations below what fish can tolerate. In conclusion, aquaculture has a variety of detrimental effects on the ecosystem, all of which can be summed up by the following: Depletion of wild stock, introduction of inactive species, habitat loss, nutrient pollution, and chemical pollution (Jeanson et al., 2022).

9.4 Lack of Aquaculture Policies and Framework

It needs effective laws, rules, and governance frameworks to strike a balance between aquaculture's financial advantages and social and environmental factors. Policies and regulations have not been adequately enforced in developing nations such as Nigeria. Although Nigeria lacks national legislation specifically on aquaculture (Adeleke et al., 2020), the Minister in charge of fisheries matters is empowered to oversee the licensing of enclosures, such as pens and cages, under the 1992 Inland Fisheries Decree (FAO, 2022). According to the current policy, the private sector drives aquaculture's output at every stage of the value chain, while the government fosters favorable conditions (Adeleke et al., 2020). Because of this problem, a number of fish farmers have resorted to shrewd tactics to increase their yield and profit margin which is detrimental over time to fish, end users, and aquatic microorganisms in particular (Abe & Agbugui, 2023). If policies were appropriately placed to increase fish output, then policymakers should have the ability to execute them effortlessly (Abe & Agbugui, 2023).

9.5 Lack of Agricultural Insurance

One of the noteworthy ways that farmers can share or transfer the risks and uncertainties related to their farming enterprises is through agricultural insurance policies. These policies encourage farmers to increase their production investments, boost their confidence in attempting new and improved farming practices, make it easier for them to obtain credit from financial institutions by providing additional collateral, and ultimately give them financial support in the form of indemnity that guarantees the continuation of their farming enterprise (Avadí et al., 2022). The purpose of agriculture insurance is to cover monetary losses brought on by a decline in anticipated agricultural product output. Crops, cattle, fisheries, and forestry are all covered by agricultural insurance policies (PUNCH, 2022). The term "aquaculture insurance" describes all the various types of insurance that would normally be used to protect an aquaculture business operation. For a reasonably large aquaculture company, this would include insurance protection for buildings and equipment, employees, stock, livestock, liabilities, motor vehicles, vessels and divers, odds in transit, and other insurable interests (FAO, 2023c).

9.6 Lack of Improved Fish Seeds

There is a pressing need to address the challenges of inadequate production of high-quality brood stock and seed encountered by the aquaculture industry in Nigeria where the demand-supply gap for fish seed is considerably huge (Palma & Viegas, 2022) as Nigeria currently requires at least 1 billion fingerlings/seeds annually to meet market demand for table-size fish while it barely produces about 55 million fingerlings from all available sources (Palma & Viegas, 2022). Many technical problems arise in the production of fish seed either in the pond or hatchery system. Principal among these are the lack of and poor management of broodstock, especially feeding and handling; and the poor record keeping of all activities regarding induced spawning, care of eggs, fry, feeding, and general management of fingerlings. Many erstwhile enthusiastic farmers have suffered colossal failures due to poor management skills. Aquaculture involves the risk of fish seed loss due to wide pH fluctuations, oxygen depletion, parasites, water pollution, disease, predators, flooding, and vandalism (Adelesi & Baruwa, 2022).

9.7 Market Availability

Marketing is closely related to all endeavors. Fish marketing in Nigeria is hinged on some basic questions: What do consumers want? Which species? What price, size, form, quality, quantity, and grading? Fish marketing is a complex venture to an extent. From harvesting to transportation and processing over a wide geographical range by a large number of traders, the sector needs the support of all. The major constraints facing fish marketing in Nigeria are pronounced in the various pre-harvesting methods of fishery products in Nigeria. According to Avadí et al. (2022), there is a lack of cooperation for the enhancement of a reliable marketing system for the country. Also, there is a lack of development of large-scale post-harvest technologies and the improvement of transportation systems that can support the delivery of fish products to the market.

9.8 Lack of Modern Infrastructure

Several authors have attested to the fact that there is a lack of infrastructural facilities in the coastal waters of Nigeria where artisanal fisheries are mainly practiced (Kadi, 2024). The lack of capital for investment in modern infrastructure does not encourage fish farming, thereby forcing a huge number of fish farmers to abandon the enterprise in search of other means of livelihood. Some of the numerous consequences of this include rural-urban migration, an increase in crime rate poor income from fishing and poor investment capacity (Kadi, 2024).

The inadequacy of power infrastructure, credit facilities, water and sanitation systems, and post-technological infrastructure such as cold rooms, and smoking kilns in the coastal communities further compound the heavy spoilage, thus making fresh fish unaffordable to an average Nigerian. The creeks are generally difficult terrain, thus making it inaccessible for extension workers and fishers to operate because of inadequate road infrastructure. This accounted for the high post-harvest losses (Oboh, 2022).

9.9 Insufficient Government Funding for the Sector

The government of Nigeria pointed out that the fish production deficit stands at about 2.5 million tons, hence importation of the commodity continues, which has drained foreign exchange. This is majorly due to the lack or inadequate government funding of the sector. In this regard, the total demand for fish is 3.6 million tons annually while Nigeria is producing 1.1 million tons from all sources (aquaculture, artisanal, and industrial sectors) leaving a deficit of about 2.5 million tons to be supplemented by importation (NT, 2021). Restricted access to credit is a common challenge in fisheries especially in artisanal fisheries, hence the need for preferential allocation of loan-able funds by the government.

9.10 Lack of Affordable Nutritionally Balanced Fish Feed

Affordable, nutritionally balanced fish feed has been hard to come by (Olapade & Saboleh, 2022), particularly when it comes to the production of fingerlings for Nile tilapia (*Oreochromis niloticus*) and African catfish (*Clarias gariepinus*). Due to their excellent culture prospects, both fish species are significant in global aquaculture, particularly in Africa (FAO, 2024). Regretfully, the fact that fish feed is being imported at a price that ordinary Nigerian farmers cannot afford is disheartening. Fish producers in the nation are compelled to continue feeding their fish with crude feed substitutes like termites, rice bran, and domestic garbage due to the prohibitive cost of imported feed. The industry performs poorly because of this discordant feeding technique. The absence of more affordable and high-quality feeds, as reported by Palma and Viegas (2022), is a major constraint to the development of aquaculture in developing countries. The economic feasibility of aquaculture development in Nigeria has been severely hampered by the exorbitant and skyrocketing cost of fish feed and fish feed ingredients. The production of high-quality aquaculture feed has also been a major constraint hindering the growth of the aquaculture industry in Nigeria. Due to currency devaluation, unstable exchange rates, and inflation, feed costs currently make up about 80% of production costs. This means that prospective investors must carefully plan if they hope to succeed in the business. Due to low local production and competitive ingredient use with other animal feeds, Nigeria's aquaculture business is mostly dependent on imports of both manufactured feeds and feed ingredients. The biggest obstacle facing Nigeria's aqua-feed production sector is a shortage of feed ingredients, particularly since nutrition is crucial in fish farming (aquaculture), where feed accounts for almost half of the variable production costs. The price of raw materials for aqua-feeds is thought to have suffered significantly because of the naira's indirect depreciation. Moreover, there is a limited amount of corn and soybeans, and both humans and animals dispute them. Subpar output is the reason feed ingredients are unavailable.

9.11 Poor Extension Services and Low Level of Technical Know-How

Dan-Awoh (2023) states that a significant problem in fish culture is the low level of acceptance of fishing technology because of a lack of extension services, which is one of the causes that contribute to low productivity in the industry. Because most farmers in Nigeria have low levels of education, there is poor contact between the extension workers and the farmers. The majority of farmers' educational standing needs to be improved in order to increase the yield of fish produced through aquaculture. The government should regularly require farmers to attend training programs in order to keep them up to date on practical skills that will be very beneficial. By taking these actions, Nigerian aquaculture would be able to flourish in a sustainable and profitable manner by increasing fish production. One of the main issues facing Nigeria's aquaculture business is a shortage of skilled workers in the fields of machine operators, feed formulation and processing, and aquaculture nutrition (Dan-Awoh, 2023).

10 Future Prospects and Opportunities for Growth in the Aquaculture Sector in Nigeria

Nigeria has a vast potential for aquaculture due to its abundant natural resources (aquatic and human), therefore, aquatic animal production is forecasted to grow another 14% by 2030. It is vital this growth is sustainable and goes hand in hand with safeguarding ecosystems, reducing pollution, protecting biodiversity, and ensuring social equity (Omeje et al., 2021). As Nigeria's population increases from 196 million in 2018 to 263 million in 2030, Nigeria would need 752,000 tons more fish to maintain 11.2 kg of per capita fish consumption and so would need 3.14 million tons more fish to reach 20.3 kg global per capita fish consumption by 2030 (Omeje et al., 2021). Furthermore, a more recent estimate states that the population of Nigeria is expected to rise from 206 million in 2020 to 401 million in 2050 (FAO, 2022). The Federal Department of Fisheries (FDF) (Ogunji & Wuertz, 2023) states that the Nigerian coastline has enormous potential for inland aquaculture and offshore cage culture, observed also that the maximum sustainable yield (MSY) of Nigerian marine waters, noting that Nigeria has about 853 km of coastline and maritime water of 210,900 km^2, including the exclusive economic zone (EEZ). Jeanson et al. (2022) and FAO (2024) noted that the figures for coastline and marine water are about 900 km, 217,313 km^2, 960 km, and 309,120 km^2, respectively. These findings show that there are significant commercial prospects for sustainable fish farming in Nigeria (Okoye et al., 2022). Recently, in an interview with the News Agency of Nigeria (NAN), the Director of the West African Region of the World Aquacultural Society (WAS) unequivocally stated that the untapped potential of Nigeria's blue economy can be worth 286 billion dollars which will significantly boost the overall GDP of the country. With the brand-new Ministry of the Blue Economy, Nigeria can be on the brink of unlocking new opportunities in its aquacultural sector. With the right goals, objectives, and policy plans of the new ministry, it has the opportunity to ensure food security, job creation, growth, and poverty eradication (Dan-Awoh, 2023).

11 Conclusion and Prospects

Considering the wide gap between the national annual fish demand and supply, the aquaculture industry is a viable ground for investment. According to NIMASA 2024, opportunities exist in nearly all aspects of the fishing industry and some of the areas for possible investment include:

Production of eco-friendly fishing equipment
Establishment of modern fish farms
Shrimp fishing
Establishment of an ice-melting plant
Fish processing for export
Fish fingerling production
Table-size fish production
Brood stock production
Fish seed production
Ornamental fish farming
Production system design and construction
Processing and preservation
Capacity building in project management
Fabrication of fish handling equipment
Packaging and branding

Acknowledgments The authors of this manuscript extend their sincere gratitude to all the lab technologists of the Edo State University Uzairue for the support given to the successful completion of this article.

We also appreciate the support provided by the management of the Edo State University Uzairue whose resources and expertise greatly facilitated this work. Additionally, we are grateful to our colleagues for their constructive criticism and collaboration.

References

Abe, G. O., & Agbugui, M. O. (2023). Reoccurring constraints in fish culture and fish nutrition in Nigeria. *Journal of Fisheries and Livestock Production, 11*, 439. https://doi.org/10.4172/2332-2608.1000439

Adeleke, B., Robertson-Andersson, D., Moodley, G., & Taylor, D. (2020). Aquaculture in Africa: A comparative review of Egypt, Nigeria, and Uganda Vis-À-Vis South Africa. *Reviews in Fisheries Science & Aquaculture, 29*(2), 167–197. https://doi.org/10.1080/23308249.2020.1795615

Adelesi, O. O., & Baruwa, O. I. (2022). Profitability analysis of smallholder aquaculture farms: The case of Lagos State, Nigeria. *Journal of Agriculture and Rural Development in the Tropics and Subtropics, 123*, 109–120.

Agbugui, M. O., Emawunegbe, L. I., & Yaro, A. C. (2020). Indigenous fish feed resources in Nigeria. *Journal of Bio Innovation, 9*(6), 1339–1353.

Ahmed, N., & Turchini, G. M. (2021). Recirculating aquaculture systems (RAS): Environmental solution and climate change adaptation. *Journal of Cleaner Production, 297*, 126604. Available at: https://doi.org/10.1016/j.jclepro.2021.126604.

Ajayi, O., Akinrinlola, A., Usman, A., Muhammed, A., & Van der Knaap, M. (2022). Aquaculture development in Nigeria and FAO's role. In FAO aquaculture news (Ed.), *International year of artisanal fisheries and aquaculture* (pp. 21–24). FAO. Special Issue.

Akinnao, S. O. (2023). Eutrophication: Causes, consequences, physical, chemical and biological techniques for mitigation strategies. *Environmental Challenges, 12*, 100733. Available at: https://doi.org/10.1016/j.envc.2023.100733.

Ali, B. A., & Mishra, A. (2022). Effects of dissolved oxygen concentration on freshwater fish: A review. *International Journal of Fisheries and Aquatic Studies, 10*(4), 113–127. Available at: https://doi.org/10.22271/fish.2022.v10.i4b.2693.

Avadí, A., Cole, S. M., Kruijssen, F., Dabat, M. H., & Mungule, C. M. (2022). How to enhance the sustainability and inclusiveness of smallholder aquaculture production systems in Zambia? *Aquaculture, 547*, 737494. https://doi.org/10.1016/j.aquaculture.2021.737494

Dan-Awoh, D. (2023). *Aquaculture in Nigeria holds 296 billion dollars annual revenue potential.* World Aquaculture Society.

FAO. (2020). *The state of world fisheries and aquaculture-sustainability in action.* FAO.

FAO. (2022). *Aquaculture growth potential in Nigeria: WAPI factsheet.* FAO.

FAO. (2023a). *Nigeria GLOBEFISH Market Profile.* https://www.fao.org/3/cc5640en/cc5640en.pdf. Accessed 12 Oct 2023.

FAO. (2023b). The role of aquaculture in improving food security and nutrition. *Committee on World Food Security.* CFS: 2003/6.

FAO. (2023c). *Unlocking the potential of sustainable fisheries and aquaculture in Africa, the Caribbean and the Pacific—Nigeria; FISH4ACP factsheet.* FAO. Available online: https://www.fao.org/in-action/fish-4-acp/where-we-work/africa/nigeria/

FAO. (2024). *(ND) Fisheries and aquaculture.* Accessed 22 June 2024 at https://www.fao.org/fishery/en/topic/16064

Jeanson, A. L., Gotzek, D., Mam, K., & Torres, Y. (2022). Inland fisheries management-Case studies of inland fish. In *Encyclopedia of inland waters.* https://doi.org/10.1016/B978-0-12-819166-8.00170-5

Kadi, S. (2024). *Integrated fish farming: Principles, advantages and disadvantages.* Notes on Zoology.

Nigerian Tribune. (2021). Nigeria imports 2.2m tonnes of fish annually. *Nigerian Tribune.* Available online: https://tribuneonlineng.com/nigeria-imports-2-2m-tons-of-fish-annually/.

Oboh, A. (2022). Diversification of farmed fish species: A means to increase aquaculture production in Nigeria. *Reviews in Aquaculture, 14*, 2089–2098.

Ogunji, J., & Wuertz, S. (2023). Aquaculture development in Nigeria: The second biggest aquaculture producer in Africa. *Water, 15*, 4224. https://doi.org/10.3390/w15244224

Okai, E. K. (2020). Nigeria launches new aquaculture initiatives. *West Africa Aquaculture Correspondent.*

Okoye, L. I., Kalu, C. U., Uzonwanne, M. C., & Ezenekwe, R. U. (2022). Small-scale fish farming and poverty reduction in some selected states in Southeast, Nigeria. *Socialscientia Journal of the Social Sciences and Humanities, 7*(4), 84–100.

Olapade, O. J., & Saboleh, P. (2022). Development of fish feeds for African catfish (*Clarias gariepinus* Burchell 1822) farming in Sierra Leone, West Africa. *International Journal of Fisheries and Aquaculture, 14*(2), 15–21. https://doi.org/10.5897/IJFA2022.0837

Omeje, J. E., Achike, A. I., Arene, C. J., Faleke, S. A., Manuwuike, Q. C., & Usman, G. A. (2021). Socioeconomic determinants of net income in fish farming in Kainji lake basin, Nigeria. *Global Journal of Agricultural Sciences, 20*(2021), 53–61. https://doi.org/10.4314/gjass.v20i1.8

Palma, M., & Viegas, I. (2022). Aquaculture: Farming our food in water. In *Encyclopedia of the UN sustainable development goals (ENUNSDG)* (pp. 44–52).

PUNCH. (2022). *2.4 million metric tonnes fish imports depleting Nigeria's forex—FG.* https://punchng.com/2-4-million-metric-tonnes-fish-imports-depleting-nigerias-forex-fg/.

Nematophagous Fungi: Biocontrol of Parasites and Biotech Advances

Madhuri Kaushish Lily, Koushalya Dangwal, Kamlesh Kumar Bhatt, and Suneel Kumar Singh

Abstract Nematophagous fungi present an eco-friendly solution to managing parasitic nematodes in ruminants, addressing critical challenges in livestock health and productivity. These fungi, which naturally prey on nematodes, offer an alternative to chemical anthelmintics, reducing the risks associated with resistance caused by their overuse. This chapter examines the potential of nematophagous fungi, focusing on their mechanisms of action, including trapping, parasitizing, and producing nematotoxic compounds. Key species such as *Arthrobotrys oligospora*, *Duddingtonia flagrans*, and *Pochonia chlamydosporia* have demonstrated significant efficacy in reducing nematode populations both in vitro and in vivo.

Field trials incorporating fungal spores into animal feed show promising outcomes, with reduced nematode burdens and improved livestock health. Integrating these fungi into grazing systems supports sustainable livestock management practices, aligning with global efforts to minimize chemical inputs and enhance animal welfare.

Beyond parasite control, nematophagous fungi exhibit immense biotechnological potential. Their enzymatic capabilities, particularly in degrading nematode cuticles, can be utilized to develop novel biopesticides, enhance soil health, and promote sustainable agriculture. Advances in fungal genomics and biotechnology open avenues for applications as biocontrol agents and biofertilizers, further broadening their scope in various industries.

This chapter highlights the multifaceted benefits of nematophagous fungi, emphasizing their role in sustainable agriculture and biotechnology. Continued research and innovation are essential to fully harness their potential for addressing agricultural challenges and advancing biotechnological applications.

Keywords Biocontrol · Biotechnology · Integrated pest management (IPM) · Nematophagous fungi · Ruminant parasitic nematodes (RPNs)

M. K. Lily (✉) · K. Dangwal · K. K. Bhatt · S. K. Singh
Department of Science, Modern Institute of Technology, Rishikesh, Uttarakhand, India

A. Gupta (ed.), *Prospects of Fungal Biotechnologies for Livestock Volume 1*, Fungal Biology, https://doi.org/10.1007/978-3-031-90994-8_8

1 Introduction

Ruminant parasitic nematodes (RPNs) represent a significant challenge to global livestock health and productivity, primarily affecting cattle, sheep, and goats. These parasites cause gastrointestinal disorders, reduced feed efficiency, stunted growth, and decreased reproductive performance. These impacts translate into substantial economic losses in the livestock industry, estimated to amount to billions of dollars annually worldwide (Barone et al., 2020; Mendoza-de Gives et al., 2022; Sharma et al., 2023). The burden of these nematodes is particularly severe in regions where livestock production plays a vital role in food security and rural livelihoods.

Traditionally, the control of RPNs has relied heavily on chemical anthelmintics. These drugs have been widely adopted due to their efficacy, ease of application, and immediate impact on parasite burdens. However, the extensive and often indiscriminate use of anthelmintics has contributed to the rapid emergence of anthelmintic resistance among nematode populations. This phenomenon has been reported globally, posing a critical threat to the sustainability of chemical control measures (Berhanu et al., 2024; Sander & Neumann, 2024). In some regions, resistance has reached alarming levels, rendering commonly used drugs ineffective. This resistance crisis has necessitated a paradigm shift toward alternative and more sustainable parasite control strategies.

To address these challenges, integrated parasite management (IPM) has gained attention as a holistic approach. IPM combines practices such as rotational grazing, selective breeding for parasite-resistant livestock, and biological control methods. Among these, biological control using nematophagous fungi has emerged as a promising strategy, offering both efficacy and environmental sustainability (Berhanu et al., 2024; Filippe et al., 2024). Nematophagous fungi, including *Arthrobotrys oligospora*, *Duddingtonia flagrans*, and *Pochonia chlamydosporia*, have demonstrated their ability to reduce nematode populations through diverse mechanisms. These fungi can trap nematodes using specialized hyphal structures, parasitize eggs and larvae, and produce nematotoxic compounds that impair nematode viability (Céspedes-Gutiérrez et al., 2021; Mendoza-de Gives et al., 2022; Fonseca et al., 2022; Pérez-Anzúrez et al., 2022; Tian, 2023; Hao et al., 2024; Khan & Mohiddin, 2023; Rahman et al., 2023; Rodrigues et al., 2024).

Integrating nematophagous fungi into livestock management systems presents a sustainable and ecologically sound alternative to conventional chemical treatments. These fungi not only reduce the risk of resistance development, but also align with increasing consumer and regulatory demands for environmentally friendly agricultural practices (Li et al., 2022). Furthermore, their potential extends beyond parasite control, offering opportunities for biotechnological innovation in areas such as soil health, nutrient recycling, and environmental remediation.

This chapter explores the biology, ecological roles, and mechanisms of action of nematophagous fungi. It further examines their practical applications in managing RPNs, evaluates their efficacy in field conditions, and discusses their expansive potential as biotechnological tools in sustainable livestock production systems. By understanding the multifaceted capabilities of these fungi, this work aims to contribute to the development of integrated and innovative solutions to combat RPNs and enhance livestock productivity globally.

2 Ruminant Parasitic Nematodes (RPN)

Ruminant parasitic nematodes (RPNs) are a critical concern in livestock management, posing a direct threat to the health, productivity, and overall welfare of ruminants, including cattle, sheep, and goats. These parasites primarily inhabit the gastrointestinal tract, where they disrupt normal physiological functions and compromise the nutritional status of their hosts. The effects of RPN infections manifest in various clinical symptoms, ranging from reduced appetite and feed intake to impaired digestion and nutrient absorption. These disruptions result in significant economic losses, as infected animals experience diminished growth rates, lower weight gain, reduced milk yield, and poor wool quality, ultimately affecting their market value and long-term productivity (Sharma et al., 2023).

In more severe cases, heavy infestations can lead to life-threatening conditions characterized by anemia, diarrhea, and a weakened immune system, increasing susceptibility to secondary infections and even mortality. The impact of these parasites extends beyond individual animal health, as entire herds may suffer productivity declines, creating financial challenges for farmers, particularly in regions where livestock farming constitutes a primary source of livelihood. Furthermore, the stress and discomfort caused by RPNs raise ethical concerns about animal welfare, underscoring the need for effective and sustainable management strategies.

Addressing the challenges posed by RPNs requires an integrated approach that not only mitigates their direct effects on livestock but also considers the broader implications for food security, rural economies, and sustainable agricultural practices. This necessitates exploring innovative control measures that balance efficacy with environmental and ethical considerations.

2.1 Economically Important RPN (Table 1)

The most economically significant parasitic nematodes affecting ruminants belong to the genera *Haemonchus*, *Ostertagia* (also known as *Teladorsagia*), *Trichostrongylus*, *Cooperia*, and *Nematodirus*. These genera are widespread in grazing livestock systems and are responsible for substantial productivity losses worldwide due to their pathogenicity and impact on host health (Céspedes-Gutiérrez et al., 2021; Fonseca et al., 2022; Pérez-Anzúrez et al., 2022; Berhanu et al., 2024). Below is a detailed overview of the primary species within these genera and their pathological effects:

Table 1 Nematophagous fungi used to control ruminant parasitic nematodes (RPNs) (Céspedes-Gutiérrez et al., 2021; Fonseca et al., 2022; Mendoza-de Gives et al., 2022; Pérez-Anzúrez et al., 2022; Rahman et al., 2023; Tian, 2023; Berhanu et al., 2024; Hao et al., 2024; Rodrigues et al., 2024)

Fungus	Mode of action	Target nematodes	Details
Duddingtonia flagrans	Trapping	*Haemonchus contortus*, *Ostertagia* spp.	Traps and digests nematode larvae in feces/soil
Arthrobotrys oligospora	Trapping	*Cooperia* spp., *Trichostrongylus* spp.	Captures nematodes with adhesive networks
Monacrosporium thaumasium	Trapping	*Haemonchus contortus*	Traps and kills nematodes in pastures
Pochonia chlamydosporia	Egg-parasitic	*Haemonchus contortus*, *Cooperia* spp.	Infects nematode eggs, reducing larvae
Lecanicillium lecanii	Entomopathogenic	Various nematodes	Kills nematodes through contact, used in pest management
Clonostachys rosea	Antagonistic	*Haemonchus contortus*	Competes with nematodes for resources
Paecilomyces lilacinus	Egg-parasitic	*Haemonchus contortus*, *Trichostrongylus* spp.	Parasitizes eggs, preventing larvae from hatching
Hirsutella rhossiliensis	Larvae-parasitic	Various nematodes	Infects nematode larvae, applied to pastures
Nematoctonus robustus	Trapping	*Ostertagia* spp., *Cooperia* spp.	Traps nematodes in varied environments
Purpureocillium lilacinum	Egg-parasitic	*Haemonchus contortus*, *Trichostrongylus* spp.	Degrades nematode eggs in soils

2.1.1 *Haemonchus contortus*

Commonly referred to as the barber's pole worm, *H. contortus* is one of the most pathogenic gastrointestinal nematodes, particularly in small ruminants such as sheep and goats. Its blood-feeding behavior results in severe anemia, hypoproteinemia, and weight loss. In cases of heavy infection, it can lead to sudden death, especially in young or immunocompromised animals. The worm's high fecundity and rapid life cycle make it challenging to control, as it can quickly build up large populations on pastures, leading to reinfection. *H. contortus* is notorious for its ability to develop resistance to anthelmintics, further complicating management efforts.

2.1.2 *Ostertagia ostertagi*

Known as the brown stomach worm, *O. ostertagi* primarily affects cattle, causing the disease ostertagiasis. This condition is characterized by diarrhea, reduced appetite, and significant weight loss. The worm targets the gastric glands of the abomasum, disrupting their function and causing damage that gives the abomasal lining a nodular appearance, commonly referred to as "Moroccan leather." The resulting disruption of digestive processes and protein absorption significantly affects animal growth and productivity.

2.1.3 *Trichostrongylus* spp.

Trichostrongylus species are smaller nematodes that reside in the small intestine of ruminants. They cause trichostrongylosis, which is associated with diarrhea, poor nutrient absorption, and impaired growth rates. Infections are particularly problematic in young animals, where malnutrition and weight loss can have long-term consequences on productivity.

2.1.4 *Cooperia* spp.

Cooperia species are often found alongside other gastrointestinal nematodes in mixed infections. Although individually less pathogenic, their combined presence can exacerbate the overall parasitic burden, compounding the negative effects on livestock health. Symptoms typically include mild diarrhea and reduced feed efficiency, which can significantly impact productivity in suboptimal grazing conditions.

2.1.5 *Nematodirus* spp.

Nematodirus species, particularly *Nematodirus battus*, pose a significant threat to young lambs. These nematodes are known for their synchronized hatching patterns, which often lead to outbreaks of severe diarrhea and dehydration during spring grazing. The high mortality rates associated with *N. battus* infections make it one of the most concerning parasites in sheep production systems, especially in temperate regions.

The combined impact of these nematode genera underscores the need for effective control strategies tailored to specific species and their ecological contexts. Understanding their life cycles, pathogenic mechanisms, and interactions with the host and environment is critical for developing integrated management approaches that minimize economic losses while safeguarding animal health and welfare.

3 Biology and Mechanisms of Action of Nematophagous Fungi

Nematophagous fungi, a unique group of microorganisms with the ability to capture, kill, and digest nematodes, play a pivotal role in biological control strategies for parasitic nematodes. These fungi are classified based on their modes of action into three main categories: predatory fungi, endoparasitic fungi, and fungi that form nematode-trapping structures (Gustavo & Lopes Dias, 2023; Rahman et al., 2023; Berhanu et al., 2024; Filippe et al., 2024) (Table 1). Each group employs distinct mechanisms to target and neutralize nematodes, making them effective tools for sustainable parasitic management.

3.1 Predatory Fungi

Predatory fungi, such as species of *Arthrobotrys*, are renowned for their ability to produce intricate structures that physically ensnare nematodes. These fungi utilize adhesive networks, adhesive knobs, and constricting rings on their hyphae to capture nematodes. The trapping structures are highly specialized and are triggered by the presence of nematodes in the environment.

A prominent example is *Arthrobotrys oligospora*, a well-studied predatory fungus that forms complex three-dimensional adhesive networks. When the nematode contacts these structures, it becomes immobilized, allowing the fungus to initiate its invasion process. Hyphae penetrate the nematode's cuticle and secrete hydrolytic enzymes such as proteases and chitinases, which degrade the nematode's tissues. This enzymatic digestion releases nutrients that the fungus absorbs for growth and reproduction (Berhanu et al., 2024). The effectiveness and adaptability of predatory fungi highlight their potential for integration into parasite management systems.

3.2 Endoparasitic Fungi

Endoparasitic fungi, including *Hirsutella rhossiliensis* and *Harposporium anguillulae*, employ a direct and invasive approach to target nematodes. These fungi produce infective propagules such as conidia or spores that adhere to the nematode's cuticle upon contact. Once attached, the spores germinate and penetrate the nematode's body using mechanical pressure and enzymatic degradation.

After breaching the cuticle, the fungi colonize the internal tissues, disrupting vital physiological processes and ultimately causing the death of the nematode. The fungal life cycle continues as the nematode cadaver serves as a substrate for sporulation, releasing new infective propagules into the environment. The production of specific enzymes, such as proteases and chitinases, is crucial for breaking down the nematode's structural barriers, ensuring successful infection and nutrient acquisition (Pérez-Anzúrez et al., 2022). The specificity and effectiveness of endoparasitic fungi make them valuable assets in targeted nematode management.

3.3 Nematode-Trapping Fungi

Nematode-trapping fungi are among the most versatile and adaptive groups of nematophagous fungi. These fungi produce specialized trapping structures, such as adhesive knobs, sticky nets, and constricting rings, in response to nematode activity in the environment.

A well-documented example is *Duddingtonia flagrans*, which forms thick-walled chlamydospores capable of surviving in harsh environmental conditions. In the presence of nematodes, these chlamydospores germinate, producing trapping structures that immobilize nematode larvae. The traps may function mechanically, such as constricting rings that tighten upon contact, or adhesively, with sticky surfaces that bind nematodes. Adhesive traps are the most common and effective structures, ensuring rapid immobilization of the nematodes (Filippe et al., 2023; Rahman et al., 2023; Berhanu et al., 2024; Hao et al., 2024).

The adaptability of nematode-trapping fungi to environmental conditions and their ability to persist in soil and manure highlight their potential for long-term biological control applications. Their use in integrated management systems could significantly reduce reliance on chemical anthelmintics while enhancing sustainable livestock productivity.

The classification and diverse mechanisms of action of nematophagous fungi underline their ecological and biotechnological importance in managing nematode infestations. Predatory, endoparasitic, and nematode-trapping fungi each contribute unique attributes that can be harnessed to control parasitic nematodes effectively. Understanding their biology and ecological roles is critical to optimizing their application in sustainable livestock management systems.

4 Application of Nematophagous Fungi in Controlling RPN

The application of nematophagous fungi in controlling RPNs is garnering significant interest due to their specificity, environmental compatibility, and safety for non-target organisms. These fungi offer an innovative, sustainable alternative to chemical anthelmintics, addressing the challenges of resistance and environmental contamination associated with traditional methods (Mendoza-de Gives et al., 2022; Gustavo & Lopes Dias, 2023; Khan & Mohiddin 2023; Rahman et al., 2023; Berhanu et al., 2024; Filippe et al., 2024; Sander & Neumann, 2024) (Table 1). Below is an overview of their applications, ranging from field deployment to integration into pest management frameworks.

4.1 Field Applications

Integrating nematophagous fungi into grazing systems involves the strategic application of fungal spores through animal feed or direct soil treatment. These methods have demonstrated significant reductions in nematode populations in pastures, leading to improved animal health and productivity.

4.1.1 Incorporation into Livestock Feed

A prominent application method involves mixing fungal spores, such as those of *Duddingtonia flagrans*, into livestock feed. Upon ingestion, the spores pass through the animal's digestive system and are excreted in the feces, where they germinate and trap nematode larvae. This targeted approach reduces pasture contamination, thereby lowering reinfection rates. Field trials have reported up to 80% reductions in infective larvae on pastures when *D. flagrans* was used as a feed additive (Céspedes-Gutiérrez et al., 2021; Tian, 2023; Hao et al., 2024).

4.1.2 Direct Soil Application

Applying fungal spores directly to pasture soil is another effective strategy. This method aims to establish a persistent fungal presence in the environment, allowing continuous trapping of nematode larvae. Field studies have shown that soil applications of nematophagous fungi can reduce nematode populations by over 70%, offering a long-term solution to nematode control. This approach is particularly useful in rotational grazing systems, where treated pastures can maintain low infection pressures.

4.2 Laboratory Studies and Efficacy

Extensive laboratory studies have provided strong evidence for the efficacy of nematophagous fungi in reducing RPN populations under controlled conditions. These studies serve as a foundation for optimizing field applications and understanding the fungi's biological mechanisms.

4.2.1 Efficacy in Controlled Environments

In laboratory experiments, fungi such as *Arthrobotrys oligospora* and *Duddingtonia flagrans* have demonstrated the ability to significantly reduce nematode larvae in soil and fecal samples. For instance, experiments with *D. flagrans* showed an over

80% reduction in larval counts in treated feces. Such results highlight the fungi's potential as a reliable biological control agent (Pérez-Anzúrez et al., 2022; Tian, 2023).

4.2.2 Optimization of Growth Conditions

Research has identified critical environmental factors influencing the efficacy of nematophagous fungi. These include:

(i) *Soil pH*: Optimal fungal activity occurs in soils with a pH range of 5.5–7.5.
(ii) *Moisture Levels*: Adequate moisture is essential for forming nematode-trapping structures.
(iii) *Organic Matter*: High organic content enhances fungal growth and activity.

These findings underscore the importance of tailoring field conditions to maximize the effectiveness of fungal applications.

4.3 Integrated Pest Management (IPM)

The use of nematophagous fungi aligns seamlessly with integrated pest management (IPM) principles, emphasizing sustainability and the combination of multiple control strategies.

4.3.1 Synergy with Other Strategies

Incorporating nematophagous fungi into IPM frameworks involves combining biological control with rotational grazing, pasture management, and strategic use of chemical anthelmintics. This multi-faceted approach reduces reliance on anthelmintics, mitigates resistance development, and enhances overall pest control efficacy.

4.3.2 Soil Health and Ecosystem Benefits

IPM strategies can also include practices that promote soil health, such as planting cover crops and applying organic amendments. These practices create favorable conditions for nematophagous fungi, enhancing their activity and effectiveness in suppressing nematode populations (Berhanu et al., 2024). By fostering a balanced ecosystem, IPM frameworks ensure long-term sustainability and resilience against parasitic infestations.

The application of nematophagous fungi represents a promising advancement in RPN control, offering specific, safe, and environmentally friendly solutions. From field applications to integration into IPM systems, these fungi provide a sustainable alternative to conventional chemical treatments. Ongoing research and optimization of deployment strategies will be crucial for maximizing their potential and addressing the global challenge of parasitic nematodes in livestock production.

5 Advantages and Challenges of Nematophagous Fungi in Controlling RPNs (Szewc et al., 2021; Mendoza-de Gives et al., 2022; Khan & Mohiddin, 2023)

5.1 Advantages

5.1.1 Specificity and Safety

Nematophagous fungi target nematodes with high specificity, sparing non-target organisms such as soil microbiota, beneficial insects, and other fauna. This characteristic ensures minimal ecological disruption compared to chemical treatments, which often have broad-spectrum activity and unintended consequences for the environment. Their safety profile makes them a suitable choice for sustainable agriculture and integrated pest management systems.

5.1.2 Reduced Resistance Development

Chemical anthelmintics have faced widespread resistance issues due to overuse and misuse, leading to reduced efficacy in controlling RPNs. Nematophagous fungi, being biological control agents, act through diverse and complex mechanisms, such as trapping structures and enzymatic degradation, which reduce the likelihood of resistance development in nematode populations. This makes them a valuable tool for long-term sustainable pest management.

5.1.3 Environmental Sustainability

Nematophagous fungi integrate seamlessly into existing livestock management systems without significant disruptions. Unlike chemical anthelmintics that persist as residues in the environment, these fungi naturally decompose, leaving no harmful residues. Their application helps mitigate the contamination of pastures, water bodies, and soil, contributing to environmentally friendly livestock practices.

5.2 Challenges

5.2.1 Environmental Variability

The activity and survival of nematophagous fungi are highly influenced by environmental factors, such as temperature, moisture, soil pH, and organic matter content.

(i) *Moisture Dependency*: For example, *Duddingtoniaflagrans* requires adequate moisture to form effective nematode-trapping structures. This dependency limits its applicability in arid or semi-arid regions.
(ii) *Temperature Sensitivity*: Extreme temperatures can inhibit fungal growth and sporulation, reducing field efficacy. Site-specific conditions must be carefully assessed to optimize fungal performance.

Developing strains or formulations that are resilient to diverse environmental conditions remains a critical area of research.

5.2.2 Production and Formulation

Scaling up the production of nematophagous fungi for commercial use presents significant technical and economic challenges.

(i) *Production Costs*: Cultivating fungal spores at a large scale while maintaining viability and effectiveness can be expensive.
(ii) *Formulation Stability*: Developing formulations that ensure the spores remain viable during storage, transportation, and field application is crucial. For instance, granular or pelletized formulations are being explored to improve ease of application and enhance shelf life.

Advancements in biotechnology and fermentation techniques are needed to lower costs and improve the scalability of fungal production.

5.2.3 Regulatory and Adoption Barriers

The adoption of nematophagous fungi is hindered by regulatory and sociocultural factors.

(i) *Regulatory Complexity*: Biological control agents often face lengthy approval processes due to stringent safety and efficacy evaluations. Regulatory frameworks need to adapt to facilitate the timely introduction of eco-friendly pest control solutions.
(ii) *Farmer Awareness and Acceptance*: Many farmers are unfamiliar with the use of biological control agents and may be skeptical of their efficacy compared to

conventional chemical treatments. Demonstrating cost-effectiveness, ease of use, and reliability through extension services, field trials, and education campaigns is essential to build confidence and encourage adoption.

Nematophagous fungi offer significant advantages in the sustainable management of RPNs, including specificity, safety, and reduced resistance development. However, environmental dependencies, production challenges, and regulatory hurdles must be addressed to fully realize their potential. Strategic investments in research, development, and outreach will be crucial for overcoming these barriers and promoting the adoption of nematophagous fungi as a cornerstone of sustainable livestock management.

6 Expanding Biotechnological Horizons of Nematophagous Fungi

Beyond their established role in controlling ruminant parasitic nematodes (RPNs), nematophagous fungi exhibit significant potential in diverse biotechnological fields. Their versatility stems from their unique biological properties, which enable them to address challenges in agriculture, industry, and medicine (Szewc et al., 2021; Adil et al., 2023; Filippe et al., 2023; Rahman et al., 2023; Berhanu et al., 2024; Pokhare et al., 2024; Sander & Neumann, 2024).

6.1 Biocontrol of Plant-Parasitic Nematodes

Nematophagous fungi are gaining recognition as effective biocontrol agents for plant-parasitic nematodes, which inflict severe damage on crops and contribute to significant agricultural losses worldwide. These fungi reduce nematode populations in the soil, improving plant health and increasing crop yields.

(i) *Case Studies*: *Pochonia chlamydosporia* has shown remarkable efficacy in controlling root-knot nematodes (*Meloidogyne* spp.) in vegetable crops. Field trials have demonstrated reduced nematode infestations, leading to healthier root systems, improved plant growth, and higher yields.
(ii) *Integration into Farming Practices*: As part of integrated pest management (IPM) systems, nematophagous fungi can reduce the dependency on chemical nematicides, promoting sustainable agriculture and minimizing environmental harm.

6.2 Enzyme Production

Nematophagous fungi produce an array of enzymes, such as proteases, chitinases, and lipases, which facilitate the degradation of nematode cuticles. These enzymes have diverse applications across industries:

(i) *Bioremediation and Waste Management*: Chitinases derived from these fungi can degrade chitinous waste, such as crustacean shells, aiding in waste recycling and bioremediation efforts.
(ii) *Detergent Industry*: Fungal proteases enhance cleaning efficiency by breaking down proteinaceous stains, offering an eco-friendly alternative to synthetic additives.
(iii) *Agricultural Pest Control*: Enzymes like chitinases have the potential to manage insect pests, complementing the fungi's nematode-targeting properties to provide broader biocontrol applications.

6.3 Genetic Engineering

Genetic engineering has unlocked new possibilities for optimizing the functionality and resilience of nematophagous fungi:

(i) *Enhanced Biocontrol Efficacy*: Advances in genetic modification allow for the manipulation of fungal genes to improve pathogenicity and environmental tolerance. For instance, the genetic engineering of *Duddingtoniaflagrans* has increased its production of chlamydospores, enhancing its persistence in diverse field conditions.
(ii) *Bioactive Compound Production*: Modifying metabolic pathways enables these fungi to produce novel bioactive compounds with antimicrobial, antifungal, or insecticidal properties. Such compounds hold promise for pharmaceutical development and sustainable pest control strategies.

6.4 Pharmaceutical and Agricultural Innovations

The secondary metabolites and enzymes produced by nematophagous fungi are of interest in both pharmaceutical and agricultural sectors:

(i) *Pharmaceutical Applications*:

  (a) *Antimicrobial Agents*: Novel antibiotics derived from these fungi could address the growing challenge of antimicrobial resistance.
  (b) *Anticancer Properties*: Compounds with anticancer activities offer the potential for developing innovative therapies for cancer treatment.

(ii) *Agricultural Benefits*:

(a) *Soil Health Management*: The enzymatic activity of nematophagous fungi contributes to nutrient cycling and organic matter decomposition, enhancing soil fertility.
(b) *Crop Resilience*: By improving soil health and suppressing nematode populations, these fungi bolster crop resilience against biotic stressors.

The versatility of nematophagous fungi positions them as invaluable assets beyond nematode control. Their applications in biocontrol, enzyme production, genetic engineering, and pharmaceutical innovations underscore their potential to address critical challenges in agriculture, industry, and healthcare. Continued research and development will expand their practical applications, paving the way for sustainable solutions in multiple sectors.

7 Conclusion and Prospects

Nematophagous fungi offer a groundbreaking, eco-friendly solution for controlling ruminant parasitic nematodes, addressing critical challenges posed by chemical anthelmintic resistance. Their unique biological mechanisms, including trapping, parasitizing, and producing nematotoxic compounds, make them a powerful tool for sustainable livestock management. Beyond parasite control, their versatility extends to biotechnological applications such as biopesticides, biofertilizers, and soil health enhancement, solidifying their importance in sustainable agriculture and environmental stewardship.

Despite the challenges in large-scale application and commercialization, advancements in fungal genomics, biotechnology, and delivery systems hold the promise of overcoming these barriers. Continued investment in research and development will be pivotal in refining their efficacy, affordability, and integration into existing agricultural practices. By harnessing the full potential of nematophagous fungi, we can not only revolutionize parasite management but also contribute to global efforts toward sustainability, reduced chemical dependency, and improved animal welfare, marking a transformative step in both agriculture and biotechnology.

Acknowledgments We sincerely thank all those who contributed to this chapter. Special thanks to the Director, Modern Institute of Technology (MIT), Rishikesh, Uttarakhand for providing the resources for our research. We are also grateful to the editors and the anonymous reviewers for their constructive feedback.

References

Adil, A., Bhat, A., Shakeel, S., Waqar, S., Zafar, A., Handoo, A., & Khan, A. (2023). Microbes vs. nematodes: Insights into biocontrol through antagonistic organisms to control root-knot nematodes. *Plants (Basel, Switzerland), 12*(3), 451. https://doi.org/10.3390/plants12030451

Barone, C. D., Wit, J., Hoberg, E. P., Gilleard, J. S., & Zarlenga, D. S. (2020). Wild ruminants as reservoirs of domestic livestock gastrointestinal nematodes. *Veterinary Parasitology, 279*, 109041. https://doi.org/10.1016/j.vetpar.2020.109041

Berhanu, M., Gebeyaw, D. T., Kefale, D., & Kang, Y. (2024). Overview of nematophagous fungi, isolation techniques, and their role in biological control of helminthic parasites: A literature review. *Acta Entomology and Zoology, 5*(1), 133–143. https://doi.org/10.33545/27080013.2024.v5.i1b.133

Céspedes-Gutiérrez, E., Aragón, D. M., Gómez-Álvarez, M. I., Cubides-Cárdenas, J. A., & Cortés-Rojas, D. F. (2021). Survival of the nematophagous fungus *Duddingtonia flagrans* to in vitro segments of sheep gastrointestinal tract. *Experimental Parasitology, 231*, 108172. https://doi.org/10.1016/j.exppara.2021.108172

Filippe, E., de Freitas Soares, J., Moreira Marques Ferreira, J. P., Genier, H. L., & Aguilar-Marcelino, L. (2023). Biological control 2.0: Use of nematophagous fungi enzymes for nematode control. *Journal of Natural Pesticide Research, 4*, 100025. https://doi.org/10.1016/j.napere.2023.100025

Filippe, E., de Freitas, S. J., Aguilar-Marcelino, L., & Ribeiro Braga, F. (2024). Editorial: Nematophagous fungi as nematode control agents. *Frontiers in Fungal Biology*. https://doi.org/10.3389/ffunb.2023.1353132

Fonseca, J. D. S., Ferreira, V. M., Freitas, S. G., Vieira, Í. S., & Araújo, J. V. (2022). Efficacy of a fungal formulation with the nematophagous fungus *Pochonia chlamydosporia* in the biological control of bovine nematodiosis. *Pathogens (Basel, Switzerland), 11*(6), 695. https://doi.org/10.3390/pathogens11060695

Gustavo, D., & Lopes Dias, R. C. (2023). Nematophagous fungi and their influences on parasitic control in sheep. *Research, Society and Development, 12*(4), e21512441126. https://doi.org/10.33448/rsd-v12i4.41126

Hao, L., Wang, X., Gao, M., Liu, T., Ma, Y., Zhang, Y., Li, Q., Wang, R., & You, X. (2024). Preparation and application of biocontrol formulation of nematode-trapping fungus *Duddingtonia flagrans*. *Veterinary Parasitology*. https://doi.org/10.1016/j.vetpar.2024.110119

Khan, M. R., & Mohiddin, F. A. (2023). Biocontrol strategies for nematode management, an overview. https://doi.org/10.1007/978-981-99-2893-4_5

Li, S., Wang, D. C., Gong, J., & Zhang, Y. (2022). Individual and combined application of nematophagous fungi as biological control agents against gastrointestinal nematodes in domestic animals. *Pathogens, 11*(2), 172. https://doi.org/10.3390/pathogens11020172

Mendoza-de Gives, P., Braga, F. R., & Araújo, J. V. (2022). Nematophagous fungi, an extraordinary tool for controlling ruminant parasitic nematodes and other biotechnological applications. *Biocontrol Science and Technology, 32*(7), 777–793. https://doi.org/10.1080/09583157.2022.2028725

Pérez-Anzúrez, G., Olmedo-Juárez, A., von Son de Fernex, E., Alonso-Díaz, M., Delgado Núñez, E., López-Arellano, M., González-Cortazar, M., Zamilpa, A., Ocampo-Gutiérrez, A., Paz-Silva, A., & Gives, P. (2022). *Arthrobotrys musiformis* (Orbiliales) kills *Haemonchus contortus* infective larvae (Trichostrongylidae) through its predatory activity and fungal culture filtrates. *Pathogens, 11*. https://doi.org/10.3390/pathogens11101068

Pokhare, S., Mhatre, P. H., Manjunatha, D. T., Patil, N., & Agarrwal, R. (2024). Fungal biocontrol agents for nematode management in organic agriculture. *Sustainability in Plant and Crop Protection*, 183–206. https://doi.org/10.1007/978-3-031-52557-5_7

Rahman, M. U., Chen, P., Zhang, X., & Fan, B. (2023). Predacious strategies of nematophagous fungi as bio-control agents. *Agronomy*. https://doi.org/10.3390/agronomy13112685

Rodrigues, J. A., Feitosa, T. F., & Vilela, V. L. R. (2024). A systematic review on products derived from nematophagous fungi in the biological control of parasitic helminths of animals. *World Journal of Microbiology and Biotechnology, 40*(7), 224. https://doi.org/10.1007/s11274-024-04036-5

Sander, C., & Neumann, S. (2024). Nematophagous fungi as biological control agents of parasitic nematodes in soils of wildlife parks. *International Journal for Parasitology. Parasites and Wildlife, 26*, 101033. https://doi.org/10.1016/j.ijppaw.2024.101033

Sharma, A., Sharma, S., Kour, S., Avatsingh, A. U., Perveen, K., Alsulami, J. A., & Singh, N. (2023). Gastrointestinal nematodes and protozoa in small and large ruminants from rural agro-climatic regions of Northern India. *Diversity, 15*, 1131. https://doi.org/10.3390/d15111131

Szewc, M., de Waal, T., & Zintl, A. (2021). Biological methods for the control of gastrointestinal nematodes. *Veterinary Journal, 268*, 105602. https://doi.org/10.1016/j.tvjl.2020.105602

Tian, Y. (2023). Biological control of gastrointestinal nematodes in horses fed with grass in association with nematophagous fungi *Duddingtonia flagrans* and *Pochonia chlamydosporia*. *Biological Control, 182*, 105219. https://doi.org/10.1016/j.biocontrol.2023.105219

Pathways for Human and Animal Health Using Legume Natural Products

Abel Inobeme, John Tsado Mathew, Elijah Yanda Shaba, Kshitij R. B. Singh, Esther Bernard, Ranjita Saikia, Charles Oluwaseun Adetunji, Oritseseyigbemi Gregory Edema, Jonathan Inobeme, Mathew Folaranmi Olaniyan, Adamu Aliyu, Mary Ohunene Yakubu, Vware Manawa, Igechi Gabriel Ozekhome, Okewu Liberty Apeh, Alexander Ikechukwu Ajai, Abdullahi Mann, and Temidayo Olanrewaju Aduloju

Abstract Legumes have been recognized as the second most popularly consumed food after cereals. The role of legumes in human and animal nutrition cannot be overemphasized. Legumes are well known as a vital component of several healthy dietary patterns. Their outstanding proximate compositions, absence of cholesterol, and experimentally proven deficiency in saturated fat have accounted for their wide consumption as an alternative to other protein sources. The presence of numerous biologically active compounds and vital natural products has further revealed their use in medicinal supplements and as indispensable pathways for human and animal health. More recent studies have further reported the presence of numerous antioxidant-containing natural products that are valuable in defending the body

A. Inobeme (✉)
Department of Chemistry, Edo State University Uzairue, Iyahmo, Edo State, Nigeria
e-mail: inobeme.abel@edouniversity.edu.ng

J. T. Mathew
Department of Chemistry, Ibrahim Badamasi Babangida University, Lapai, Lapai, Niger, Nigeria

E. Y. Shaba · A. I. Ajai · A. Mann
Department of Chemistry, Federal University of Technology, Minna, Minna, Niger, Nigeria

K. R. B. Singh
Department of Chemistry, Govt, PG. Autonomous College, Durg, Chhattisgarh, India

E. Bernard
Department of Chemical Engineering, Nassarawa State University, Keffi, Keffi, Nigeria

R. Saikia
Material Science Unit, Asian Institute of Technology, Klong Luang, Thailand

C. O. Adetunji
Department of Microbiology, Edo State University Uzairue, Iyahmo, Edo State, Nigeria

A. Gupta (ed.), *Prospects of Fungal Biotechnologies for Livestock Volume 1*, Fungal Biology, https://doi.org/10.1007/978-3-031-90994-8_9

from various notable degenerative diseases such as osteoporosis, stroke, heart disease, cancer, Arteriosclerosis amongst others. Empirical evidence have shown that this class of food materials has significant potential in the prevention and treatment of several health conditions. This chapter therefore focuses on the role of legume natural products in human and animal health. It gives a general overview of legumes and their nutritional value; it highlights the various bioactive compounds present in legumes. It also discusses comprehensively the role of legumes in animals and human nutrition, with recent reports on studies on natural products from legumes as a pathway for human and animal health. Finally, the future direction of research in this dimension is also attempted.

Keywords Antioxidants · Legumes · Natural products · Photochemical · Phytotherapy

1 Introduction

Legumes are a well-known category of plants that belong to a family called Fabaceae also called Leguminosae. They are popular for their remarkable nutritional and ecological values. The role in human and animal diets has been well documented for long in numerous cultures. They contain high amounts of proteins, minerals, vitamins, dietary fiber, and carbohydrates (Müller et al., 2022). Typical examples of

O. G. Edema
Department of Physics with Electronics, Federal Polytechnic, Auchi, Edo State, Nigeria

J. Inobeme
Department of Geography, Ahmadu Bello University, Zaria, Nigeria

M. F. Olaniyan
Department of Medical Laboratory Science, Edo State University Uzairue, Iyahmo, Edo State, Nigeria

A. Aliyu
Department of Medicinal Plant Research and Traditional Medicine, NIPRD, Idogwari, Faidna, Nigeria

M. O. Yakubu
Department of Chemistry, Nigerian Defence Academy, Kaduna, Nigeria

V. Manawa
Department of Biochemistry, Edo State University Uzairue, Iyamho, Edo State, Nigeria

I. G. Ozekhome
Department of Chemistry, Ignatius Ajuru University, Port Harcourt, River State, Nigeria

O. L. Apeh
Department of Polymer and Textile Technology, Yaba College of Technology, Lagos, Nigeria

T. O. Aduloju
Department of Quantity Surveying, Faculty of Environmental Science, Lagos State University, Lagos, Nigeria

legumes include chickpeas, lentils, beans, and peas. They are well-cultivated in different countries across the globe. One of the outstanding characteristics of this plant is its potential to fix nitrogen in the atmosphere through symbiotic relations with the bacteria present in its root noodles, a process that aids in the enhancement of the fertility of soils and boosts sustainable farming practices (Pérez-Jiménez et al., 2021).

The nutritional values of legumes have been attributed to the previous of various active compounds or natural products such as phenolic acids, flavonoids, polyphenols, and saponins. These compounds are responsible for the numerous therapeutic potentials of this plant such as antioxidant, anti-inflammatory, and anticancer potentials (Siddique et al., 2021; Mathew et al., 2014a, b). They also aid in the improvement of the regulation of the blood sugar balance, and cholesterol lowering. This has further enhanced the shift in focus to legumes amongst the various foods because of rising cases of diseases that are linked to lifestyles (Rosato et al., 2020).

Legumes are also well known as proteins of high quality for animals due to the fact that they provide the needed amino acids that help in reproduction, growth, and overall animal health. The efficiency of most commercial feeds available in the market is now enhanced by the introduction of legumes thereby bringing about a general rise in the productivity of farm animals (Borges et al., 2021). Also, the potential of legumes in the aspect of nitrogen fixation promotes sustainability in animal feed formulation, since there is no more need for synthetic fertilizer to be used in farming (Silva et al., 2020).

This chapter focuses on the role of legumes in animals and human nutrition, with a primary focus on the various natural products present in legumes which is responsible for nutritional potentials (Fig. 1). The chapter introduces the types of legumes, cultivation and distribution of legumes, and proximate compositions. It further unravels the various natural products present in legumes and their roles in human and animal health.

2 Types of Legumes

Agricultural legumes are cultivated mainly for human use, animal feed and silage, and green manure that improves soil condition. Legumes come in over 16,000 different types with a wide range of color, flavor, and texture. Among the well-known legumes are peanuts, chickpeas, key beans, green peas, lima beans, black beans, soya beans, lentils, pinto beans, black-eyed peas, fava beans, etc.

2.1 Peanuts

Sativum is the Latin term for pea. In the Middle Ages, the pea, Pisum sativum, was frequently used in Pease pottage. The Middle Ages saw widespread usage of the pea, or Pisum sativum, in Pease pottage. There are two main types of peas: field

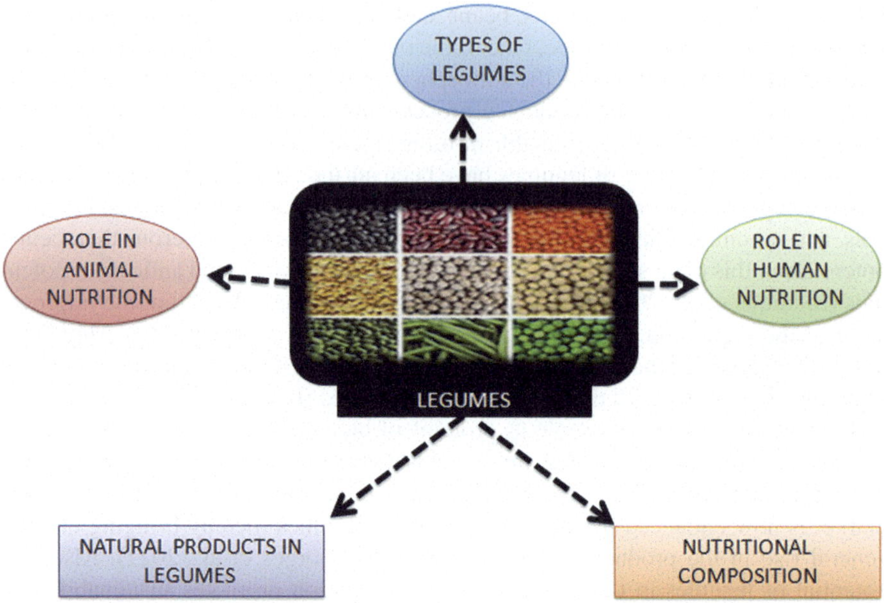

Fig. 1 Schematic representation of the chapter

peas and sweet or green peas. The domesticated Pisum sativum is what is commonly referred to as field peas, cowpeas, or gray peas. For of the color of the seed and the flour they provide when ground, these are known as "gray." This is not your garden pea; they are the horticultural equivalent of peas. Drying occurs after field peas are picked from the vine. Field peas tend to keep shorter, stronger vines than wild peas do (Maria & Zhao, 2020).

In actuality, peanuts are legumes rather than nuts, with over 30 different vitamins and minerals and an astounding 7 g of protein per serving, they are also extremely beneficial for human consumption. Regular peanut consumption can decrease cholesterol and cut the risk of heart disease. After roasted and salted, peanuts make a great snack and also work well in a wide range of cooked foods (Allamine et al., 2023).

2.2 *Chickpeas*

A cool-season legume crop farmed for food all over the world is the chickpea. The primary edible component of the plant is the seed. Bengal gram or garbanzo gram are other names for it. The two different varieties of chickpeas, referred to as "desi" (also known as microsperma) and "kabuli" (also known as macrosperma), have different plant types and distribution patterns. The Mediterranean region is home to the Kabuli varieties, whilst Central Asia and the Indian subcontinent are home to Desi types. Most Kabuli kinds are taller than the Desi types. The majority of chickpeas are consumed as seeds. It is pounded and used as flour (besan) to produce various treats on the Indian subcontinent, while chickpea is consumed in boiled, salted, roasted, and fermented forms across Asia and Africa and used in stews, soups, and salads (Jukanti et al., 2012).

2.3 *Kidney Beans*

Kidney beans (Phaseolus vulgaris) hold a significant place among the world's legume crops and their high percentage of protein (20–25%), multifaceted carbohydrate content (50–60%), an excellent supply of vitamins, minerals, and polyunsaturated fatty acids, makes them an important part of human nutrition. Kidney bean seeds are made up of starch which contains up to 25–45% of polysaccharides. Starch is the primary form of stored carbohydrates found in plants, the supply of glucose is directly provided by it and accounts for 50–70% of the energy in the human diet. Starch often finds its uses in food items, including sauces, soups, candies, ice cream, sugar syrups, snack foods, meat products, infant feeds, and fat substitutes. Kidney beans hold up nicely in soups, stews, and other recipes that require lengthy cooking times because of their meaty texture and mild, slightly sweet flavor. Since it takes on the tastes of other foods, this kind of bean is also very adaptable. Kidney beans offer an astounding 15 g of protein per cup and are a great source of vitamins, minerals, antioxidants, and dietary fiber (Kaur, 2023).

2.4 *Green Peas*

Green pea, or Pisumsativum, is a significant legume that has its place in the Fabaceae family. The plant is a yearly vegetable that grows best in the winter and is therefore primarily found in temperate regions. Pisumsativum is consumed extensively worldwide and has a variety of culinary uses. It can be dried, boiled, or used in salads and soups. They are further processed and canned for later use because of their great demand. Green peas are prized for their nutritional value, much like any other legume. The plant also contains calcium, iron, potassium, magnesium, phosphorus, and a number of amino acids, according to earlier studies. Green peas can be used to treat heart and stomach ailments, the protein in green peas can also assist in meeting your body's daily needs for protein. Additionally, it might aid in managing weight and diabetes (Jasper et al., 2019).

2.5 *Lima Beans*

Phaseolus lunatus L., commonly called butter bean or lima bean, is a tropical legume crop that grows predominantly in temperate and subtropical parts of Southeast Asia and Africa. It falls under the Fabaceae family (Jayalaxmi et al., 2020). It adjusted to the more humid areas' highly leachable infertile soils, having a high degree of genetic variety across the globe. As a result, lima beans from different regions may have different qualities from one another. The limited production and tough texture of dry beans have prevented this legume from being widely consumed up to this point. One potential benefit of this high protein content is that it can help address nutritional issues caused by low protein consumption by making nutrient-dense food products. It is believed that lima bean seeds are an excellent source of nutrients. They are regarded as a reliable source of proteins, ranging in concentration from 14.24% to 24.92%, and abundant in vital amino acids (Ibeabuchi et al., 2019). Since lima beans have a healthy macronutrient composition and contain slow-release carbohydrates, lima beans have a low glycemic index, which makes them advantageous for health. Nevertheless, lima beans, like other legumes, also have antinutritional elements such as trypsin inhibitors, phenolic compounds, phytic acid, poisonous cyanogenic glycoside chemicals, and saponin. Antinutritions reduce the legumes' nutritional value. However, some of these substances may be beneficial to the body as antioxidants, anticancer agents, and other health advantages (Brawijaya et al., 2022).

2.6 Soya Beans

Soya beans are made for human consumption, while oil-extracted soy meal is used as a nutritious livestock feed.

Soybeans are a prominent agricultural crop that provides both vegetable oil and high-quality concentration protein (Glycine max (L.). For many generations, people in Asia and other areas of the world have utilized soybean seeds to produce a wide range of fresh, fermented, and dried dishes. Nutritious food items made from soya beans such as miso, tofu, soy milk, and soy sauce, are made for the purpose of human feeding, while oil-extracted soy meal is used as a nutritious livestock feed. Besides being used domestically, insecticides, papers, paints, chemicals, plastics, inks, varnishes, and cosmetics can all be made from the use of soybean oil. Utilizing soy oil to create biodiesel has recently created a new avenue for industrial applications for renewable energy sources. Since soybeans are a legume crop and can use atmospheric nitrogen through biological nitrogen fixation, they are less reliant on synthetic nitrogen fertilizer (Brawijaya et al., 2022). Insecticides, papers, paints, chemicals, plastics, inks, varnishes, and cosmetics can all be made from the use of soybean oil.

2.7 Lentils

Lentils are an essential component of traditional meals and belong to the legume family. The interest in lentils as functional foods has increased significantly as a result of the presence of bioactive components, antioxidants, high nutritional content, and other phytochemicals that give lentils health benefits. Research suggests that eating lentils may assist in lowering the risk of emerging lingering ailments like tumors, obesity, heart disease, and diabetes. One of the finest sources of prebiotics is lentils, which also contain nutritionally substantial levels of prebiotic carbohydrates such as sugar alcohols and fructooligosaccharides (FOS). Lentils are a major source of prebiotic carbohydrates; over 13 g (12.3–14.1 g) of prebiotics may be contained in about 100 g serving of lentils, according to the total prebiotic carbohydrate concentrations of lentils. Also due to their reputation as excellent sources of protein, lentils and pulses are important sources of nutrition for those living in developing nations and low-income areas (Montejano-ram & Valencia-cantero, 2024).

2.8 Pinto Beans

Common beans, also known as dry beans (Phaseolus vulgaris L.), are considered almost ideal due to their high vitamin B, fiber content, prebiotic, protein, and chemically varied micronutrient makeup. Additionally, dry beans can be produced in a range of eco-agricultural zones and sold in a number of ways, such as whole, canned goods, unprocessed seeds, mixes, and as a gluten-free alternative to wheat flour. Dry beans are therefore utilized all over the world and account for 50% of the grain legumes that people eat. Nearly about 8–10% of the daily required intake of protein is provided by dry beans for each 100 g plateful of cooked beans; the only essential amino acid that is missing is methionine. As a result of the high fiber content of dry beans, they are classified as vegetables. Other prebiotics found in dry beans include fructooligosaccharides including stachyose and raffinose as well as resistant starch (RS). These substances act as substrates for bacterial fermentation in the human gut, which affects gut metabolism and the microbial ecology of the GI tract. Furthermore, dry beans have amounts of calcium and iron that, respectively, fulfill 2–6% of the daily reference intake (DRI) for a 2000-calorie (kcal) diet. Dry beans have an especially high concentration of vitamin B. Folic acid (vitamin B9) is included in around 74% of a cup of cooked beans (225 g).This vitamin aids in the synthesis, maintenance, and methylation of DNA among other biological processes. It also serves as a cofactor in numerous other processes. For both adults and children to avoid anemia, folic acid is necessary for the normal synthesis of red blood cells (Urrea, 2015).

2.9 *Fava Beans*

Fava bean is a fellow of the vetch genus in the Fabaceae household. In colder climates, fava beans are grown as a warm-season crop and in warm temperate as a winter annual. The ideal range of soil temperatures for germination is 60–65 °F. Temperatures beneath 40 °F or directly above 76 °F will not cause germination, yet cultivars may vary in this regard. The plant's pods, beans, and shoots can be eaten; the portion(s) consumed varies depending on the area and culture. Fava beans are consumed either dried, fresh, or roasted as a snack food. Dry fava beans have a nutritional profile that includes 50% carbohydrate, 2% fat, 24% protein, and approximately 700 calories contained in a cup. Fava beans have high levels of L-DOPA, an amino acid that is utilized in treating Parkinson's disease and dopaminedystonia. Fava bean is an excellent source of forage, and it can either be grazed or used for hay and silage. In Egypt and Sudan, fava bean straw is utilized as a cash crop (Dhull et al., 2022).

3 Nutritional Composition of Legumes

3.1 *Protein*

They are known to be a good source of plant-derived proteins, with a composition of 20–25% protein in terms of dry weight (Borges et al., 2021; Inobeme et al., 2014). They also act as a reliable source of numerous essential amino acids, although the amount and types of amino acids do vary amongst different types of legumes. When they are combined with other food materials like corn and rice, they provide a remarkable profile of protein.

3.2 *Carbohydrates*

There is a predominance of complex carbohydrates in leguminous plants. The most common are starch and non-starch such as dietary fibers and oligosaccharides. Oligosaccharides such as stachyose and raffinose are well known for their improvement of gut health by enhancing the growth of the bacteria present in the gut (Chia et al., 2019).

3.3 *Vitamins*

Numerous essential vitamins are found in legumes most especially folate and other B vitamins. There are also numerous mineral elements such as zinc, potassium, magnesium, and iron. Folate is very important in the health of humans especially in pregnant women because of its role in the synthesis of DNA (Silva et al., 2020).

3.4 *Dietary Fiber*

About 30% of dietary fibers are present in legumes and this plays a vital role in the process of digestion by aiding the movement of the bowel. The presence of high fiber also counters the effect of cholesterol by the reduction of cholesterol levels. Fiber also helps in the regulation of the sugar levels in the blood (García-Amoedo et al., 2021).

4 Natural Products in Legumes

There are numerous active compounds present in legumes, with the major ones being saponins, flavonoids, and polyphenols, which have antioxidant potentials (Fig. 2). These compounds have been connected to various therapeutic values (Siddique et al., 2021).

4.1 *Phytosterols*

They have structural similarities that are close to those of cholesterol. They have been reported to have the potential to reduce cholesterol levels in the blood and promote the condition of health. Typical examples include stigmasterol and sitosterol (Singh & Kumar, 2020).

4.2 *Flavonoids*

Flavonoids are abundant, a subgroup of polyphenols in legumes. This class of compounds is well-known for their anti-inflammatory and antioxidant properties, and plays a vital role in the reduction of chances of chronic disease occurring. Some of

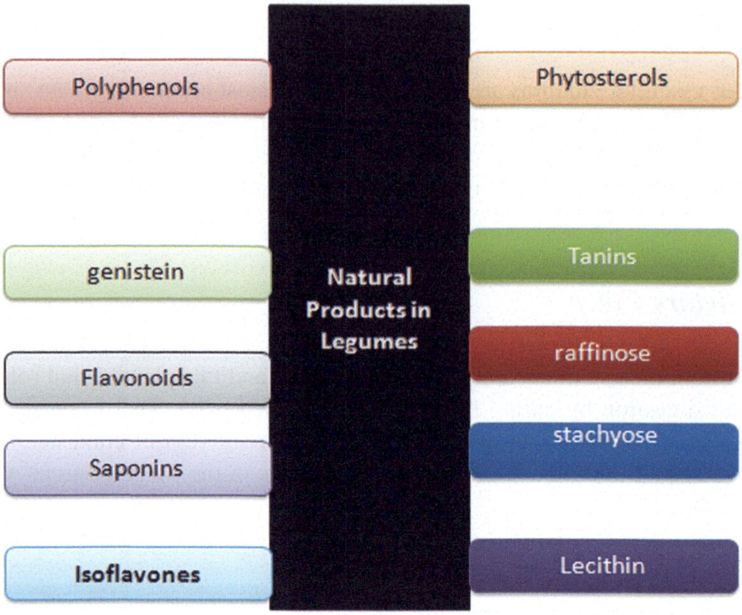

Fig. 2 Natural products in legumes

the most common flavonoids that are found in leguminous plants include catechin, kaempferol, and quercetin. They are also well-recognized for their role in reducing the occurrence of cancer and improvement of the cardiovascular system (Choudhury & Sinha, 2020; Mathew et al., 2014a, b).

4.3 Polyphenols

They are compounds that occur naturally and have a remarkable antioxidant potential. There are various kinds of polyphenols that are found in legumes which involve tannins, phenolic acids, and flavonoids (Ebeed & Aref, 2019).

4.4 Saponins

These are glycosides with a remarkable ability to form very stable forms in aqueous media. They have been reported in various types of legumes, most especially in chickpeas and soybeans. They are well known for their hypocholesterolemic properties, anticancer potential, and immune-modulating activity. They have also been documented to have antioxidant and anti-inflammatory effects making them suitable for combating diseases (Feldman & Pomeranz, 2018).

4.5 *Tannins*

Their concentrations vary among various species of legumes. They inhibit the absorption of some kinds of nutrients and also have astringent characteristics. They have also been linked to the lowering of cancer occurrence (Bhanja & Raikwar, 2021).

Other natural products in legumes include genistein, raffinose, stachyose, isoflavones, and lecithin. The distribution and concentrations vary amongst different types of legumes.

5 Metabolic Routes of Natural Products in Leguminous Plants

5.1 *Biosynthesis of Saponin*

Saponins found in legumes are very complex triterpenoid glycosides. Their synthesis is through the methylerythritol phosphate or mevalonate pathway. Beta amyrin synthase and squalene are enzymes that encode genes and play a vital role during the biosynthesis of saponin (Brawijaya et al., 2022).

5.2 *Flavonoids Biosynthesis*

The synthesis of flavonoids in legumes is through the phenylpropanoids pathways which involves the conversion of phenylalanine to flavonoids structural backbone. The formation of numerous flavonoid compounds is through the enzymatic reaction involving chalcone synthase, hydroxylases, and chalcone isomerase (Allamine et al., 2023).

6 Role of Legumes in Animal Diets

Legumes are vital materials in animal nutrition, due to their numerous functions which are presented as follows:

6.1 Feed Formulation

They are the primary source of plant-based proteins in the formulation of feeds for various animals as a result of their high protein content. They have also been identified as a replacement for some expensive sources of protein source during the commercial formulation of highly economical feeds (Adeola et al., 2020; Jayanegara et al., 2022).

6.2 Digestion of Food

Ruminants find the high fiber contents of legumes to be of immense benefits, due to the promotion of the overall digestion system by the enhancement of gut health. This invariably aids their health performance by boosting the efficiency of nutrient absorption (Jayanegara et al., 2022; Beltran et al., 2020).

6.3 Sustainability

The cultivation of legumes is a promising pathway for the retention and enhancement of soil fertility. The bacteria present in the root nodules of these plants help in the fixation of atmospheric nitrogen into soil nitrogen, a process that promotes sustainability as there is no need for introducing artificial or synthetic fertilizers with their inherent environmental impact (Bhanushali et al., 2022).

7 Benefits of Legumes in Human Health

7.1 Regulation of Weight

The high fiber content combines with the high protein in legumes to promote a feeling of fullness, thereby enhancing the prevention of obesity and the management of weight (Pérez-Jiménez et al., 2021).

7.2 Anticancer Potentials

Recent studies have documented that some of the natural products present in legumes have the potential to lower the risk of cancer in humans. They also aid the detoxification functioning of the liver and influence the expression of genes (Beltran et al., 2020). Epidemiological assessments have also reported the frequent consumption of legumes can aid in the reduction of the risk of breast and colorectal cancers (Gil et al., 2018).

7.3 Stabilization of Blood Sugar Content

The presence of a low glycemic index in legumes aids in the maintenance of the stability of the sugar levels in the blood. This is of paramount importance to people suffering from type 2 diabetes (Rosato et al., 2020; Chia & Tan, 2019).

7.4 Improvement of Cardiovascular Health

Increased consumption of leguminous plants has been connected to improvement in cardiovascular health. This is achieved through the reduction in cholesterol levels of the blood as well as lowering the risk of high blood pressure. These protective potentials are further enhanced by the high fibers and other phytochemicals (Bhanushali et al., 2022).

7.5 Other Medicinal Values

Legumes have been long employed in traditional medicines in numerous cultures due to their potential to promote health. For example, soybeans have been used in the enhancement of hormonal balance due to the presence of the active compound known as isoflavones. Also, Trigonella fenum-graecum (fenugreek seeds) has been employed in Ayurvedic medicine as a result of its anti-diabetic potential (Gil et al., 2018).

8 Prospect of Legumes in Development of Functional Foods

There is a continuous rise in the introduction of legumes into numerous food products in the food industry which is a result of their nutritional benefits. There has been a rise in the utilization of plant-based proteins obtained from legumes as a result of the rising quest for vegan alternatives and vegetarian (Jayanegara et al., 2022).

Furthermore, there is an increase in the incorporation of legumes into functional food. At present, there are numerous products such as pasta, dairy alternatives, and snacks that are legume-based (Bui et al., 2020; Müller et al., 2022).

9 Conclusion and Future Studies

Legumes are indispensable food sources for humans and animals especially in an era characterized by food scarcity and the quest for rich sources of proteins and other nutrients. The varieties of active ingredients and natural products found in these food materials further account for their role in animal and human nutrition, as they aid in boosting overall health. They contain a rich amount of proteins, vitamins, minerals, and various natural products which are capable of supporting health. Their role will continually increase, with the increasing demand for sustainable food sources for animals and humans. The current research in the area of legumes should therefore explore novel routes for the recommendations and development of functional food.

Future studies on legumes should focus on the in-depth characterization and the nutritional profiling of various species of legumes, as well as the overall bioavailability of major nutrients and how different methods of processing affect the retention of nutrients. There is also a need for more studies on the functional characteristics of natural products that are obtained from legumes and the effects of geographical parameters and varying species of legumes. There is a current dearth of studies exploring the perception of consumers and barriers to the consumption of legumes. Such research should therefore focus more on strategies for increasing awareness on the health and nutritional benefits of legumes.

Acknowledgments The authors are sincerely grateful to their various institutions for their support in creating a conducive atmosphere for research.

References

Adeola, O., Aregheore, E. M., & Ipema, A. H. (2020). Legumes in animal nutrition: A review on inclusion limits in diets for livestock. *Animal Feed Science and Technology, 271*, 114718.

Allamine, H. M., Roumane, M., Allamine, Y. M., & Nazal, A. M. (2023). Biochemical characterization of peanuts and cowpeas consumed in the Sahelo-Saharan zone of Chad. *Nutrition and Food Science International Journal, 12*(4), 10–13. https://doi.org/10.19080/NFSIJ.2023.12.555842

Beltran, C., Rojas, S., & Ochoa, M. (2020). The role of legumes in cancer prevention: An overview of research. *Nutrition Research Reviews, 33*(1), 120–139.

Bhanja, S. K., & Raikwar, M. (2021). Nutraceutical aspects of legumes: Exploring health benefits of bioactive compounds. *Food Research International, 140*, 110–116. https://doi.org/10.1016/j.foodres.2020.110116

Bhanushali, A., Shrivastava, N., & Sharma, D. (2022). Cardiovascular benefits of legumes: A meta-analysis of clinical studies. *Journal of Cardiovascular Nutrition, 11*(1), 1–10.

Borges, M. C., Ferreira, D. J., & Cazals, L. (2021). Nutritional composition of common legumes in the Americas. *Food Science and Nutrition, 9*(3), 1345–1356.

Brawijaya, U., Veteran, J., Brawijaya, U., Veteran, J., Brawijaya, U., Veteran, J., Yudarta, U., & Yudharta, J. (2022). Physicochemical and protein characterization of lima bean. *Food Research, 6*(1), 168–177.

Bui, T. A., Watanabe, A., & Nishimura, T. (2020). Nutritive value of legumes: Economic and health benefits. *Food Reviews International, 36*(8), 792–817.

Chia, S. Y., & Tan, L. H. (2019). The role of oligosaccharides in gut health. *Journals of Nutrition and Metabolism*, 6825164.

Choudhury, A., & Sinha, A. (2020). Novel applications of leguminous natural products in drug discovery. *Journal of Herbal Medicine, 20*, 100286. https://doi.org/10.1016/j.hermed.2020.100286

Dhull, S. B., Rose, P. K., Kidwai, M. K., Noor, R., & Chawla, P. (2022). A review of nutritional profile and processing of faba bean (Vicia faba L.). *Legume Science, 4*(8), 1–13. https://doi.org/10.1002/leg3.129

Ebeed, H. T., & Aref, A. S. (2019). Polyphenols in legumes: Sources and health effects. *Current Trends in Nutraceuticals and Health, 1*(1), 11–20. https://doi.org/10.21599/ctnh.2019.1.1

Feldman, M., & Pomeranz, Y. (2018). Nutritional properties of legumes as functional foods. *International Journal of Food Sciences and Nutrition, 69*(6), 698–710. https://doi.org/10.1080/09637486.2018.1469825

García-Amoedo, L. H., Martinez-Urrea, L. M., & Corral, J. (2021). Dietary fibers from legumes: Implications for nutrition and health. *Nutrients, 13*(5), 1527.

Ibeabuchi, J. C., Okafor, D. C., Ahaotu, N. N., Eluchie, C. N., & Agunwah, I. M. (2019). Effect of dehulling on proximate composition and functional properties of Lima bean (Phaseolus lunatus) grown in Enugu State. *Journal of Food Research, 8*(2), 116–121. https://doi.org/10.5539/jfr.v8n2p116

Inobeme, A., Nlemadim, A. B., Obigwa, P. A., Ikechukwu, G., & Ajai, A. I. (2014). Determination of proximate and mineral compositions of white cowpea beans (Vigna unguiculata) collected from markets in Minna, Nigeria. *International Journal of Scientific and Engineering Research, 5*, 502–504.

Jasper, I., Of, C., & State, B. (2019). No title. *International Journal of Agriculture, Environment and Bioresearch, 4*(06), 308–316.

Jayalaxmi, B., Vijayalakshmi, D., Usha, R., Revanna, M. L., & Chandru, R. (2020). Effect of different processing methods on proximate, mineral and antinutrient content of lima bean (Phaseolus lunatus) seeds effect of different processing methods on proximate, mineral and antinutrient content of lima bean (Phaseolus lunatus) seeds. *Legume Science, 39*(4), 6. https://doi.org/10.18805/lr.v0iOF.7108

Jayanegara, A., Rumenos, J., & Goh, F. M. (2022). Nutritional potentials and benefits of legumes for ruminant nutrition: A review. *Animal Health Research Reviews, 23*(1), 85–94.

Jukanti, A. K., Gaur, P. M., Gowda, C. L. L., & Chibbar, R. N. (2012). Nutritional quality and health benefits of chickpea (Cicer arietinum L.): A review. *British Journal of Nutrition, 108*, 1–44.

Kaur, J. (2023). Kidney beans (Phaseolus vulgaris L.) its nutrient profile, health benefits, value-added products and anti-nutritional properties. *Legume Science, 12*(7), 1524–1528.

Maria, S., & Zhao, T. (2020). Phenotypic diversity of pea (Pisum sativum L.) varieties and the polyphenols, flavonoids, and antioxidant activity of their seeds. *Ciencia Rural, 50*(5), 1–16.

Mathew, J. T., Adamu, A., Inobeme, A., Muhammed, S. S., Otori, A. A., & Salihu, A. B. (2014a). Comparative nutritional values of Bambara nut obtained from major markets in Minna Metropolis, Niger State, Nigeria. *Pure and Applied Chemistry, 72*, 25701–25703.

Mathew, J. T., Ndamitso, M. M., Otori, A. A., Shaba, E. Y., Inobeme, A., & Adamu, A. (2014b). Proximate and mineral compositions of seeds of some conventional and non conventional fruits in Niger state, Nigeria. *Academic Research International, 5*(2), 113.

Montejano-ram, V., & Valencia-cantero, E. (2024). The importance of lentils : An overview in. *Agriculture, 14*, 1–15.

Müller, T., Moller, H. B., & Kristensen, T. (2022). The role of legumes in sustainable agriculture: Nitrogen fixing and soil health. *Agronomy for Sustainable Development, 42*(1), 20.

Pérez-Jiménez, J., Negrón, C., & Decker, E. A. (2021). The impact of legumes on satiety and weight management: A systematic review. *Obesity Reviews, 22*(1), e13184.

Rosato, V., et al. (2020). Legumes and the risk of chronic diseases: A meta-analysis. *Journal of Nutrition, 150*(7), 1722–1730.

Siddique, K. H. M., et al. (2021). Phytochemical characterization of legumes: Implications for health and disease resistance. *Frontiers in Nutrition, 8*, 645806.

Silva, G. S., et al. (2020). Folate content in legumes and its biological significance across life stages: A review. *Journal of Dietary Supplements, 17*(5), 529–547.

Singh, S., & Kumar, D. (2020). Phytochemical and pharmacological properties of legumes: A comprehensive review. *Phytochemistry Reviews, 19*(4), 859–888. https://doi.org/10.1007/s11101-020-09630-y

Urrea, C. A. (2015). Pinto beans (Phaseolus vulgaris L.) as a functional food: Implications on human health. *Agriculture, 3*(12), 90–111. https://doi.org/10.3390/agriculture3010090

Disease Management Strategies for Sustainable Aquaculture in Nigeria: Enhancements, Innovations, and Future Directions

Marian Onwude Agbugui, Abel Inobeme, Fran Ehizokhale Abhulimen, Elizabeth Princess Okhamafe, John Chiwuzulum Odozie, John Temitope Ogbiti, and Habibat Joy Ozemoka

Abstract Aquaculture in Nigeria has experienced significant growth in recent years, contributing substantially to the economy and food security. However, disease outbreaks pose a major threat, impacting productivity and sustainability. The paper reviews prevalent diseases affecting key aquaculture species such as tilapia, catfish, and shrimp, emphasizing the economic losses incurred and the ecological implications. It discusses traditional approaches such as antibiotics and vaccines, alongside emerging methods including probiotics, immunostimulants, and genetic selection for disease resistance. Furthermore, the abstract examines the role of enhanced biosecurity measures and environmental management practices in preventing disease outbreaks. It underscores the importance of education and training for aquaculturists to implement effective disease surveillance and control strategies.

M. O. Agbugui (✉) · E. P. Okhamafe
Department of Biological Sciences, Edo State University Uzairue, Uzairue, Edo State, Nigeria
e-mail: agbugui.marian@edouniversity.edu.ng

A. Inobeme
Department of Chemistry, Edo State University Uzairue, Uzairue, Edo State, Nigeria

F. E. Abhulimen
North Somerset Council, Weston-super-Mare, UK

J. C. Odozie
Department of Agricultural and Development Economics, Ajayi Crowther University, Oyo Town, Oyo, Nigeria

J. T. Ogbiti
Department of Computer Science, Edo State University Uzairue, Uzairue, Edo State, Nigeria

H. J. Ozemoka
Department of Animal and Environmental Biology, University of Benin, Benin City, Edo State, Nigeria

A. Gupta (ed.), *Prospects of Fungal Biotechnologies for Livestock Volume 1*, Fungal Biology, https://doi.org/10.1007/978-3-031-90994-8_10

Keywords Environmental management · Fish diseases · Health · Nutrition · Productivity · Sustainability

1 Overview of Fish Diseases in Nigerian Aquaculture

Fish farming in Nigeria centred primarily on species such as catfish and tilapia, serves as a significant source of protein for its population. While aquaculture holds substantial potential for economic growth and food security, it is concurrently challenged by various disease outbreaks that compromise productivity (Olaifa et al., 2022). Diseases in aquaculture facilities manifest through pathogens including bacteria, viruses, and parasites, each capable of causing severe health issues among fish stocks and leading to devastating financial losses. Prevalent diseases in Nigerian aquaculture include bacterial infections such as those caused by Aeromonas species, parasitic infestations like *Ichthyophthirius multifiliis* (commonly known as Ich), and various viral diseases, including infectious pancreatic necrosis virus (IPNV). These diseases have significant economic and productivity impacts, leading to reduced growth rates, increased mortality, and higher operational costs due to the need for treatments and interventions. For example, bacterial infections can cause ulcerations and septicemia, drastically reducing market value and overall yield (Khor et al., 2024).

2 Importance of Disease Management

The control and management of diseases are paramount to the sustainability of fish farming in Nigeria. Without effective disease management, outbreaks can result in high mortality rates, diminished growth rates, and escalating costs related to treatments and loss of stock (Scarfe & Palić, 2020). Implementing robust disease management strategies is essential for reducing the incidence and impact of infectious diseases. This includes developing integrated approaches that combine accurate diagnostics, preventative measures, and the judicious use of medications and vaccines, aiming to sustain not only the health of the fish but also the environmental and economic viability of the aquaculture industry.

3 Common Diseases Affecting Farmed Fish in Nigeria

3.1 Key Pathogens and Affected Species

Aquaculture in Nigeria, which predominantly includes the farming of species like catfish (*Clarias gariepinus*) and tilapia (*Oreochromis niloticus*), is significantly impacted by a variety of pathogens that can cause severe diseases, leading to economic losses (Khor et al., 2024). These pathogens include bacterial, fungal, viral, and parasitic agents, each contributing differently to the morbidity and mortality rates in aquaculture operations.

3.2 Bacterial Diseases

Among the bacterial pathogens, *Aeromonas hydrophilia* and *Edwardsiella tarda* are particularly notorious for their impact on Nigerian fish farms. *Aeromonas hydrophilia* is known for causing hemorrhagic septicemia, leading to high mortality rates, especially in catfish farms. This bacterium thrives in poor water quality conditions, making fish more susceptible to infections that manifest as ulcers, tail rot, and abdominal swelling (Semwal et al., 2023). On the other hand, *Edwardsiella tarda* is responsible for Edwardsiellosis, which presents as jaundice, anemia, and hemorrhagic lesions, significantly reducing the marketability of the affected fish (Pakingking & Nguyen, 2022) (Figs. 1 and 2).

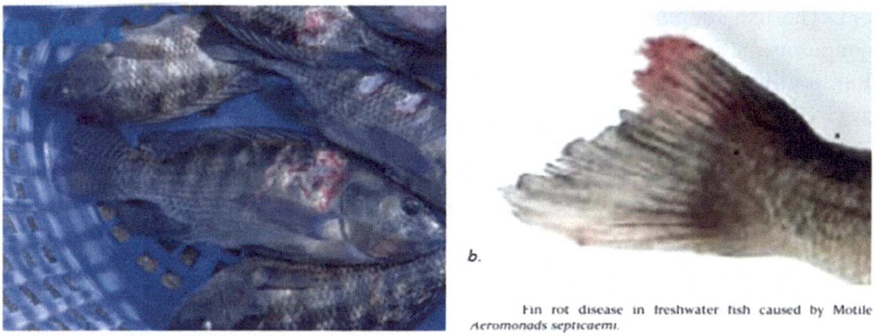

Fin rot disease in freshwater fish caused by Motile *Aeromonads septicaemi.*

a. Aeromonas hydrophilia
fish bacterial infection causing skin ulceration in Tilapia fish

Fig. 1 Bacterial infection on the skin and tail of fish

a

F. *columnare* elicited gill lesions in rainbow trout (*Oncorhynchus mykiss*) fry (operculum removed). In young fish, the disease is mostly acute and the gill is the major site of damage. The lesions are exhibited by pale necrotic areas (arrow, bar ~ 1 cm).

b

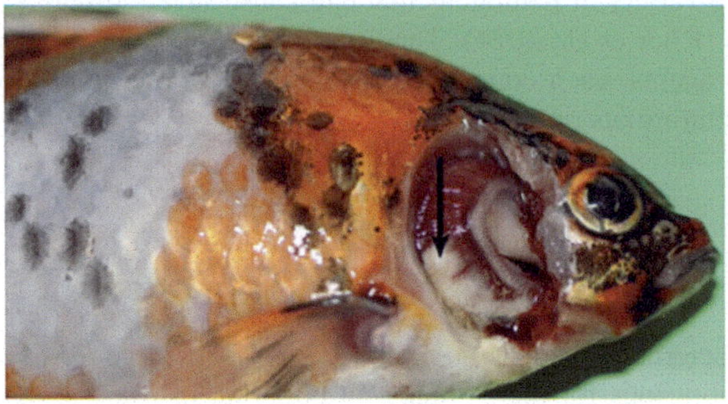

Gill lesions in a shubunkin (*Carassius auratus*) (operculum removed) caused by *F. columnare*, yelloish white areas of degeration are visible in the next part of the first gill arch (arrow). when F.columnare infection spreads rapidly throughout the gill lamellae, the fish may die in a short period of time without another apparent lesions. respiratory distress caused by damage to the gills appers to be the cause of death. bar=1cm

Fig. 2 Bacterial infection on the gills of fish

3.3 Fungal Diseases of Fish

Fungal infections in fish often occur when a fish's immune system is already weakened by illness or injury, and the fungus takes advantage of the change in the system. The fish fungus attaches superficially and stays on the skin and scales. The symptoms include white to brown fluffy patches on the body, eyes, and fins in addition to gills. Increased respiration, decreased appetite, and sudden death in severe cases (Pakingking & Nguyen, 2022).

Fungal infections usually arise due to poor water quality, stress from aggressive tank mates, injuries, tank cleanliness and hygiene, dead fish in the tank, injured or molded fish, and sometimes overcrowding. The three common fungal infections in fish include and are shown in Fig. 3:

1. The gill/mouth rot: Typical gill or mouth rot in fish is caused by the highly aggressive *Branchiomyces* spp.
2. Egg fungus: Opportunistic water mold can attach itself to various dead tissues, including scales and unfertilized eggs.
3. Systemic fungus: These infections are commonly found in warmer water with high-density tanks.

a. Gill and mouth rot causes by *Branchiomyces spp.*

b. note the blood streaked fins, head and mouth regions eroded with white cotton wool-like growths.

c. same as b above.
Off-white cotton wool-like tufts develop around the tail, and body

d. note the blood regions around the fins the fish becomes motionless, lack appetie, no eating becomes thin or emaciated.

Fig. 3 Fungal infections in fish

3.4 *Viral Diseases*

Viral infections present significant challenges to aquaculture in Nigeria, particularly impacting primary farmed species like catfish *Clarias gariepinus* and tilapia *Oreochromis niloticus*. Among the notable viral threats, Tilapia Lake Virus (TiLV) and Channel Catfish Virus (CCV) are predominant, each causing severe outbreaks that can lead to high mortality rates.

TiLV in tilapia is especially concerning due to its rapid spread and lethal impact, primarily affecting younger fish populations. Symptoms include abdominal swelling, skin lesions, and neurological signs, which can swiftly lead to significant losses in productivity (He et al., 2023). For catfish, Channel Catfish Virus (CCV), a type of herpesvirus, leads to Catfish Disease, characterized by similar clinical signs including congestion and organ failure, abdominal distension, severely affecting juvenile stocks (He et al., 2023) (Fig. 4).

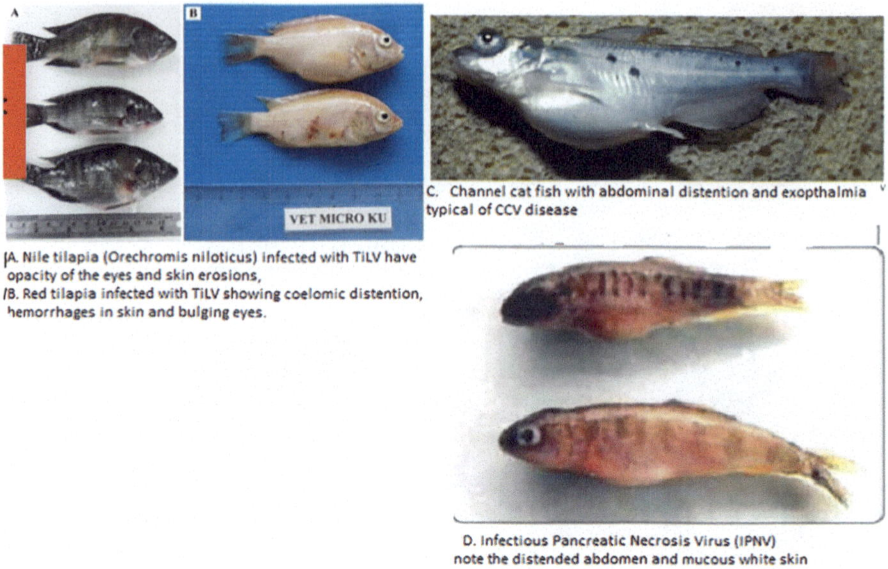

Fig. 4 Viral diseases in the Nie tilapia (**a**), the red tilapia (**b**), in Channel catfish (**c**), and mugil fish (**d**)

3.5 Parasitic Diseases

Parasitic infections, particularly by protozoans like *Ichthyophthirius multifiliis*, commonly known as Ich, cause significant issues in both wild and cultured fish populations. Ich manifests as white spots on the skin, gills, and fins, and is highly contagious and stressful to infected fish, often leading to massive die-offs if not promptly managed (Von Gesdorff et al., 2022) (Fig. 5). The life cycle of this parasite and its ability to form cysts under the skin of fish makes it difficult to eliminate once established in a farm setting.

4 Impact on Fish Growth and Farm Yields

The impact of these diseases on fish growth and farm yields is profound. Infected fish often exhibit stunted growth due to impaired physiological functions and high-energy diversion to fight infections. Chronic disease presence in farms leads to reduced feed conversion efficiency, thereby escalating feed costs and reducing the overall profitability of farm operations. Furthermore, the presence of diseases can lead to entire stock losses in severe outbreak scenarios, devastating the economic stability of aquaculture operations (Abdella et al., 2024).

For instance, farms afflicted by *Aeromonas hydrophilia* may experience up to 30–70% mortality, drastically reducing yield and requiring significant investments

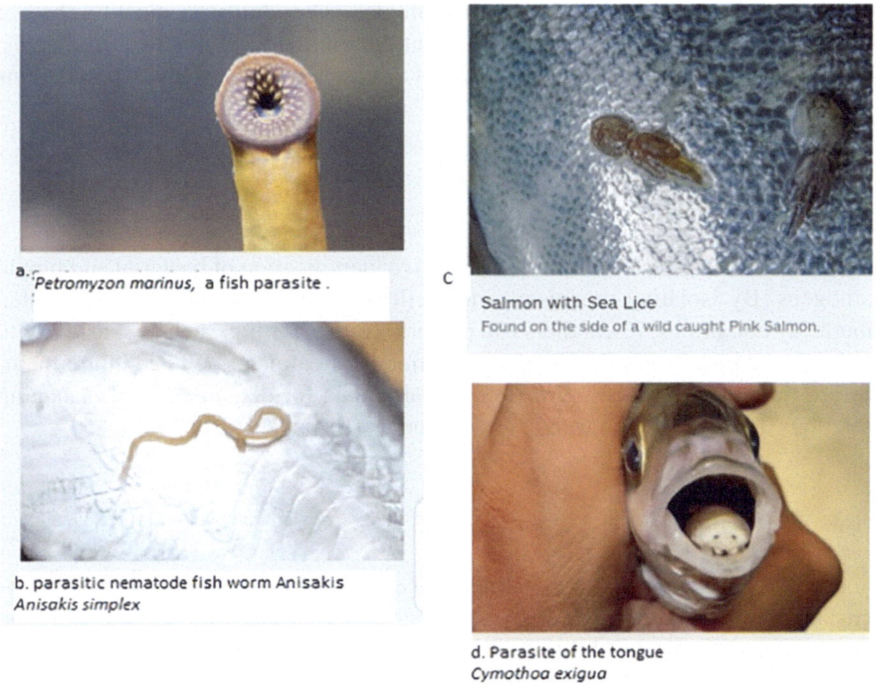

Fig. 5 Parasitic diseases of fish. Both endo and ectoparasites are of importance

in disease management and stock replenishment. The economic impact is compounded by market rejection of visibly infected or poor-quality fish, further straining financial returns (Abdella et al., 2024).

5 Diagnostic Methods for Fish Diseases

5.1 Conventional Diagnostics

Conventional diagnostic methods form the cornerstone of initial disease screening and pathogen identification in aquaculture, employing a range of techniques such as microscopy, culture methods, and biochemical tests.

5.1.1 Microscopy

Microscopy offers a direct and immediate means to visually identify pathogens within samples taken from infected fish. This approach is vital for recognizing morphological features of bacterial, fungal, and parasitic agents and for observing histopathological changes within tissue samples. The ability to quickly observe these details allows for the rapid assessment and initiation of treatment protocols, which are critical in preventing the spread of disease within aquaculture populations (Knoll et al., 2023).

5.1.2 Culture Methods

Culture methods are essential for the definitive identification of bacterial and fungal pathogens. By isolating organisms and facilitating their growth under controlled laboratory conditions, researchers can perform further testing, such as sensitivity to antibiotics, which is crucial for selecting effective treatments. While culture methods are time-intensive, they provide invaluable data for diagnosing and managing infectious diseases in fish farming environments (Evans & Riley, 2021).

5.1.3 Biochemical Tests

Biochemical tests analyze the metabolic and enzymatic activities of cultured organisms, helping to differentiate between pathogen species and strains. These tests can determine pathogenic capabilities, resistance profiles, and other critical factors that influence the choice of treatment strategies. Understanding these aspects is essential for managing disease in a targeted and effective manner (Spiga et al., 2021).

5.1.4 Molecular Diagnostics

Advancements in molecular diagnostics have greatly enhanced the sensitivity and specificity of pathogen detection, utilizing techniques such as PCR and ELISA to identify disease-causing agents at a molecular level.

5.1.5 Polymerase Chain Reaction (PCR) Techniques

PCR is instrumental in amplifying trace amounts of DNA or RNA from pathogens, making it possible to detect infections that might be missed by conventional methods. This technique is particularly valuable for identifying and quantifying viral pathogens, as well as providing early detection of bacterial and parasitic DNA in tissue samples. Its rapid processing time allows for timely decisions on quarantine and treatment, which are critical in managing disease outbreaks in aquaculture settings (Lei et al., 2021).

5.1.6 Enzyme-Linked Immunosorbent Assay (ELISA)

ELISA is used to measure specific antigens or antibodies in fish samples, offering quantitative data on pathogen presence or the immune response of the fish. This information is crucial for understanding the severity of an infection and monitoring the effectiveness of treatment regimens. ELISA can also be adapted to screen large numbers of samples efficiently, making it an excellent tool for routine surveillance and disease prevention in aquaculture operations (Abdella et al., 2024).

5.1.7 Recent Advances

Recent technological advancements have introduced more sophisticated methods such as Clustered Regularly Interspaced Short Palindromic Repeats (CRISPR) based diagnostics, which enhance the capability to diagnose diseases quickly and accurately.

6 Disease Prevention and Control Measures in Aquaculture

Effective disease management requires accurate diagnosis, achieved through visual inspections and laboratory tests. Visual inspections involve observing fish behavior and physical signs such as lesions, discoloration, or abnormal swimming patterns (Evans & Riley, 2021). Laboratory tests, including histopathology, microbiological cultures, and molecular diagnostics, provide precise identification of pathogens (Evans & Riley, 2021).

Treatment options vary depending on the disease but commonly include antibiotics for bacterial infections, which must be used judiciously to prevent resistance. Herbal treatments, such as extracts from neem and garlic, have shown promise in treating parasitic and bacterial infections due to their antimicrobial properties (Khor et al., 2024). Improved management practices, such as maintaining optimal water quality and reducing stress factors, are critical for disease control (Benzie, 2024).

6.1 Bio-security Measures

Biosecurity is critical in aquaculture to minimize disease outbreaks and ensure the health and productivity of aquatic species. Effective biosecurity measures encompass several stringent practices aimed at preventing the introduction and spread of infectious agents (Benzie, 2024).

6.2 Quarantine

Quarantine protocols are the first line of defense in protecting farm populations from new or reintroduced individuals potentially harboring pathogens. Isolating these individuals for a specified duration allows for observation and health assessments, which are crucial in identifying any signs of disease before these fish are allowed to interact with the main stock. This proactive measure significantly reduces the risk of disease transmission, providing a buffer that helps maintain the health integrity of aquaculture operations (Sharan et al., 2023).

6.3 Sanitation Practices

Sanitation is paramount in disrupting the pathways through which infectious diseases can spread within an aquaculture system. Regular and thorough disinfection of equipment, tanks, and water systems prevents the accumulation and spread of infectious agents. This includes protocols for cleaning nets, tanks, and water pumps, and the use of disinfectants that are effective against a broad spectrum of pathogens without harming the aquatic species. Effective sanitation practices are critical in maintaining an environment that is inhospitable to pathogens while safe for fish (Li et al., 2021).

6.4 Farm Management Strategies

Comprehensive farm management strategies integrate daily operations with proactive health management to minimize disease outbreaks. This includes maintaining proper records of stock health, monitoring mortality rates, and adjusting farming practices based on observed health issues. Strategic planning also involves the segregation of populations by age and health status to prevent cross-contamination and manage outbreaks more effectively. These integrated approaches require consistent training for farm staff on the latest health management protocols and the importance of rapid response to potential health issues (Li et al., 2021).

6.5 Nutritional Management

Nutrition plays a critical role in enhancing the immune response and overall health of farmed fish, making it a fundamental aspect of disease prevention in aquaculture (Abe & Agbugui, 2023).

6.6 Role of Diet in Disease Prevention

Nutritionally balanced diets are tailored to the specific needs of the species and the life stage of the fish, providing all necessary nutrients in optimal ratios. A well-formulated diet supports immune function by supplying essential vitamins, minerals, and energy sources, which collectively enhance the fish's resistance to diseases. Nutritional adequacy is thus directly correlated with improved survival rates and reduced incidences of disease (Labh, 2023).

6.7 Specific Nutrients That Enhance Immune Responses

Certain nutrients are specifically known for their roles in immune modulation. For instance, vitamins C and E are potent antioxidants that protect cellular integrity and boost immune functions. Trace elements like zinc and selenium are crucial for the proper functioning of various enzymatic processes that support immune defenses. Omega-3 fatty acids, particularly EPA and DHA, are critical for maintaining cell membrane fluidity, which enhances cellular immune responses and barriers against pathogens (Moraes & De Almeida, 2020).

6.8 Environmental Management

Environmental management in aquaculture involves the manipulation and control of various physical and chemical parameters to optimize conditions for aquatic life, thereby reducing stress and susceptibility to disease (Abe & Agbugui, 2023).

6.9 Water Quality Management

Maintaining optimal water quality is perhaps the most crucial environmental factor in disease prevention. Parameters such as temperature, pH, oxygen levels, and nitrogenous waste must be rigorously controlled. Fluctuations in these parameters can stress aquatic organisms, making them more susceptible to diseases. Regular monitoring and adjustments are essential to ensure that water quality is within the species-specific optimal range, thereby fostering a stress-free environment conducive to health (Li et al., 2021).

6.10 Waste Disposal

Effective waste management is essential to prevent the accumulation of organic matter and pathogens. Techniques such as sedimentation, filtration, and the use of bioreactors not only reduce the load of organic waste but also control the proliferation of pathogens. Properly designed waste management systems ensure the removal of harmful substances and maintain the overall health of the aquatic environment (Tom et al., 2021).

6.11 Control of Farm Density

Regulating stocking densities is vital for disease management. Overcrowding can lead to increased stress, higher incidences of physical injuries, and greater transmission rates of infectious diseases. Managing farm densities thus plays a direct role in controlling stress and reducing the outbreak potential of diseases, making it a critical aspect of environmental management in aquaculture (Li et al., 2021).

7 Use of Vaccines and Medication in Fish Health Management

7.1 Vaccination Strategies

Vaccination is a pivotal aspect of health management in aquaculture, serving as a preventative measure against prevalent diseases. Efficient vaccination strategies enhance the overall health status of fish populations, reducing the need for medications and minimizing outbreak impacts.

7.2 Types of Vaccines Used

In aquaculture, the two primary types of vaccines used are inactivated (killed) and live attenuated vaccines. Inactivated vaccines consist of pathogens that have been killed. They are safe since they cannot cause disease, but they often require adjuvants to enhance immune response and multiple doses to achieve effective immunity. Live attenuated vaccines, in contrast, use a weakened form of the pathogen that can still replicate but does not cause disease in healthy recipients. These vaccines typically elicit a stronger and longer-lasting immune response with fewer doses. The choice between these vaccines depends on the specific disease, the life stage of the fish, and the risk of vaccine-induced disease (Bedekar et al., 2022).

7.3 Medications

Medications are an integral component of disease control in aquaculture, used to treat bacterial, viral, and parasitic infections. However, their use is complicated by efficacy, resistance, and food safety issues.

7.4 Commonly Used Medications

The most commonly used medications in fish farming include antibiotics and anti-parasitics. Antibiotics are used to treat bacterial infections, and anti-parasitics are employed against a range of protozoan and metazoan parasites. The choice of medication, dosage, and treatment duration depend on the specific pathogen, the severity of the outbreak, and the pharmacokinetics of the drug in an aquatic environment (Hoque et al., 2022).

7.5 Issues with Resistance

Resistance to medications, particularly antibiotics, has become a significant challenge in aquaculture. The misuse or overuse of antibiotics can lead to the development of resistant bacterial strains, which not only renders the antibiotics ineffective but also poses a risk to human health and the environment. This challenge necessitates stringent management practices, including the judicious use of antibiotics and regular monitoring for resistance patterns (Larsson & Flach, 2022).

7.6 Regulatory and Safety Considerations

Regulatory frameworks governing the use of vaccines and medications in aquaculture are essential to ensure the safety and efficacy of these interventions. In Nigeria, these regulations are designed to protect both aquatic animal health and consumer safety.

7.7 Overview of Regulations

In Nigeria, the use of drugs and vaccines in aquaculture must comply with regulations set forth by veterinary authorities and health agencies. These regulations ensure that all veterinary products are safe, effective, and of good quality. They also dictate specific guidelines for the registration, importation, manufacturing, distribution, and use of veterinary drugs and vaccines (Limbu, 2020).

Regulatory bodies also enforce safety standards to prevent environmental contamination and ensure that residues in fish products do not exceed safe levels. Compliance with these regulations is monitored through surveillance and periodic inspections of aquaculture facilities (Limbu, 2020).

7.8 Adoption of Recirculating Systems

Recirculating Aquaculture Systems (RAS) is a sustainable technology that addresses many environmental concerns associated with traditional aquaculture. In RAS, water is filtered and reused within the system, significantly reducing the volume of water needed and the discharge of pollutants. This system minimizes the impact on water quality and conserves water resources, making it a viable solution for sustainable aquaculture practices (Ahmed & Turchini, 2021).

8 Comprehensive Management of Water Quality and Fish Health in Nigerian Aquaculture Systems

The sustainability and productivity of aquaculture in Nigeria are critically dependent on maintaining optimal water quality and ensuring fish health. Poor water quality can lead to a range of issues, including reduced growth rates, increased susceptibility to diseases, and higher mortality rates, ultimately impacting the economic viability of aquaculture operations (Manyise et al., 2024). Fish health is intrinsically linked to water quality, as suboptimal conditions can stress fish, making them more prone to infections and diseases. This chapter covers essential parameters for optimal water quality, such as temperature, pH, and dissolved oxygen, discusses water quality monitoring techniques, explores methods for mitigating water pollution in aquaculture systems, and provides strategies for disease management and fish health maintenance.

8.1 Parameters for Optimal Water Quality

8.1.1 Temperatures

Temperature is a crucial parameter for the optimal growth and health of aquaculture species such as catfish and tilapia, which are predominant in Nigerian aquaculture. For catfish (*Clarias gariepinus*), the ideal temperature range is between 27 and

30 °C, while tilapia (*Oreochromis niloticus*) thrives best at 25–32 °C (Kasihmuddin et al., 2024). Temperature directly influences physiological processes, including growth rate, feed conversion efficiency, and immune response. Optimal temperatures enhance metabolic rates, leading to improved growth and better feed utilization (Khieokhajonkhet et al., 2022). Conversely, suboptimal temperatures can suppress the immune system, making fish more susceptible to diseases (Ma et al., 2023).

To manage temperature effectively, several strategies can be employed. Pond depth control is critical; deeper ponds tend to have more stable temperatures due to thermal stratification (Ships & Ports, 2021). During periods of extreme heat, providing shade using floating vegetation or shade cloth can help reduce water temperature and prevent heat stress (Ships & Ports, 2021).

8.1.2 pH

The pH level of water is another vital parameter for aquaculture, with an optimal range typically between 6.5 and 9.0 for most species. pH imbalances can cause significant stress to fish, impairing growth and increasing susceptibility to diseases. Low pH levels (acidic conditions) can lead to acidosis, affecting respiration and osmoregulation, while high pH levels (alkaline conditions) can cause ammonia toxicity (Bozorg-Haddad et al., 2021). Managing pH involves regular monitoring and adjustments. Lime (calcium carbonate) can be added to increase pH in acidic waters, while acidifiers such as phosphoric acid can be used to lower pH in alkaline conditions (Yilmaz et al., 2022).

8.1.3 Dissolved Oxygen

Dissolved oxygen (DO) is essential for the metabolic processes and overall health of fish. It is critical for respiration and energy production, with optimal DO levels for most aquaculture species being above 5 mg/L (Bozorg-Haddad et al., 2021). Insufficient DO can lead to hypoxia, causing stress, reduced growth rates, and increased mortality (Mallat et al., 2022). Enhancing DO levels can be achieved through several methods. Mechanical aerators, such as paddle wheels and diffused air systems, are commonly used to increase pond oxygenation (Yilmaz et al., 2022). Water agitation techniques, such as splashing or cascading water, also help increase oxygen levels. Additionally, managing aquatic plants in ponds can assist in maintaining DO, as plants produce oxygen through photosynthesis during daylight hours (Ramesh et al., 2024).

8.1.4 Ammonia

Ammonia is a critical parameter in aquaculture due to its toxicity to fish at high concentrations. Ammonia exists in two forms: un-ionized ammonia (NH_3), which is highly toxic, and ionized ammonia (NH_4^+), which is less harmful. Elevated levels of ammonia can cause gill damage, reduce oxygen-carrying capacity, and impair fish growth and immune function (Randall & Tsui, 2002). Regular monitoring and maintaining low ammonia levels are essential. Biological filtration using nitrifying bacteria can convert ammonia to less toxic compounds, thereby mitigating its harmful effects (Jain et al., 2022). Effective feed management to minimize waste and regular water changes are also crucial strategies for controlling ammonia levels.

8.1.5 Nitrite

Nitrite is an intermediate product in the nitrification process and is also toxic to fish. High nitrite levels can lead to "brown blood disease," where nitrite binds with hemoglobin, forming methemoglobin, which impairs oxygen transport in fish (Yilmaz et al., 2022). This condition can cause respiratory distress and mortality if not managed properly. Maintaining optimal biological filtration is critical, as nitrifying bacteria convert nitrite to nitrate, a less toxic end product (Bozorg-Haddad et al., 2021). Regular water quality monitoring and adequate water exchange are necessary to prevent nitrite accumulation and ensure a healthy aquatic environment.

8.1.6 Nitrate

Nitrate is the final product of the nitrification process and is less toxic than ammonia and nitrite but can still be harmful at high concentrations. Chronic exposure to elevated nitrate levels can lead to reduced growth rates, compromised immune responses, and reproductive issues in fish (Yilmaz et al., 2022).

Nitrate accumulation can be controlled through regular water changes and the use of denitrifying biofilters, which convert nitrate to nitrogen gas, thereby removing it from the water column (Luo et al., 2020). Maintaining a balanced feeding regimen and ensuring efficient waste removal are also essential practices to keep nitrate levels within safe limits for aquaculture species.

8.1.7 Salinity

Salinity refers to the concentration of dissolved salts in water and is a crucial parameter for species-specific aquaculture systems. Most freshwater fish require salinity levels of less than 0.5 ppt, while marine species need significantly higher levels (Seale et al., 2024). Sudden changes in salinity can cause osmotic stress, affecting fish metabolism, growth, and immune function (Yang et al., 2022).

Monitoring salinity levels and gradually acclimating fish to any changes are vital for maintaining their health (Seale et al., 2024). Additionally, managing water sources and ensuring proper mixing of salt in aquaculture systems help maintain stable salinity conditions conducive to fish health and growth.

8.1.8 Hardness

Water hardness, primarily determined by the concentrations of calcium and magnesium, is important for various physiological processes in fish, including bone formation and enzyme activation. Optimal hardness levels vary among species but generally, a range of 50–200 mg/L as $CaCO_3$ is recommended for freshwater aquaculture. Low hardness can lead to poor growth and increased susceptibility to diseases, while excessively high hardness can cause scaling and other operational issues in aquaculture systems (Ajayi et al., 2022). Regular monitoring and adjusting hardness using calcium or magnesium supplements help maintain optimal conditions for aquaculture species.

8.2 Water Quality Monitoring Techniques

8.2.1 Regular Water Testing

Frequent water quality assessments are critical for the sustainability of aquaculture systems. Regular monitoring allows for the early detection of water quality issues, which can prevent significant losses due to poor water conditions or disease outbreaks. Key parameters that require regular testing include chemical parameters such as pH, ammonia, nitrite, and nitrate; physical parameters like temperature and turbidity; and biological parameters including pathogen presence (Hoque et al., 2022). Standard testing procedures involve the use of colorimetric kits for parameters like pH and ammonia, which provide quick and accurate measurements. Digital meters are highly recommended for their precision and ease of use, particularly for monitoring dissolved oxygen and temperature. Regular testing, ideally every week, ensures that any deviations from optimal conditions can be swiftly addressed, maintaining the health and productivity of the aquaculture system (Lindholm-Lehto, 2023).

8.2.2 Remote Sensing and Automation

The advent of modern technologies such as sensors and automated systems has revolutionized water quality monitoring in aquaculture. These technologies enable real-time monitoring of key water quality parameters, providing continuous data collection and early detection of potential problems. Automated systems can measure and report on parameters such as dissolved oxygen, pH, and temperature, reducing the need for manual testing and thereby lowering labor costs (Garrido-Momparler & Peris, 2022). The use of remote sensing technologies, such as satellite and drone-based sensors, allows for the monitoring of larger aquaculture operations with greater efficiency. Successful implementations in Nigerian aquaculture include the deployment of automated monitoring systems in catfish farms, which have shown significant improvements in water quality management and fish health outcomes (Anani et al., 2022). These systems facilitate prompt corrective actions, ensuring optimal conditions are maintained continuously.

8.2.3 Data Analysis and Interpretation

Collecting and analyzing water quality data is essential for informed decision-making in aquaculture management. Data collection involves recording measurements from regular testing and automated systems, and storing this data in a structured format for analysis. Interpretation of data trends helps identify patterns and potential issues, allowing for proactive management decisions. For example, consistently low dissolved oxygen levels might indicate the need for improved aeration techniques (Ali & Mishra, 2022). Analytical tools such as statistical software and spreadsheets can be used to process and visualize data, making it easier to identify trends and outliers.

Case studies from Nigerian aquaculture farms illustrate the impact of effective water quality management. For instance, a study on a tilapia farm demonstrated that regular monitoring and data-driven management decisions led to a significant increase in fish growth rates and a significant reduction in disease incidence (WorldFish, 2021). Another study highlighted the success of using automated monitoring systems in catfish farming, resulting in better water quality control and higher survival rates (Anani et al., 2022). These examples underscore the importance of systematic data analysis and interpretation in achieving sustainable aquaculture practices.

8.3 Mitigation of Water Pollution in Aquaculture Systems

8.3.1 Sources of Pollution

Primary pollutants in aquaculture systems include excess feed, fish waste, and chemicals used for treatments. Excess feed and fish waste contribute to the accumulation of organic matter, which decomposes and releases harmful substances such as ammonia, nitrite, and nitrate into the water. These substances can significantly degrade water quality, leading to hypoxia, increased susceptibility to diseases, and reduced growth rates in fish. Chemicals from treatments, such as antibiotics and disinfectants, can also accumulate and disrupt the aquatic ecosystem, affecting both the farmed species and surrounding environments (Zhang et al., 2020).

8.3.2 Pollution Control Measures

Effective pollution control measures are essential to minimize the impact of these pollutants on water quality and fish health. Best practices include optimizing feed management to reduce waste by using high-quality, appropriately sized pellets and feeding strategies that match the fish's nutritional requirements (Pailan & Biswas, 2022). Regular cleaning of ponds and proper disposal of dead fish are critical to preventing the build-up of organic matter and reducing the risk of disease outbreaks (Kunjiraman et al., 2024).

The introduction of biofilters and sedimentation ponds can naturally treat water by promoting the growth of beneficial bacteria that break down organic matter and convert harmful substances into less toxic forms (Sinha et al., 2022). Additionally, Recirculating Aquaculture Systems (RAS) are highly effective in maintaining water quality by continuously filtering and recycling water, thereby minimizing the need for water exchange and reducing the discharge of pollutants into the environment (Ahmed & Turchini, 2021).

8.3.3 Rehabilitation of Polluted Water

Rehabilitation of polluted water involves techniques such as bioremediation and chemical treatments. Bioremediation uses beneficial bacteria to degrade organic pollutants and neutralize toxins, effectively restoring water quality. For instance, the application of nitrifying bacteria can convert harmful ammonia and nitrite into nitrate, which is less toxic and can be utilized by aquatic plants. Chemical treatments, such as the use of oxidizing agents, can also neutralize toxins and improve water quality (Thombre, 2023).

9 Conclusion and Prospects

The success of aquaculture in Nigeria hinges critically on maintaining optimal water quality and ensuring robust fish health. Water quality parameters such as temperature, pH, and dissolved oxygen are vital for the physiological well-being and growth of aquaculture species. Poor water quality can lead to increased stress, disease susceptibility, and mortality, which significantly hampers productivity and economic viability. Effective disease management, through biosecurity measures, vaccination programs, and improved nutrition, is equally crucial to mitigate the adverse effects of bacterial, parasitic, and viral infections on fish populations. Effective disease management is paramount in enhancing the sustainability and economic viability of aquaculture in Nigeria. By minimizing the incidence and severity of disease outbreaks, aquaculture operations can achieve higher productivity and stability, resulting in increased profitability and reduced economic losses. The implementation of robust disease control measures, including advanced diagnostic methods, vaccination, and effective treatment strategies, plays a critical role in maintaining healthy fish populations. These practices ensure the longevity of aquaculture enterprises and safeguard public health by preventing the spread of zoonotic diseases and reducing antibiotic resistance. Looking ahead, the future of aquaculture disease management in Nigeria holds promising opportunities for further research and development. Continued advancements in genetic engineering, vaccine development, and environmentally sustainable practices are expected to significantly improve disease resistance in aquaculture species. Additionally, the integration of digital technologies such as artificial intelligence and machine learning into disease surveillance and management could revolutionize the field, enhancing precision and efficiency. By adopting technologies like RAS and environmentally friendly inputs, aquaculture can significantly reduce its impacts on local ecosystems and contribute positively to global food security. Embracing these innovations will be crucial for the sector's adaptation to global challenges, ensuring its sustainability and growth in the coming decades. To ensure sustainable growth in Nigeria's aquaculture sector, it is imperative to adopt and consistently implement best practices in water quality management and disease prevention.

Acknowledgments The authors of this manuscript extend their sincere gratitude to all the lab technologists of the Edo State University Uzairue for the support given to the successful completion of this article.

We also appreciate the support provided by the management of the Edo State University Uzairue whose resources and expertise greatly facilitated this work. Additionally, we are grateful to our colleagues for their constructive criticism and collaboration.

References

Abdella, B., Shokrak, N., Abozahra, N. A., Elshany, Y. M., Kadira, H. I., & Mohammed, R. A. (2024). Aquaculture and *Aeromonas hydrophila*: A complex interplay of environmental factors and virulence. *Aquaculture International, 32*(6), 7671–7681. https://doi.org/10.1007/s10499-020-01535-y

Abe, G., & Agbugui, M. O. (2023). Recurring constraints in fish culture and fish nutrition in Nigeria. *International Journal of Livestock Production, 11*, 1–5.

Ahmed, N., & Turchini, G. M. (2021). Recirculating aquaculture systems (RAS): Environmental solution and climate change adaptation. *Journal of Cleaner Production, 297*, 126604. Available at: https://doi.org/10.1016/j.jclepro.2021.126604.

Ajayi, O., Akinrinlola, A., Usman, A., Ahmed, M., & Van der Knaap, M. (2022). Aquaculture development in Nigeria and FAO's role. 65, FAO *Aquaculture News*, pp. 21–24.

Ali, B. A., & Mishra, A. (2022). Effects of dissolved oxygen concentration on freshwater fish: A review. *International Journal of Fisheries and Aquatic Studies, 10*(4), 113–127. Available at: https://doi.org/10.22271/fish.2022.v10.i4b.2693.

Anani, O. A., Adetunji, C. O., Olugbemi, O. T., Hefft, D. I., Wilson, N., & Olayinka, A. S. (2022). IoT-based monitoring system for freshwater fish farming: Analysis and design. In *AI, edge and IoT-based smart agriculture* (pp. 505–515). Elsevier. Available at: https://doi.org/10.1016/B978-0-12-823694-9.00026-8.

Bedekar, M. K., Kole, S., & Marappan, M. (2022). Types of vaccines used in aquaculture. In M. Makesh & K. V. Rajendran (Eds.), *Fish immune system and vaccines* (pp. 45–63). https://doi.org/10.1007/978-981-19-1268-9_3

Benzie, J. A. H. (2024). Farmer perspectives on desired catfish attributes in aquaculture systems in Nigeria. An exploratory focus group study. *Aquaculture, 588*, 740911. Available at: https://doi.org/10.1016/j.aquaculture.2024.740911.

Bozorg-Haddad, O., Delpasand, M., & Loáiciga, H. A. (2021). Water quality, hygiene, and health. In *Economical, political, and social issues in water resources* (pp. 217–257). Elsevier. Available at: https://doi.org/10.1016/B978-0-323-90567-1.00008-5.

Evans, T. J., & Riley, P. A. (2021). Principles of microscopy, culture and serology-based diagnostics. *Medicine, 49*(10), 648–653. https://doi.org/10.1016/j.mpmed.2021.07.009

Garrido-Momparler, V., & Peris, M. (2022). Smart sensors in environmental/water quality monitoring using IoT and cloud services. *Trends in Environmental Analytical Chemistry, 35*. https://doi.org/10.1016/j.teac.2022.e00173

He, T., Zhang, Y.-Z., Gao, L.-H., Miao, B., Zheng, J.-S., Pu, D.-C., Zhang, Q.-Q., Zeng, W.-W., Wang, D.-S., Su, S.-Q., & Zhu, S. (2023). Identification and pathogenetic study of tilapia lake virus (TiLV) isolated from naturally diseased tilapia. *Aquaculture, 565*, 739166. Available at: https://doi.org/10.1016/j.aquaculture.2022.739166

Hoque, F., Adhikari, S., Mandal, N. R., Hussan, A., Chattopadhyay, D., Paul, N. B., Das, A., Chakrabarti, P. P., & Swain, K. S. (2022). A retrospective study on the incidence of fish diseases and use of therapeutants in aquaculture farms of Monya, the fisheries hub of West Bengal, India. *Exploratory Animal and Medical Research, 12*(2), 195–204. https://doi.org/10.52635/eamr/12.2.195-204

Jain, P., Dhanker, R., Bhardwaj, A., Singhla, G., Saxena, K., & Hussain, T. (2022). Aerobic and anaerobic ammonia-oxidizing bacteria: A resilient challenger or innate collaborator. In *Development in wastewater treatment research and processes* (pp. 195–222). Elsevier. Available at: https://doi.org/10.1016/B978-0-323-91901-2.00011-5.

Kasihmuddin, S. M., Cob, Z. C., Noor, N. M., & Das, S. K. (2024). Effect of different temperature variations on the physiological state of catfish species: A systematic review. *Fish Physiology and Biochemistry, 50*(2), 413–434. Available at: https://doi.org/10.1007/s10695-024-01323-8.

Khieokhajonkhet, A., Sangphrom, S., Aeksiri, N., Tatsapong, P., Wuthijaree, K., & Kaneko, G. (2022). Effects of long-term exposure to high temperature on growth performance, chemical composition, haematological and histological changes, and physiological responses in hybrid catfish [♂*Clarias gariepinus* (Burchell, 1822) ×♀C. macrocephalus (Günther, 1864)]. *Journal of Thermal Biology, 105*, 103226. https://doi.org/10.1016/j.jtherbio.2022.103226

Khor, L., Bodunde, O. A., Wills, R., Hanson, L., Adeyemo, O. K., Aina, O. O., Alarape, S. A., Delamare-Deboutteville, J., & Chadag, V. M. (2024). Understanding aquaculture biosecurity to improve catfish disease management in Ogun and Delta states, Nigeria. *Aquaculture, 584*, 740664. Available at: https://doi.org/10.1016/j.aquaculture.2024.740664.

Knoll, M. A., Steixner, S., & Lass-Flörl, C. (2023). How to use direct microscopy for diagnosing fungal infections. *Clinical Microbiology and Infection: the Official Publication of the European Society of Clinical Microbiology and Infectious Diseases, 29*(8), 1031–1038. Available at: https://doi.org/10.1016/j.cmi.2023.05.012.

Kunjiraman, S., Singh, I. S. B., Sarasan, M., & Puthumana, J. (2024). Immobilized microbial consortia: An eco-friendly and sustainable solution for aquaculture waste management. *The Microbe, 4*, 100100. Available at: https://doi.org/10.1016/j.microb.2024.100100.

Labh, S. N. (2023). Applications of biotechnology in aquaculture nutrition and fisheries. In *Frontiers in aquaculture biotechnology* (pp. 163–170). Elsevier. Available at: https://doi.org/10.1016/B978-0-323-91240-2.00007-5.

Larsson, D. G. J., & Flach, C. F. (2022). Antibiotic resistance in the environment. *Nature Reviews Microbiology, 20*(5), 257–269. Available at: https://doi.org/10.1038/s41579-021-00649-x.

Lei, S., Chen, S., & Zhong, Q. (2021). Digital PCR for accurate quantification of pathogens: Principles, applications, challenges and future prospects. *International Journal of Biological Macromolecules, 184*, 750–759. Available at: https://doi.org/10.1016/j.ijbiomac.2021.06.132.

Li, L., Shen, Y., Yang, W., Xu, X., & Li, J. (2021). Effect of different stocking densities on fish growth performance: A meta-analysis. *Aquaculture, 544*, 737152. Available at: https://doi.org/10.1016/j.aquaculture.2021.737152.

Limbu, S. M. (2020). Antibiotics use in African aquaculture: Their potential risks on fish and human health. In A. L. K. Abia & G. R. Lanza (Eds.), *Current microbiological research in Africa* (pp. 203–221). Springer International Publishing. Available at: https://doi.org/10.1007/978-3-030-35296-7_8.

Lindholm-Lehto, P. (2023). Water quality monitoring in recirculating aquaculture systems. *Aquaculture, Fish and Fisheries, 3*(2), 113–131. Available at: https://doi.org/10.1002/aff2.102.

Luo, G., Xu, J., & Meng, H. (2020). Nitrate accumulation in biofloc aquaculture systems. *Aquaculture, 520*, 734675. https://doi.org/10.1016/j.aquaculture.2019.734675

Ma, S., Lv, Y., Hou, L., Jia, Z., Lin, S., Wang, S., He, X., & Hou, J. (2023). Effect of acute temperature stress on energy metabolism, immune performance and gut microbiome of largemouth bass (Micropterus salmoides). *Aquaculture and Fisheries*, S2468550X23001168. https://doi.org/10.1016/j.aaf.2023.10.001

Mallat, J., Rahman, N., Hamed, F., Hernandez, G., & Fischer, M. O. (2022). Pathophysiology, mechanisms, and management of tissue hypoxia. *Anaesthesia, Critical Care & Pain Medicine, 41*(4). https://doi.org/10.1016/j.accpm.2022.101087

Manyise, T., Basiita, R. K., Mwema, C. M., Oyesola, O., Siriwardena, S., Fregene, B., Cole, S. M., Dompreh, E. B., Lam, R. D., R. D. Lam., Lozano, D. L., & Rossignoli, C. M., John A.H. Benzie. (2024). Farmer perspectives on desired catfish attributes in aquaculture systems in Nigeria. An exploratory focus group study, Aquaculture, Volume 588, 740911, ISSN 0044–8486, https://doi.org/10.1016/j.aquaculture.2024.740911.

Moraes., G., & de Almeida, L. C. (2020). Nutrition and functional aspects of digestion in fish, In (Eds):Baldisserotto, B., Urbinati, E.C., Cyrino, J.E.P. Biology and Physiology of Freshwater Neotropical Fish, Academic Press, Pages 251–271. https://doi.org/10.1016/B978-0-12-815872-2.00011-7.

Olaifa, E., Osabuohien, E., & Haruna, I. (2022). Enhancing fish production for food security in Nigeria. Materials Today: Proceedings, 65, 2208–2214. https://doi.org/10.1016/j.matpr.2022.06.243

Pailan, G. H., & Biswas, G. (2022). Feed and Feeding Strategies in Freshwater Aquaculture. In: Lama, T., Burman, D., Mandal, U.K., Sarangi, S.K., Sen, H. (eds) Transforming Coastal Zone for Sustainable Food and Income Security. Springer, Cham. https://doi.org/10.1007/978-3-030-95618-9_35

Pakingking, R. V., & Nguyen, V. V. (2022). Edwardsiellosis. In *Aquaculture pathophysiology* (pp. 345–356). Elsevier. Available at: https://doi.org/10.1016/B978-0-12-812211-2.00026-3.

Ramesh, P., Jasmin, A., Tanveer, M., Roshan, R. U., Ganeshan, P., Rajendran, K., Roy, S. M., Kumar, D., Chinnathambi, A., & Brindhadevi, K. (2024). Optimizing aeration efficiency and forecasting dissolved oxygen in brackish water aquaculture: Insights from paddle wheel aerator. *Journal of the Taiwan Institute of Chemical Engineers, 156*, 105353. Available at: https://doi.org/10.1016/j.jtice.2024.105353.

Randall, D. J., & Tsui, T. K. N. (2002). Ammonia toxicity in fish. Marine Pollution Bulletin, Vol. 45, (1–12)17–23. https://doi.org/10.1016/S0025-326X(02)00227-8. https://www.sciencedirect.com/science/article/pii/S0025326X02002278

Scarfe, A. D., & Palić, D. (2020). Aquaculture biosecurity: Practical approach to prevent, control, and eradicate diseases. In *Aquaculture health management* (pp. 75–116). Elsevier. Available at: https://doi.org/10.1016/B978-0-12-813359-0.00003-8.

Seale, A. P., Cao, K., Chang, R. J. A., Goodearly, T. R., Malintha, G. H. T., Merlo, R. S., Peterson, T. L., & Reighard, J. R. (2024). Salinity tolerance of fishes: Experimental approaches and implications for aquaculture production. *Reviews in Aquaculture, 16*(3), 1351–1373. Available at: https://doi.org/10.1111/raq.12900.

Semwal, A., Kumar, A., & Kumar, N. (2023). A review on pathogenicity of Aeromonas hydrophilia and their mitigation through medicinal herbs in aquaculture. *Heliyon, 9*(3), e14088. Available at: https://doi.org/10.1016/j.heliyon.2023.e14088.

Sharan, M., Vijay, D., Yadav, J. P., Bedi, J. S., & Dhaka, P. (2023). Surveillance and response strategies for zoonotic diseases: A comprehensive review. *Science in One Health, 2*, 100050. Available at:. https://doi.org/10.1016/j.soh.2023.100050

Ships and Ports. (2021). *Nigeria imports 2.2m tonnes of fish annually*. https://shipsandports.com.ng/nigeria-imports-2-2m-tonnes-of-fish-annually/.

Sinha, S., Chugh, P., Abiha, U., & Singh, R. (2022). Biofiltration: An emerging and promising technology for the treatment of water and air pollutants. In *An innovative role of biofiltration in wastewater treatment plants (WWTPs)* (pp. 459–476). Elsevier. Available at: https://doi.org/10.1016/B978-0-12-823946-9.00010-3.

Spiga, L., Jimenez, A. G., Santos, R. L., & Winter, S. E. (2021). How microbiological tests reflect bacterial pathogenesis and host adaptation. *Brazilian Journal of Microbiology, 52*(4), 1745–1753. Available at: https://doi.org/10.1007/s42770-021-00571-7.

Thombre, N. V. (2023). Oxidation in water and used water purification. In J. Lahnsteiner (Ed.), *Handbook of water and used water purification* (pp. 1–23). Springer International Publishing. Available at: https://doi.org/10.1007/978-3-319-66382-1_174-1.

Tom, A. P., Jayakumar, J. S., Biju, M., Somarajan, J., & Ibrahim, M. A. (2021). Aquaculture wastewater treatment technologies and their sustainability: A review. *Energy Nexus, 4*, 100022. Available at: https://doi.org/10.1016/j.nexus.2V021.100022.

Von Gesdorff, J. L., Puspasari, K., & Insariani. (2022). Infection by Ichthyophthirius multifiliis. In *Aquaculture pathophysiology* (pp. 493–503). Elsevier. Available at: https://doi.org/10.1016/B978-0-12-812211-2.00040-8.

WorldFish. (2021). *Nigeria fish futures. Aquaculture in Nigeria: Increasing income, diversifying diets and empowering women. Report of the scoping study*. Available at: https://worldfishcenter.org/publication/nigeria-fish-futures-aquaculture-nigeria-increasing-income-diversifying-diets.

Yang, Y.H.C., Briant, L.J.B., Raab, C.A., Mullapudi, S.T., Maischein, H.M., Kawakami, K., Stainier, D.Y.R, (2022). Innervation modulates the functional connectivity between pancreatic endocrine cells. eLIFE 11: (Journal) ZDB-PUB-220405–22.

Yilmaz, S., Yilmaz, E., Dawood, M. O. A., Ringø, E., Ahmadifar, E., & Abdel-Latif, H. M. R. (2022). Probiotics, prebiotics, and synbiotics used to control vibriosis in fish: A review, Aquaculture, Vol. 547, 737514. https://doi.org/10.1016/j.aquaculture.2021.737514.

Zhang, H., Tang, W., Chen, Y., & Yin, W. (2020). Disinfection threatens aquatic ecosystems. *Science, 368*(6487), 146–147. Edited by J. Sills. Available at: https://doi.org/10.1126/science.abb8905.

Glossary

Adenosine Derivatives fungal metabolites involved in energy metabolism
Adjuvants aiding or assisting
Aflatoxins a type of mycotoxin that can cause liver damage and other health issues in animals
Amylolytic Fungi fungal species that break down starch in feed digestion
Animal Protein protein sources derived from animal
Anthelmintics fungal-derived compounds used to combat parasitic worm infections in animals
Antibiotic Resistance development of fungal-based alternatives to traditional antibiotics
Antimicrobial Agents fungal-derived compounds that inhibit the growth of harmful microorganisms
Antioxidants compounds in fungi that help reduce oxidative stress and protect animal health
Aquaculture this is the breeding, raising, and harvesting of fish, shellfish, and aquatic plants. It is also referred to as farming of organisms in water
Aquaponics a system of aquaculture in which the waste produced by farmed fish or other aquatic creatures supplies the nutrients for plants grown hydroponically
Aspergillus Oryzae a fungal species used in animal feed to produce enzymes that enhance digestion
Autolytic Enzyme Activity fungal enzymes that break down dead cells in feed substrates
Bacterial Pathogens bacteria that can cause pathogens
Beta-glucans polysaccharides found in fungal cell walls, known for their immunomodulatory and gut health benefits
Biodegradation breakdown of organic matter by fungal enzymes
Bioinformatics computational tools and techniques used to analyze fungal genomes and discover new bioactive compounds

A. Gupta (ed.), *Prospects of Fungal Biotechnologies for Livestock Volume 1*, Fungal Biology, https://doi.org/10.1007/978-3-031-90994-8

Bioremediation this refers to a managed or spontaneous process in which chemical reactions mediated by biological organisms degrade or transform contaminants to less toxic or nontoxic forms thereby remedying or eliminating environmental contamination

Bio-security the prevention of disease-causing agents entering or leaving any place where they can pose a risk to farm animals, other animals, humans, or the safety and quality of a food product

Brackish water that is saltier than freshwater but not as salty as saltwater. Usually a mix of both waters

Cellulases enzymes produced by fungi that break down cellulose into simpler sugars, aiding in fiber digestion.

Chitin a fibrous component in fungal cell walls that has antimicrobial and immune-boosting properties

Community Network Analysis (CNA) a computational approach to study interactions within fungal communities and their ecological roles

Cytokine Modulation by Fungi how fungal additives influence immune signaling molecules

Cytokine Signaling Proteins help control inflammation in the body

Diets the kind of food that an individual (organism) habitually eats

Disease Management a system of coordinated health care interventions and communications for defined patient populations with conditions where care efforts can be implemented

Dissolved Oxygen this refers to the concentration of oxygen gas incorporated in water

Ecdysteroids from Fungi steroid-like fungal compounds that support growth and development

Eco-friendly Harvesting sustainable methods of collecting fungi without harming ecosystems

Ecological Balance the role fungi play in nutrient cycling, soil fertility, and maintaining ecosystem stability

Endoglucanases a specific cellulase enzyme that degrades plant cell walls in animal feed

Environmental Contamination these are chemicals that accidentally or deliberately enter the environment often as a result of human activities

Enzymes biological catalysts from fungi, such as cellulases and proteases, used in industrial and pharmaceutical applications

Epigenetic Modulation by Fungal Feed how fungal bioactives influence gene expression in livestock

Ergosterol a sterol unique to fungi, serving as a precursor for vitamin D synthesis

Exopolysaccharides (EPS) fungal-derived bioactive compounds with health benefits

Feed Conversion Ratio (FCR) a measure of an animal's efficiency in converting feed into body mass

Feeding this is the introduction of any food substance into water for consumption by the fish species

Fermentation the process by which fungi break down organic substances, often used in improving the nutritional value of animal feed

Fish Culture the process of raising desirable fish species in captivity and managing them and their environment to improve growth and reproduction

Fish Diseases unhealthy conditions or state of fish

Fish Meal a commercial product made from ground-dried fish and fish by-products

Fish Processing producing any substance or article from fish by any method. This includes cutting up, dismembering, cleaning, sorting, ioining, freezing, canning, salting, preservation, and reduction of fish

Fishery the occupation of rearing of fish

Forestry this is the science and art of creating, managing, planting, using, conserving, and repairing forests and woodlands for associated resources for humans and environmental benefits

Freshwater naturally occurring water that is not salty and is suitable for consumption if clean or processed

Fry baby fish, newly hatched fish from eggs

Fungal Adaptogens in Livestock stress-resistance compounds that improve animal resilience

Fungal Antioxidants compounds that prevent oxidative stress and cell damage

Fungal Beta-Mannans bioactive polysaccharides influencing nutrient absorption

Fungal Bioremediation use of fungi to clean up pollutants and toxins

Fungal Biotechnology the application of fungi in industries such as medicine, agriculture, and environmental sustainability

Fungal Diversity various fungal species and their ecological and medical roles

Fungal Exosomes in Nutrition nano-vesicles from fungi that transport bioactive compounds

Fungal Genomics study of fungal genes to explore bioactive compounds

Fungal Hydrolases in Rumen Fermentation enzymes aiding microbial digestion in ruminants

Fungal Infections fungal infection or any disease you get from fungus

Fungal Lectins sugar-binding proteins that regulate immune responses

Fungal Metabolites bioactive chemicals produced by fungi with pharmaceutical potential

Fungal Nutritional Supplements dietary products derived from fungi

Fungal Pathogenesis study of disease-causing mechanisms in fungi

Fungal Pharmacology study of fungal-derived drugs and their effects

Fungal Probiotics in Livestock beneficial fungi that promote gut health in animals

Fungal Proteins proteins derived from fungi, utilized as alternative nutritional sources in food and feed industries

Fungal RNA Interference (RNAi) in Animal Feed gene silencing effects that regulate metabolic pathways

Fungal Secondary Metabolites bioactive compounds produced by fungi that contribute to antimicrobial, anti-inflammatory, and growth-promoting effects

Fungal Siderophores in Mineral Absorption compounds that enhance iron uptake in animals

Fungal-Based Veterinary Medicine use of fungi in animal healthcare treatments

Fungal-Derived Biopeptides small protein fragments that enhance digestion and immunity

Gastrointestinal Epithelium Protection the role of fungal bioactives in maintaining gut lining integrity

Gastrointestinal Hormone Regulation by Fungi the impact of fungal bioactives on digestive hormones

Gene Clusters in Fungi groups of genes responsible for bioactive compound synthesis

Genomics the study of fungal genetic material to understand biodiversity and potential applications

Gizzard part of a bird's stomach in which food is broken

Glucuronidation Pathways in Fungal Additives detoxification mechanisms influenced by fungal feed

Glycoproteins in Fungal Additives protein-carbohydrate molecules enhancing gut function

Gut Microbiome the community of microorganisms in the digestive system that influences digestion, immunity, and overall health

Habit a way of life of fishes; their usual behavior that enables survival

Habitat the natural home or environment of fishes, e.g., freshwater

Hemicellulases fungal enzymes that break down hemicellulose, improving fiber digestion

Hemicellulose Degradation the process by which fungi break down complex carbohydrates

Hepatoprotective Effects of Fungal Feed the role of fungi in protecting the liver from toxins

Histopathology the diagnosis and study of disease of the tissues

Hybridization the process of breeding on fish species with another compatible fish species, e.g., Clarias and Heterobranchus spp.

Hydrophobin Proteins fungal proteins that aid in nutrient transport and bioavailability

Hypoxia low levels of oxygen in your body tissues

Idiopathic Dilated Cardiomyopathy a primary myocardial disease of unknown cause characterized by left ventricular or biventricular dilatation and impaired myocardial contractility

Immunoadjuvant Properties of Fungifungal compounds that boost vaccine efficiency in livestock

Immunomodulators substances from fungi that influence immune system function, either enhancing or suppressing responses

Immunomodulatory Properties the ability of fungal feed additives to regulate and enhance the immune system

Immunostimulant a substance that increases the stability of the immune system to fight infection and disease

Industrial Fermentation large-scale fungal cultivation for bioactive compound production

Inland Fish Decree this decree provides for the licensing of fishing crafts and regulates fishing on the inland waters of Nigeria
Integrated Fish Farming a system of producing fish in combination with other agricultural farming operations
Juveniles young immature fish, not sexually mature
Laccases fungal enzymes involved in biodegradation and biofuel production
Lipases enzymes that break down fats into glycerol and fatty acids, improving fat metabolism in animals
Mannan-Oligosaccharides (MOS) prebiotic molecules from fungi that support gut microbiota
Metabolic Pathways in Fungi biochemical routes used by fungi to produce bioactive compounds
Metabolomic Profiling of Fungal Additives analyzing fungal bioactive compounds for nutritional benefits
Microscopy technical field of using microscopes to view samples and objects that cannot be seen with the unaided eye
Mitochondrial Biogenesis in Livestock how fungal metabolites improve energy production
Mitochondrial Respiratory Chain Modulation fungal metabolites improving cellular respiration
Molecular Docking of Fungal Metabolites computational screening of fungal bioactives for feed applications
Mushroom-Based Feed Additives edible fungi used as nutritional supplements in livestock feed
Mycobiome the fungal component of an animal's gut microbiome
Mycogut Interaction the relationship between fungal species and the animal gut environment
Mycoremediation in Animal Feed using fungi to detoxify contaminated feed
Mycorrhizal Fungia type of fungi that forms symbiotic relationships with plant roots to enhance nutrient uptake
Mycotoxin-Binding Yeasts fungi that detoxify feed by binding mycotoxins
Mycotoxins toxic secondary metabolites produced by certain fungi, which can contaminate animal feed, animals, and humans, and cause health issues
Mycovitamins vitamins synthesized by fungi that contribute to animal health
Nanotechnology in Fungal Research use of nanotech to enhance fungal-based therapies
Neuroprotective Effects of Fungal Compounds the potential of fungi in reducing stress-induced neurodegeneration
Nitrogen Fixation Enhancement by Fungi the role of fungi in improving nitrogen assimilation in animal feed
Nutraceuticals bioactive fungal products that provide health benefits beyond basic nutrition
Organic Acids compounds produced by fungi that can help prevent pathogenic bacteria growth in animal feed
Parasitic Infection disease condition from parasites

pH (potential of hydrogen) level of acidity or alkalinity in water
Plant Protein protein derived from plant sources
Pollutants substances that are introduced into the environment that have undesired effects on the usefulness of the resource
Polyculture the simultaneous growth of two or more compatible fishes, e.g., *Clarias gariepinus* and Tilapia
Polyphenolic Compounds from Fungi antioxidants that reduce oxidative stress in animals
Polysaccharides complex carbohydrates found in fungi, known for their role in immunomodulation and health benefits
Postbiotic Fungal Metabolites health-beneficial compounds produced by fungal probiotics
Prebiotics non-digestible compounds that promote the growth of beneficial gut bacteria
Probiotics beneficial microorganisms, such as certain yeasts, fungi, and bacteria that support gut health in animals
Proteases fungal enzymes that help break down proteins into amino acids for easier absorption
Proteomics analysis of fungal proteins for biomedical applications
Quorum Sensing in Fungi the fungal ability to regulate behavior through chemical signaling
Regulatory Challenges the legal and safety hurdles associated with the commercial use of fungal-derived products
Resistance the inability to cause a change or the ability not to be affected by something
Saccharomyces Cerevisiae a widely used yeast species in animal feed additives, known for its probiotic properties
Saltwater naturally occurring salty water
Sanitation the process of keeping places free from dirt, infection, diseases
Scalability Challenges difficulties in large-scale production of fungal-based products
Secondary Fungal Alkaloids bioactive fungal compounds with antimicrobial properties
Secondary Metabolites bioactive compounds produced by fungi, often with pharmaceutical and antimicrobial properties
Sustainability in Feed the concept of using fungal feed additives to reduce environmental impact and reliance on chemical additives
Sustainable Cultivation eco-friendly methods of growing fungi to reduce environmental impact and conserve biodiversity
Symbiotic Relationships the mutualistic associations between fungi and plants or animals, beneficial for both organisms
Synthetic Biology engineering fungi for biotechnological and medical applications
Toxin Sequestration the process by which certain fungi can bind and neutralize harmful mycotoxins
Transcriptomics study of fungal RNA to understand gene expression

Triterpenes in Fungal Feed fungal-derived bioactive compounds enhancing metabolic functions

Vaccination the administration of a vaccine to help the immune system develop immunity from a disease

Water Quality this refers to the chemical, physical, and biological characteristics of water based on standards of its usage

Wild Harvesting the process of collecting fungi from their natural habitats for commercial and medicinal use

Xyloglucans fungal-sourced fibers that regulate digestion in livestock

Index

A. Gupta (ed.), *Prospects of Fungal Biotechnologies for Livestock Volume 1*, Fungal Biology, https://doi.org/10.1007/978-3-031-90994-8

The manufacturer's authorised representative in the EU is Springer Nature Customer Service Centre GmbH, Europaplatz 3, 69115 Heidelberg, Germany. If you have any concerns regarding our products, please contact ProductSafety@springernature.com

Printed and bound by CPI Group (UK) Ltd, Croydon, CR0 4YY
01/10/2026
02225864-0009